Explorations in
CLASSICAL
SOCIOLOGICAL
THEORY

THIRD EDITION

Explorations in CLASSICAL SOCIOLOGICAL THEORY

SEEING THE SOCIAL WORLD

THIRD EDITION

KENNETH ALLAN

University of North Carolina at Greensboro

Los Angeles | London | New Delhi
Singapore | Washington DC

Los Angeles | London | New Delhi
Singapore | Washington DC

FOR INFORMATION:

SAGE Publications, Inc.
2455 Teller Road
Thousand Oaks, California 91320
E-mail: order@sagepub.com

SAGE Publications Ltd.
1 Oliver's Yard
55 City Road
London EC1Y 1SP
United Kingdom

SAGE Publications India Pvt. Ltd.
B 1/I 1 Mohan Cooperative Industrial Area
Mathura Road, New Delhi 110 044
India

SAGE Publications Asia-Pacific Pte. Ltd.
3 Church Street
#10-04 Samsung Hub
Singapore 049483

Acquisitions Editor: Dave Repetto
Assistant Editor: Terri Accomazzo
Production Editor: Laureen Gleason
Copy Editor: Teresa Herlinger
Typesetter: C&M Digitals (P) Ltd.
Proofreader: Gretchen Treadwell
Indexer: Michael Ferreira
Cover Designer: Anupama Krishnan
Marketing Manager: Erica DeLuca
Permissions Editor: Karen Ehrmann

Copyright © 2013 by SAGE Publications, Inc.

Printed in the United States of America

Library of Congress Cataloging-in-Publication Data

Allan, Kenneth, 1951-
Explorations in classical sociological theory: seeing the social world/Kenneth Allan.—3rd ed.

p. cm.
Includes bibliographical references and index.

ISBN 978-1-4129-9243-5 (pbk.)

1. Sociology—Philosophy—History—19th century.
2. Sociology—Philosophy—History—20th century.
I. Title.

HM445.A45 2013
301.01—dc23   2012003286

This book is printed on acid-free paper.

Certified Chain of Custody
SUSTAINABLE FORESTRY INITIATIVE
Promoting Sustainable Forestry
www.sfiprogram.org
SFI-01268
SFI label applies to text stock

12 13 14 15 16 10 9 8 7 6 5 4 3 2

Contents

Preface

Imagination is more important than knowledge.

—Albert Einstein

I came to sociology by accident. After two careers, I had decided to go back to school. It was important for me to get into a profession in which I could help people and make a difference. So I picked psychology as my major. But while taking a sociology of education class for my liberal arts requirement, I heard a sociologist speak about his involvement in Nicaragua. Nicaragua had been undergoing economic, political, and social reforms since 1979. Previously, the country had known 40 years of dictatorship under the Somoza family. But during the 1980s, the U.S. government supported a rebel group (the Contras) that plunged Nicaragua into a civil war. During this politically and militarily risky time, this professor adopted a village. Two to three times a year, he took a group of students down to build houses, dig ditches, bring medical supplies, and care for the community. Here was an individual who was making a difference. His academic knowledge wasn't sterile; it affected his life and those around him. And that was exciting to me.

Two things struck me about the presentation. First, here was a man who walked the talk. He didn't hold knowledge in an unfruitful way; what he knew made him responsible. He cared about the human condition in an obvious way, and that wasn't something I had come across at the university before. The second thing that struck me about the presentation was his view of Karl Marx. I have always hated oppression. But as a good American, I had been told to hate Marx, too; he was the father of communism, and communism was bad. (I grew up during the Cold War.) Yet this man, whom I respected because of his fight against human suffering, told us that he was motivated by the theories of Marx. That day in class, I found out that Marx told a fascinating story about economic oppression. This story takes elements from the past, along with social factors in modernity, and predicts what will happen in society. It tells how (not just why) people are economically oppressed.

I was hooked. This teacher was a man who made a difference and knew *why he had to make a difference.* I wondered if it had something to do with sociology. The professor certainly claimed that it did. For 2 weeks, I met with my professors in psychology and went door-to-door in the sociology department talking to everybody I could. At the end of those 2 weeks, I changed my major to sociology. Yep, I was hooked on sociology, but what I didn't realize at the time was that I was hooked on theory as well.

For me, theory is amazing and exciting. Think about it this way: One of the things that makes us human is the way we think, and theory is all about thinking. Theory invites us to have new thoughts, ideas we've never even imagined before. Getting caught up in an idea, even a theoretical one, is electrifying! Another thing that makes us human is that we change the world around us—not just the physical world through technology, but the social world as well. And theory can help us do that too. In a sense, then, learning theory is one of the most human things we can do (and perhaps one of the most humane as well). Seeing the world differently, seeing the people around you differently, and making a difference are what social and sociological theory is all about.

As I've worked through a few editions of my books, I've found myself becoming more caught up in the whole story of how theory came about and what its purpose is. I've also become increasingly fascinated with how these theorists came up with their amazing ideas. Obviously, being brilliant helps. But as a sociologist, you know that social context plays a role as well. I've dug around a bit, and I want to share some of these insights with you in this edition of *Explorations in Classical Sociological Theory: Seeing the Social World.* This edition starts off with a look at modernity and Auguste Comte, the guy who founded sociology. I want to get our feet firmly placed on this foundation of science and theory and what Comte thought sociology was all about. Afterward, we'll take a look at some of the main characters in the story of sociology: Spencer, Marx, Durkheim, and Weber. With each of these, I'll invite you to think about what C. Wright Mills (1959) calls *the sociological imagination.* It's a way of seeing the world that combines a person's life story with his or her society's history and dominant structures. I use these factors to try and get at these theorists' sociological imagination. Why did they imagine society the way they did? Why did Marx think the way he did? Why was it so different from Weber? The real "take home" with this use of the sociological imagination is when we begin to see how our perspectives and thoughts are linked to the intersection of personal biography, history, and social structure.

Over the years, I've also found myself fascinated with the multifaceted nature of society and the theories we use to try and understand it. In your introduction to sociology class, you were probably told that we have three perspectives: functionalism, conflict theory, and symbolic interactionism. Well, that isn't exactly true. I've been doing this theory business for a number of years, and I have yet to encounter all the perspectives in our discipline. I can tell you one thing: It's a whole lot more than three. And that excites me. I mean, imagine how boring human society would be if the story could be told through three main ideas! Thankfully, we, and our discipline, are much more complex. I've tried to capture a bit more of that complexity in this edition.

The book is actually divided into three sections. Think of the first section, Chapters 1 through 5, as "A Sociological Core." Spencer, Marx, Weber, and Durkheim are in those chapters, so it makes sense in that way. We'll start off in

Chapter 1 thinking about our "founder" Auguste Comte and his work in positivism (scientific thinking). Spencer, Marx, and Durkheim more or less fit into his vision. But then comes Weber in Chapter 5. Weber may be the most robust and complex thinker sociology has ever known. Weber is a game changer.

The second section is "Another Sociological Core." (If we have more than one perspective, then we have more than one core.) We start off in that section thinking about the person, the modern individual. While Mead and Simmel are usually thought of as "in the core," their perspective is decidedly different from Comte, Spencer, Marx, Durkheim, or Weber (though in some ways Simmel is close to Weber). This other core also includes theorists of gender and race; we'll specifically listen to the voices of Harriet Martineau, Charlotte Perkins Gilman, Frederick Douglass, and W. E. B. Du Bois, voices that have been far too quiet in our discipline. They play a significant counterpoint to the modern melody we begin in Chapter 1. The final section of the book is Chapter 9. I include it mostly because one can't understand contemporary theory very well without Parsons and the Frankfurt School. But there's more: Parsons and the Frankfurt School embody, in perhaps the clearest possible way, the tensions and diversity that are intrinsic to studying society as a whole, which is the unique contribution of sociology. As you'll see in Chapter 1, sociology is the only discipline that is interested in *everything* social. Other disciplines focus on one facet, like the economy or politics, but sociology is interested in every facet of social life. It makes sense, then, that our discipline would reflect the complexity of society as a whole.

I've also included a number of elements to guide you along the way, some of which are new. You'll find the theorist's work concisely stated in the Theorist's Digest; important terms are defined in boxes throughout the text; and I provide Theoretical Hints and Enduring Questions to clue you in to important issues and give you a sense of some overarching sociological topics. The chapters conclude with a Theory Toolbox, filled with review questions and ways to see and use theory, and I also extend a theorist's ideas to see how they look in postmodernity. In addition, I really encourage you to use the headings and subheadings I've provided. Theory can be overwhelming because of the number of ideas presented, and trying to grasp a chapter as a whole can be daunting. Take the theories a step at a time. Everything under the heading "Concepts and Theory" forms a single unit. Understand each unit before trying to pull it all together to understand the theory as a whole. Also, in Chapter 1 I tell you what theory is, what it does, and how it is built. Theory isn't like an essay, novel, or poem. It is its own sort of beast. If you take the time to fully understand theory, it will make your journey through the book a lot easier. Okay, that's all I can think of right now. Good luck. I hope you enjoy the story half as much as I enjoyed writing it. If you actually take the time to get into this book, your mind will never be the same.

Acknowledgments

Without the A-team at SAGE, this project could never have been completed. As always, I want to thank my original editors, Jerry Westby and Ben Penner; you caught the vision and made it happen. My thanks to Dave Repetto for continuing to believe, and my thanks to Laureen Gleason, Erica DeLuca, and Karen Ehrmann for taking care of the details I could never manage. I especially appreciate my copy editor, Teresa Herlinger. You always know what I mean to say far better than I ever do—and you always make it fun.

SAGE and the author gratefully acknowledge the contributions of the following reviewers: Dalia Abdelhady, Southern Methodist University; John Barnshaw, University of South Florida; Duane Gill, Oklahoma State University; Robert Hard, Albertus Magnus College; Linda Jasper, Indiana University Southeast; Wanda Rushing, University of Memphis; and Anna Wilson, Chapman University.

CHAPTER 1

Beginning to See

A Sociological Core

M ost people enjoy a story with a good plot, one with a strong beginning and clear resolve at the end. Yet it's equally true that the vast majority of stories don't start with any sort of *true* beginning. The opening of a story is determined by the story itself; it's determined by the central characters and the plot the author wants to convey. Kind of like the story of your life. Your *true* beginning lies in the distant past, when your ancestors moved from one continent to another; when your distant relatives were conceived; and those precarious moments when your grandparents and parents met, when they moved past the awkwardness of new relationships and into the bonds that brought forth children. While a lot of that may be very interesting, it's not really the story of your life—at least, not with you as the central character.

The same sort of thing is true about the book you now hold. Its true beginning goes all the way back to when the first person ever thought about social things. But telling that tale would take a very long time and in the end, it wouldn't be that interesting: too many plots and characters, and far too much going on to make any kind of sense out of it. So, I'm telling a shorter tale—but even so, I'm choosing a few characters and story lines from a vast array. I'm beginning our story with the Enlightenment and modernity. While that may seem academically dry, people in the Enlightenment asked some amazing questions, such as, what is reality? What is human nature? How can we be sure that what we know is true? Even more interesting is that this story is *your* story. Now, I admit I'm extending your life story further back than your date of birth. However, if you've learned anything about sociology thus far, you know that the social context wherein you were born is really important. For example, if you're black, it matters if you were born in the United States in 1780 or 1980; your life chances and the story you could tell about your life would be utterly different.

What the story of modernity and the Enlightenment are going to give you is what C. Wright Mills calls *the sociological imagination,* the ability to see how society, history, and personal narratives intersect. Mills (1959) tells us that the very first thing you see with the sociological imagination is the real you: "The first fruit of this imagination . . . is the idea that the individual can understand his own experience and gauge his own fate only by locating himself within his [historical] period" (p. 5). I hope that by the end of this book, you'll see that the effects of these social forces go deep; they influence how we think, feel, and act, as well as what we value and what we know.

In this book, you'll learn about various theorists and what they had to say about society. There are other themes in the book as well. Sometimes they will be explicit, but they will always be there percolating in the background, even if I don't point them out. These themes or plot lines are the story of sociology as a modern way of knowing, the story of democracy as a central theme in modernity, and the story of the modern person (that would be you). As with any good book, you're going to find that these stories or characters are not free from conflict; they aren't things that just sit there without moving or changing. Change is endemic in human affairs, and thus society and people are organic in the sense that they are never quite the same from one moment to the next. My intent in telling you these stories is to invite you to have eyes to see the social world in which you live, move, and have your being. Every character in this book—including such things as the Enlightenment and modernity— invites us to see the social world in a different way. Revising this book is always a joy for me because I invariably find new perspectives and different ways of seeing. I hope you find these stories we're about to share to be as fascinating as I do.

The Making of Modernity and the Modern Way of Knowing

The words *modern* and *modernity* are used in a number of different ways. Sometimes modern is used in the same way as contemporary or up-to-date. Other times it's used as an adjective, as in modern art or modern architecture. In the social disciplines, there has been a good bit of debate about the idea of modernity. Some argue that we are no longer modern; others that we never were; and still others that we are living in some different form of modernity, like liquid modernity. We're not going to enter into this debate directly here, but the existence of the debate is important for us. It's important because this debate has implications for the kind of person we can be and the kind of society in which we live. In this book, we're going to begin thinking about society and our place in it using a specific view of modernity, one that assumes a rational actor and an ordered world that can be directed. It's important for you to know that this approach to understanding modernity and knowledge is just one of many possibilities. So, this story of modernity is simply our beginning; it's our touchstone, the place from which we will organize our thinking. In Max Weber's terms (Chapter 5), I'm giving you an *ideal type* of modernity.

As a social, historical period, modernity began in the seventeenth century and was marked by significant social changes, such as massive movements of populations from small local communities to large urban settings, a high division of labor, high commodification and use of rational markets, the widespread use of bureaucracy, and large-scale integration through national identities. In general, the defining institutions of modernity are nation-states and mass democracy, capitalism, science, and mass media; the historical moments that set the stage for modernity are the Renaissance, Enlightenment, Reformation, the American and French Revolutions, and the Industrial Revolution. We'll be thinking further about most of these throughout our time together, but for now I'd like you to see that as a result of these factors, a new *zeitgeist*, a new "spirit of the age," came into existence. After centuries of stagnation—they weren't called the "Dark Ages" for nothing—some people began to believe that the enlightened mind could achieve anything. It was in this age that our modern beliefs in reason, empiricism, progress, and equality began.

Enduring
Issues

Epistemology—the study of how we know things—has been a thorny issue for people who think about such things for millennia. It gained even more importance due to modern positivism and science. The usual issues, as we'll see in Marx, Durkheim, and Weber, concern the way the subject of study exists (ontology) and the appropriate way to study such a thing (epistemology). Other issues became equally important as the social and behavioral sciences developed. Critical thought, for example, takes issue with one of the goals of science: control. According to critical theory, if sociology is a science, then the goal of sociological knowledge is to control people, whether intentionally or not. Some contemporary feminists argue that because science creates objective knowledge, it denies the subjective experiences of women and minorities (see P. H. Collins, 2000; D. E. Smith, 1987).

One of the earmarks of the modern era is *intentionality*. For most of human history, collective life simply happened; social groups grew and changed more by happenstance than design and were legitimated in retrospect. But modern society was intentional. It began with ideas and theories of human nature, knowledge, and what society could and should be; these ideas were then purposely used to improve the human condition. Some of the clearest examples of this purposefulness can be found in such political documents as the U.S. Declaration of Independence, and the French and U.S. Constitutions. Those documents are based on theories and philosophies of natural law and human nature; they intentionally and legally set up specific relationships among social institutions and citizens.

Institutions of Modernity

One of the most notable institutional changes concerned religion. Prior to modernity, religion occupied the central position in society as a whole and in government specifically. This dominance of religion dates back to 140 BCE in the East, with Emperor Wu of Han and Confucianism, and 313 CE in the West, with Constantine and Christianity. The modern age is marked by the intentional separation of church and state. This separation is necessary because democracy cannot function under absolute truth and legitimation, which is what religion does for government. A government that is legitimated by religion cannot be questioned; to do so would be to question God. Thus, a *theocracy* (rule by God) is the polar opposite of a *democracy* (rule by the people). In a theocracy, the power to rule goes from the top (God) down; in a democracy, the power to rule comes from the bottom (citizens) up. However, while it's true there needs to be a separation, it's also clear by looking at early social thinkers and sociologists that this separation did not necessarily mean that religion wasn't important or would go away.

Religion was separated out so that government could change. Prior to modernity, the primary form of government in Western Europe was feudalism, which was based on land tenure and personal relationships. These relationships, and thus the land, were organized around the monarchy with clear social, hereditary divisions between royalty and peasants. Therefore, the experience of the everyday person in feudal Europe was one where personal obligations and one's relationship to the land were paramount. Every person was keenly aware of his or her obligations to the lord of the land. These were seen as a kind of familial relationship, with fidelity as its chief goal. The quintessential form of government in modernity is the democratic nation-state, whose chief goal is to protect the freedom and sovereignty of the individual.

This can be seen as one of the most astonishing aspects of this period: *Modernity is founded on a specific view of and belief in the person.* One easy way to see this is through the different sorts of political actors feudalism and democracy assume. The majority of people in a feudalistic state are *subjects;* the majority in a democracy are *citizens.* The distinctions are not incidental—subjects are *subjected* to power; citizens *hold* power—nor are these distinctions merely outward. They imply something about how the individual experiences his or her life—subjects must be cared for and guided; citizens must use reason to guide society. Further, this idea of the reasoning, responsible citizen was perceived as the result of birth—not birth into a social class, but birth of a universal human nature. Notice what the U.S. Declaration of Independence says: People are "endowed by their Creator with certain unalienable Rights." Early modernity brought with it, then, this vision of the autonomous person: "the human person as a fully centered, unified individual, endowed with the capacities of reason, consciousness, and action, whose 'centre' [sic] consisted of an inner core which first emerged" at birth and continued to unfold "throughout the individual's existence" (S. Hall, 1996a, p. 597).

Further, the idea that modern society has about the person is itself an institution upon which the social system stands. It is the basis of democracy and the fountainhead of the Enlightenment: "*Enlightenment is the human being's emergence from his self-incurred minority. Minority is the inability to make use of one's own understanding without direction from another. . . . Have courage to make use of your *own* understanding!* is the motto of enlightenment" (Kant, 1784/1999, p. 17). Eighteenth century political literature is filled with references to the "enlightened citizen" and is always linked to education. Thus, the institution of the reasoning person goes hand in hand with another institutional change specific to modernity: the shift from education for the elite to education for the masses, from religious to scientific education.

Of course, education for the masses makes perfect sense in a democracy. Power in a modern state rests upon the people. The method through which democratic citizens are to exercise their power is through knowledge, which is why education is a key to the success of democracy—it's also the reason why freedom of the press is guaranteed in the U.S. Constitution; people need information in order to govern. Education is linked to democracy not simply to supply knowledge for voting; education is necessary in democracy because "A democracy is more than a form of government; it is primarily a mode of associated living, of conjoint communicated experience" (Dewey, 2009/1916, p. 73).

Yet there's something more basic and perhaps more profound implied by this link between the modern person and education: Democracy is based on an idea, a theory of government. One of the first studies of modern democracy was by Alexis de Tocqueville, who came to the United States in 1831. Tocqueville (1835–1840/2002) tells us that many who left their homelands to come to America didn't do it for economic gain or even for personal freedom: Rather, "they tore themselves away from the sweetness of their native country to obey a purely intellectual need . . . they wanted to make *an idea* triumph" (p. 32). In his 2011 State of the Union Address, President Barack Obama emphasized this same point:

> We're the home to the world's best colleges and universities, where more students come to study than any place on Earth. What's more, we are the first nation to be *founded for the sake of an idea*—the idea that each of us deserves the chance to shape our own destiny. (n.p., emphasis added)

In a way, then, democracy is conceptual; it is based in and progresses through a vision of what could be. Yet this idea isn't pure speculation; it's founded in a specific way of seeing and understanding the social world. Democracy is based in social theory—the kind of theory found in sociology.

The Birth of Sociology: August Comte

Where shall we begin the story of sociology? People have been thinking about society ever since Plato, but like going back to the proverbial first parents to tell your story, starting that far back wouldn't actually tell the story of sociology. I'm going to begin our story with Auguste Comte (1798–1857). Why Comte? Well, there are a couple of good reasons. First, Comte is the one who is given credit for coming up with the word *sociology*, and he is certainly the first to systematically organize the discipline, since Comte gave us our beginning conceptual framework. Second, many if not most professional sociologists would pick Comte as the most logical place to start. So I'm on pretty safe ground starting with him. However, a good beginning point isn't all a story needs; it also needs a plot. A literary plot is the actual organization of the story; it's the pattern of events or the story line that leads the reader along. Without a plot, a literary piece is simply a bunch of disconnected vignettes that have little in common. Although this isn't a novel, I do have a plot in mind for this book. I'm going to tell you a story, and beginning with Comte gives me a foothold.

While Comte is more or less taken for granted as a beginning point, there are some powerful implications for the kind of sociology we do. Jonathan H. Turner (1985), a well-known contemporary theorist, says, "Auguste Comte proclaimed that sociology could take its place among the sciences" (p. 24). To begin with Comte thus implies that the scientific model is to be used to understand human behavior and is the core of sociology. As straightforward as that might sound, there's a problem: Not all sociologists accept that sociology can be a science in the same way as physics or biology. Some sociologists, like many contemporary feminists, argue that the pursuit of timeless laws in society is misguided and that positivism actually works to oppress certain minority groups.

Definition

Positivism is a philosophy of science first articulated by Auguste Comte. There are three foundations of this way of understanding the universe. First is the belief in the infinite potential of the human mind and knowledge. Knowledge, according to Comte, has progressed through three distinct phases: theological, metaphysical, and positivistic. In the positivistic phase, there is virtually no limit to what human beings can discover—the limit of knowledge is set only by the boundaries of the universe. The second premise is that everything within this universe is empirical and operates according to invariant, natural laws that govern behavior in predictable ways, as with the law of gravity. The third foundation of positivism is the belief in science as the best way to improve human existence: "The positive philosophy offers the only solid basis for . . . social reorganization" (Comte, 1830–1842/1975b, p. 83).

The idea that science may not be the proper way to understand human behavior isn't new. There are a variety of critiques, but one that is salient for us comes from Max Weber (Chapter 5), generally conceded to be one of the three most important sociologists that ever lived. Weber drew inspiration for his methodology from Wilhelm Dilthey (1833–1911). Dilthey (1883/1991) characterized Comte's work as "impoverished, superficial, but analytically refined" and said that "Comte and the positivists . . . seemed to me to truncate and mutilate historical reality in order to assimilate it to the concepts and methods of the natural sciences" (p. 49). As I said, our characters are complex.

So, why do I begin with Comte? My first reason is that he gives us a clear understanding of scientific sociology and, while there are detractors, much of sociology works in this way. Second, Comte gives us insight into the original purpose of sociology. I'll have more to share about this presently. Finally, Comte gives us a point of comparison to talk about and understand the other ways of seeing social life, and we always understand things better, more clearly, if we have something to compare. For example, talking about Weber's approach to understanding society in comparison to science helps to make his position stand out in stark relief. Another example from the book is George Herbert Mead (Chapter 6). Mead is the founder of symbolic interaction, an approach that understands society as an emergent phenomenon. If we first understand that science is focused on discovering invariant laws that allow us to predict human behavior, then Mead's idea of emergent society is all the more powerful and intriguing.

For me, this *polyvocal* (many voices) approach to understanding sociology makes sense. One of the reasons is that not everything sociologists do is scientific—you can tell by looking at the field today that there are thousands of sociologists doing sociological work that isn't scientific. That could be due to the vastness of the field or the fact that we're too young a discipline to be utterly positivistic. Or, as Weber and Mead claim, it could be because there are things that people do that cannot be understood or predicted using scientific laws. Regardless of the reason, it's clear to me that acknowledging and studying the many voices that helped create sociology as a discipline is the most authentic approach. Plus, a classical theory book should give you a foundation for understanding contemporary theory and society, and these diverse perspectives are becoming more, not less, important. So I begin our story with Comte, and I will use his vision and passion for sociology as a starting point to weave a much more complex tapestry throughout the rest of the book.

Comte's Positivism

One of the things I skipped over in most of the textbooks I read as a college student was biographical information. I thought it was just fluff, and I wanted to get to the actual material because that's what mattered for the test. I'm embarrassed to say that it took me years to realize that a person's social experiences really do matter (I know, pretty slow for a sociologist). People feel, think, and act because of the society in which they live. This basic sociological insight is especially true for the kinds of people we'll meet in this book, powerful minds responding to powerful situations. So I'll start off by sharing with you some of the sociohistorical background of our theorists. Please don't make the same mistake I did. Let these life experiences speak to you and show you why people like Comte thought the way they did. The values and aspirations that guided our theorists came out of their life circumstances. Seeing that actually helps us to use the sociological imagination in our own lives.

Auguste Comte was born in southern France in 1798, and lived until 1857. In some ways, it is difficult for us to imagine the world in which Comte lived. Up until just before Comte's time, Europe had endured a long period of economic and social continuity and rule by religiously supported monarchs, but the eighteenth century changed all that, and not always peacefully. Comte was born during the French Revolution (1789–1799) to a devout royalist and Catholic family. Prior to the revolution, France had experienced severe food shortages, resulting in several food riots, plus rising inflation and a growing national debt. In response to a projected 20 percent budget shortfall, Louis XVI attempted to create a land tax that would affect even the nobles. In addition, the Third Estate, a political entity that represented about 90 percent of the population—the other two estates were composed of clerics and nobles—was pushing for equal representation.

The French Revolution came out of this caldron of social ills and gave violent birth to the First French Republic (1792–1804). The beginning years of the Republic are known as the Reign of Terror. The leading figure was Maximilien Robespierre, also known as "the incorruptible" because of his self-proclaimed purity of belief, a purity that in practice produced violent intolerance. During this 10-month terror, some 16,000 to 40,000 people lost their lives, and the guillotine became a symbol of the revolution. In 1793, both Louis XVI and Marie Antoinette (previous king and queen) were beheaded. Other European nations declared war on the new Republic in an attempt to end the revolution; this War of the First Coalition lasted from 1792 to 1797. One of the Coalition nations was Great Britain. Britain and the United States carried on maritime trade, which prompted France to attack and seize American ships, which in turn led to what's called the Franco–American war (1798–1800).

The First French Republic ended when Napoléon Bonaparte declared himself emperor; the First French Empire lasted from 1804 to 1815. During that time, Napoléon carried on an extensive military campaign, with as many as nine different "wars." Napoléon's campaign ended with his defeat at the Battle of Waterloo. Napoléon's defeat, along with the breakup of the Holy Roman Empire, left Europe in disarray. The great European powers of the time—Austria, Russia, France, the United Kingdom, and Prussia—met in Vienna to create order out of chaos. This first collective attempt at nation building virtually recast Europe and set the stage for Great Britain to become the dominant world power of the nineteenth century. In France, the monarchy was reinstated and King Louis XVIII reigned until the July Revolution of 1830, when King Louis-Philippe took control. The February Revolution of 1848 forced Louis-Philippe out and Louis-Napoléon Bonaparte (Napoléon's nephew) became president of the Second French Republic. In 1851, Louis-Napoléon staged a coup d'état and became Napoléon III, emperor of the Second French Empire (1852–1870).

Comte's time was thus riddled with political upheaval and unrest, and he responded at an early age. When he was 14, Comte broke away from his family's Catholicism and support of the monarchy, declaring himself a republican and an unbeliever. Just 8 years later, Comte (1822/1975a) characterized his society as a "system in its decline. . . . [S]ociety is hurried towards a profound moral and political anarchy" (p. 9). Yet Comte (1848/1957) saw this social disorganization as a necessary step to a greater good; the revolution was "absolutely necessary to rouse and sustain our mental efforts in the search for a new system. . . . The shock was especially necessary for the foundation of [sociology]" (pp. 65–66).

Comte found inspiration for his new science from many sources, but two men stand out specifically: Nicolas de Condorcet (1743–1794) and Henri de Saint-Simon (1760–1825). Condorcet was significant in defining the modern idea of progress. When we think of progress today, most of us see it as technical development. Condorcet saw the more basic issue out of which technology springs: the indefinite perfectibility of the human mind. While you and I may take the ideas that we can learn and grow for granted, it was revolutionary in its beginnings. In Western Europe (and in much of the world as well) people weren't generally seen as capable of independent thought and responsibility. One way to understand this is to think of traditional Catholic doctrine, which at that time held sway over much of the Western world. In Catholicism, the Church mediates salvation and spiritual life. Salvation comes not through individual decision but by keeping the sacraments, such as baptism, confession, and the Eucharist. Biblical interpretation was also seen as above the heads of most believers—Holy Writ needed to be interpreted by the priests. The first English translation of the Bible came from William Tyndale and was published in Great Britain in 1526; very few of the original 6,000 copies remain because the Church confiscated and destroyed them. Tyndale was put to death in 1536 for heresy.

What I want you to see here isn't something about Catholicism; rather, I want you to see the general conception of the person prior to the Renaissance and Enlightenment. Seeing the mind as perfectible, believing that regular people could improve their lot in life by growing intellectually, is a relatively new idea. Condorcet was one of the first to declare this concept clearly in its modern form. Condorcet saw this sort of progress linked to education and became one of the architects of the new French education system. He created the base that ultimately led to the secularization of education (previously, education had been the domain of the Catholic Church), mandatory primary education for both boys and girls (Condorcet supported full citizenship for women, arguing that gender differences are mostly based on differences in education, not biology), and the standardization of the French language (to combat class differences). Condorcet also proposed the merging of the moral and physical sciences in studying human behavior. In Comte's hands, this amalgamation became the discipline of sociology.

For a time, Comte was secretary and coauthor with Henri de Saint-Simon. Saint-Simon was one of the principal founders of Christian socialism. This was an intellectual and activist movement that combined the ethics of Christianity with the social justice concerns of socialism, especially the life chances of the poor. At the center of Christian socialism was science and industry. For Saint-Simon, the hope of equality and the end of human misery lay in the theoretical advances of science, practically applied to the production of industry. Saint-Simon believed that the purpose of Christian religion wasn't necessarily eternal salvation, but rather the extension of hope and the relief of suffering to all humankind in this world. Scientists, then, in Saint-Simon's eyes, are the priests of a new social order. From this perspective, science, technology, and capitalism are to be used ethically for the betterment of the human condition. These ideas profoundly influenced Comte, who between 1830 and 1877 published extensively on positivism, the philosophic base of scientific inquiry generally, and sociology in particular.

The Evolution of Knowledge

This new social science came into existence in due course, at a specific point in time, to furnish the theoretical knowledge upon which modern social practices could be based. Modern politics, then, "decides upon the distribution of authority and the combination of administrative institutions best adapted to the spirit of the system already determined by the theoretical labors" (Comte, 1822/1975a, p. 19). These theoretical labors are based upon positivism, a specific way of creating knowledge. The reason I said that this approach came in "due time" is because Comte saw positivism, or the scientific method, as a result of naturally evolving processes at work within both the mind and society.

For Comte, increasing human control over the natural environment marks the historical process of civilization in general. In the beginning stages of human society, we had little control over the sources of basic essentials. Hunter-gatherer groups had to continually move from one place to another so as not to deplete the food sources. As humanity figured out how to plant seeds and then to irrigate fields, we became less dependent upon nature in its pure form—we changed nature to fit our needs. Civilization marches forward, then, as we increase this span of control. This control becomes greater as our mental powers increase. According to Comte, both mind and knowledge have moved through three distinct stages—each phase defined by the relative weights of imagination and observation. Comte saw these stages as necessary, and each as building on the previous.

The first stage is *theological* and "is the necessary point of departure of the human understanding" (Comte, 1830–1842/1975b, pp. 71–72). In this stage, people sought absolute knowledge, the essential nature and ultimate cause of everything. These emphases meant that people wanted to know *why* things happen—their focus was behind the scenes, not on the thing itself. Let's say blight wiped out a farmer's crop during this stage of knowledge. His first and dominant reaction was, "Why? Why me?" This sort of question makes the farmer more concerned with the unseen forces behind the blight rather than how the blight actually works. Thus, in this phase, imagination ruled over observation: "the facts observed are explained . . . by means of invented facts" (p. 29). And these "invented facts" were seen as existing before (*a priori*) the observed ones—people made empirical observations fit in with already established beliefs. In addition to the relative weights of imagination and observation, societies, and thus the purpose or outcome of knowledge, vary on a continuum from military to industry, conquest to production. In the theological phase of the mind, "Society makes conquest its one permanent aim" (p. 52). Because knowledge is certain in the theological phase, it is certain that everyone must believe the same.

The second stage of knowledge is *metaphysical* and is a transitory phase. Comte argued that the mind couldn't make the leap straightaway from theology to the positive stage. The metaphysical stage had "a mongrel nature, connecting facts by ideas that are no longer entirely supernatural and have not yet become completely natural" (Comte, 1822/1975a, p. 29). The third phase is *positive* and is ultimately where all knowledge systems will end up in Comte's scheme. Here, observation trumps imagination; truth is judged by empirical observation, and it progresses through skepticism: the empirical assessment of all theories. Scientific theories explain *how some empirical phenomenon works.* Notice the shift in emphasis: from why something happened to how something works. Theory, then, takes the place of spiritual accounts, and as such it is the core of the positive method. Notice also the shift from faith in the unseen to skepticism and empirical examination. The basis of these different ways of knowing are poles apart: For theology, it's imagination and faith; for positivism, it's observation and doubt. Comte likened the change from one to the other to the process of growing up. Children want to know why and want to be comforted; adults want to know how something works so they can take responsibility.

Comte's law of three stages (theological → metaphysical → positive) also holds for the separate disciplines as well. Each branch of science grew to be positivistic according to how complicated the subject and how independent it was from other areas. So, for example, the first discipline to become positivistic was astronomy because, according to Comte (1830–1842/1975b), it is "made up of facts that are general, simple, and independent of other sciences" (p. 76). Then came physics, chemistry, and physiology. The last branch of study to become positivistic was the study of society, which Comte called social physics or sociology. Sociology came last because of the complexity of its subject (human society) and its dependence upon other fields of study, but Comte saw something more here as well. Think of it like an arch made out of stones. Stone arches are built by stacking one block upon another with curved stones at the top. A keystone is placed at the very top, and the tension of the curved stones against the keystone is what holds the entire arch together. The historical progress of science is like that, with each idea building upon another and each field of study based upon the achievements of the previous until the most complicated phenomena could be studied scientifically. Comte claimed that this science, sociology, would also be able to embrace and hold together the rest of the sciences, just like a keystone. Because it developed last and could encompass the other sciences, Comte dubbed sociology the "queen of the sciences."

Let's pause here for a moment and consider what Comte is telling us—this discipline, sociology, is actually pretty amazing. In order to help you see what I mean, let's look at two other disciplines, economics and biology. Economics is the study of the production, distribution, and use of goods and services; biology is the study of living organisms. It's unlikely that economists study biology or that biologists study economics. But sociologists study them both, as well as every other discipline.

The reason we study economics is pretty obvious. The economy is a social structure, and we study all social structures, which is why the economy has been at the heart of our discipline since the beginning: Both Marx and Weber were intensely interested in economic issues. But there's more to it. Sociology also sees the discipline of economics as a social factor in and of itself that creates its own effects. When the discipline of economics changes, so does society. A good example is the influence of the British economist John Maynard Keynes: Keynesian economics influenced the fiscal and monetary policies of the United States from the 1930s into the 1970s and is generally seen as the principal reason for the prosperity of the 1950s. Keynes fell out of favor for a while, but a good deal of both President George W. Bush's and President Barack Obama's economic policies are guided by Keynesian economics (including the government's response to the economic crisis of 2008).

Sociology's interest in biology may not be as apparent, but we are just as interested in biology, maybe even more so. We study biological factors just as we do economic ones. Sociobiology is a growing field. A good example of this approach is Jonathan H. Turner's recent book *On the Origins of Human Emotions: A Sociological Inquiry Into the Evolution of Human Affect* (2000). Turner draws from neurology, evolutionary biology, primatology, and many other sciences to argue that humans developed unique emotions in order to create and maintain the social relations necessary for human society to evolve. According to Turner, then, sociological factors are among the most important dynamics that pushed for human evolution.

We're also interested in how biological discoveries, such as the discoveries surrounding evolution, influence society. It's easy to see that theories of human evolution have had an enormous impact on society. More recently, the discovery of DNA fingerprinting has fundamentally changed police work, the court system, and criminology. There are a number of initiatives and organizations that are now sifting through thousands of court cases looking for places where DNA evidence might tell a different story with a different outcome; the Innocence Project is a good example (see www.innocenceproject.org/).

Besides the obvious social effects that DNA fingerprinting brings, sociology also has a deeper interest in the *discovery* of DNA fingerprinting. Notice that I put discovery in italics to emphasize it. The reason is that scientific discovery is a social process. It rarely if ever fits the model of the lone scientist in the laboratory discovering the secrets of nature. Scientific discovery is a process deeply embedded in and influenced by social interactions and the social structures surrounding scientific enterprise. In a recent article, Linda Derksen (2010) documented the social factors involved in creating DNA as scientific fact. In the end, she concludes, "The institution of science is a master at cloaking, hiding, and erasing the conditions of the production of knowledge, and it does not give up its social origins easily" (p. 234).

The point here isn't that DNA fingerprinting might not be real; sociology actually can't say anything about its reality one way or another (we're not biologists, after all). However, in order to do their work, biologists must work socially and connect to various elements in society. The social basis of biological work is something we can talk about. The point is that because humans are intrinsically social, everything we do is of interest to sociology. It truly is the only discipline that can embrace, understand, and give meaning to all others. Sociology is stunning in its scope and brilliant in its potential. Of course, I'm biased, but not without reason.

Theory

The scientific study of the natural and social worlds is based upon theory. *Theory is a logical explanation of how a given empirical phenomenon works.* Theories are always tentative because they don't claim to explain first or ultimate causes. First and final causes—the why of the universe—are theological and metaphysical knowledge that imply final answers, and final answers imply faith precisely because they can't be observed. A positive approach, on the other hand, doesn't provide final answers and it doesn't demand faith. In fact, "the study of the laws of phenomena must be relative, since it supposes a continuous progress of speculation subject to the gradual improvement of observation, without the precise reality being ever fully disclosed" (Comte, 1830–1842/1975b, pp. 220–221). We are also skeptical about our theories, and there is thus a continual back-and-forth movement (or "testing") between theory and the empirical world. This back-and-forth movement between theory and empirical observation will, according to Comte, eventually lead to a small number of laws that will account for most of the observed human actions. The discovery of these laws allows us to explain, predict, and in some ways control the natural and social worlds.

A law of nature "is a rule that is based upon an observed regularity and provides predictions that go beyond the immediate situations upon which it is based" (Hawking & Mlodinow, 2010, p. 27). The most famous example of such a law, and one that Comte noted, is Newton's law of gravity. It's pretty clear that Newton wasn't the first to notice that apples always fall to the ground; it's an empirical regularity that's been around for quite some time. Newton went past that observation and explained *how it works.* His explanation had to do with the masses of and distance between two objects. Because his explanation got to the underlying dynamics, we are able to predict with great regularity how gravity will work. Newton's law also has the quality of being universal: It reaches beyond the immediate situation (objects falling on earth) to explain how bodies in space interact. Comte (1875–1877/1975d) fully expected sociology to come up with these kinds of laws for society: "Knowledge of the principal static and dynamics laws of social existence is evidently sufficient for the purpose . . . of rendering our condition far more perfect" (p. 331).

Advantages and Goals of Positivism

Positive knowledge has at least three advantages over previous ways of knowing. First, "the positive system *discovers,* whereas other systems *invent*" (Comte, 1822/1975a, p. 47). Second, because positivism is based on empirical observation, it alone provides a basis for agreement. An old adage says, "There are three things one should never discuss in polite company: religion, politics, and sex." Part of the reason, especially for religion and politics, is that they tend to be explosive subjects. People will argue at length about religion and politics and never resolve anything because it seems like there's no basis for agreement. The only agreement is, "You believe your way and I'll believe mine."

Comte, however, utterly opposes this sentiment. For his entire life, Comte saw political disagreement so fervent that it led to thousands of deaths. His desire was to find a way out. Comte (1822/1975a) formulated sociology for the express purpose of guiding politics and social reorganization: "*scientific men ought in our day to elevate politics to the rank of a science of observation"* (p. 29). Because the workings of society are observable and thus available for all to see, they can provide the basis for agreement. The theories derived from empirical observation also provide the third advantage: relief from the seeming arbitrariness of human existence and misery. With sociological theories to guide politics, "government by measures replaces government by men" (p. 49).

For Comte, the guiding of politics and nations is too important to leave it to ideological beliefs, especially when they are used in the name of God. People were dying in Comte's time—as they are in ours—because politics based on belief will eventually be enforced through military coercion. There is no reasoning with ideology or belief because neither is based in reason—there is, then, no objective basis of agreement. However, if the workings of society and the progress of civilization are empirically available, as Comte believed they are, then a foundation for reason exists. This foundation provides a hope similar to the one we have in medical science.

Let's take an obvious example. Polio is a horrific, painful virus that can be deadly or at the very least paralyzing, and invariably leaves its victims deformed; further, the great majority of its victims are children. A global outbreak of polio began around 1840. By 1916, New York City reported over 9,000 cases and 2,343 deaths. Homes were quarantined and businesses shut down. The worst year of the epidemic in the United States was 1952, with almost 60,000 new cases reported and 3,145 deaths. Some cities prevented children under 16 from traveling without a health certificate. Fear gripped the nation and the world, as epitomized in this man's account: "The fear of polio was a fear of something you had no defense against, something that hit without logic or reason. Yesterday, it was the man down the block. Today it could be you or your children" ("Whatever Happened to Polio?" n.d.).

Thankfully, the *workings* of polio are open to reason. A vaccine became publicly available in 1955. Because scientists were able to figure out the empirical factors of the virus, the last reported case of naturally occurring polio in the United States was in 1979. Comte held that society is subject to the same sorts of empirical factors and theoretical laws. The business of sociology, then, is to discover those factors and laws that can be used to guide society in its quest for improving the human condition. If Comte is right, then the theoretical explanations of how society works will provide the basis for political agreement and a bulwark against the arbitrariness of social life.

Enduring
Issues

Hope in the ability of humankind to guide its own destiny is the heart of the Enlightenment. This point of view is rooted in the ideas of a predictable universe and the free reasoning of the human mind. Together, these forged the ideas and modern values of progress. This cluster of ideas, however, was soon questioned, initially by the Counter-Enlightenment movement of the nineteenth century and more recently by such schools of thought as postmodernism and poststructuralism.

Seeing Society

If you take another look at Comte's list of sciences—astronomy, physics, chemistry, physiology, social physics—you might notice something missing: psychology. This omission isn't because psychology wasn't around; Comte claims, and most histories of psychology confirm, that psychology had been around for some 2,000 years. Rather, Comte argued, psychology is excluded because of the logic intrinsic to scientific observation. We've seen that scientific investigation is empirical and based upon observation. According to Merriam-Webster (2002), *observation* "is an act or the faculty of observing or taking notice: an act of seeing or fixing the mind upon something" (n.p.). It is thus the mind that observes, and observation takes place when the mind is directed toward something.

The mind, however, isn't always observing; it isn't always directed. You know this from personal experience. Sometimes we act on automatic pilot, never giving a thought to what we're doing, like driving home and thinking about dinner plans, and sometimes we are so caught up in something that we experience a kind of "flow" with the activity, like at a basketball game or music show. To observe something is to be mindful of it, to intend your mind toward it—like the police officer observing you speeding while your mind is elsewhere, and like the sports reporter who watches the game in order to give an account of it. This quality of mindfulness implies objectivity, a separation from the thing being observed. Webster's (2002) defines *objective* as relating to an object, and an object as "something that is put or may be regarded as put in the way of some of the senses . . . by which the mind or any of its activities is directed" (n.p.). In Comte's (1830–1842/1975b) view, these features of objective, empirical observation are what make scientific psychology impossible: "the mind may observe all phenomena but its own" (p. 80). It's kind of like trying to make a flashlight shine its light on itself.

My purpose here isn't to discount psychology. I'm hoping, rather, to give you a clear sense of what empirical observation implies about society. Not only did scientific inquiry have to evolve to a certain point for sociology to emerge, but *a particular kind of society had to evolve* as well. Almost everybody takes the existence of society for granted. Yet society as we know and understand it today is a relatively new thing. The word *society* came into the English language from French and has a Latin base. The Latin root for society means companion or fellowship, and up until the middle of the eighteenth century, it kept this basic meaning (Williams, 1983, pp. 291–292). Society thus initially referred to a group of friends or associates, like a legal or religious society. Toward the end of the seventeenth and through the middle of the eighteenth century, the idea of society began to be used in more abstract ways to refer to something larger.

Society not only became seen as something bigger than face-to-face social interactions, but it also came to be understood as a separate entity, an entity unto itself with its own nature and laws of existence separate from the people who make it up. At the same time, this larger, more abstract entity began to be perceived as something that could exert independent influence over people. As Émile Durkheim (1895/1938) would later write, society is "external to the individual, and endowed with a power of coercion, by reason of which they control him" (p. 3). Further, society had to be seen as an entity that could be steered or guided. Though I didn't phrase it this way, *the goals of science are to explain, predict, and control*. That's how science defeated polio: Science was able to explain how it works, predict what would happen following a vaccine, and thus control it. In society, the guiding mechanism is the state: "To the extent there is something called 'society,' then this should be seen as a sovereign social entity with a nation-state at its centre that organizes the rights and duties of each citizen" (Urry, 2006, p. 168). This combination of independent existence, law-like factors and processes, and the subsequent ability to control is what gave Comte and sociology its foothold as a science.

Society, then, is an objective, dynamic system that moves along a continuum of equilibrium (social statics) and change (social dynamics). More than anything else, what we need to look at in observing society are its systematic qualities—this is "the only right way" to see society (Comte, 1830–1842/1975b, p. 228). Sociology always sees things in their context. For example, many people today are upset over the loss of "family values," and they want to bring family back to the way they believe it should be. The problem with this observation from a sociological point of view is that the institution of the family doesn't exist in a vacuum. If family has changed, it's because other social institutions have also changed, like the economy and polity (the form or system of government). Once you see this sociologically, you will realize we can't simply go back to family values without changing the entirety of society.

Sociological Methods

In order to understand this new objective entity (society), Comte proposed specific methods for creating sociological knowledge. Fundamentally, our observations as sociologists are always comparative, and, according to Comte, there are three places where we can find comparisons: experiments, different cultures, and history. There are two types of experiments. The first is the kind we usually think of, with a scientist in a white coat in a laboratory. Comte calls these direct experiments and says that they are rarely appropriate for sociology because our subject matter is too complex. Indirect or natural experiments are suitable for sociology; they occur whenever the ordinary course of life is disrupted. These disruptions often expose underlying structures and processes, social facts that are sometimes difficult to see in normal day-to-day life. A good example is Hurricane Katrina and its impact on New Orleans. As a result of the disaster, the underlying dynamics of race and the problematic relationships among federal, state, and local governmental agencies were revealed and are continuing to be studied by sociologists.

The second place we can make comparisons is across cultures and societies that currently exist. We may notice, for instance, that one culture has a high rate of eating disorders among women, whereas another culture doesn't. By comparing these cultures, we can discover the underlying social factors that create these different phenomena. Durkheim (Chapter 4) used this methodology in his famous study of suicide, demonstrating that diverse individual–group relationships create differing suicide rates. Comparisons can also be made historically, comparing states of a single society across time or the evolving path of a social structure. Max Weber (Chapter 5) used this method to trace the historical development of capitalism from traditional to rational capitalism. Weber found that modern capitalism is based on cultural conditions created by the Protestant Reformation.

I want you to notice something about this history. We usually think of history being made up of significant moments and people. So a class on the history of the United States might have a section on the Civil War and Abraham Lincoln. However, this isn't the kind of history in which sociology is interested. A history that focuses on the Civil War and President Lincoln is simply looking at the *effects* of a more important history: the history of social structures and processes. A more powerful history—the one sociology is interested in—is concerned with the cultural and socioeconomic *processes* that *created* the Civil War, and how that event in turn influenced the social system.

Comte also foresaw a potential methodological problem. If you've had a sociological methods course, you might notice something missing from Comte's description of methods. Contemporary textbooks usually divide the field into qualitative and quantitative methods. The primary difference between the two is that quantitative methods use statistical analysis and qualitative generally do not, though there are statistical packages that will quantify qualitative data. While methods books acknowledge both approaches, the majority of texts spend the bulk of their time explaining research approaches that can be analyzed statistically. This is also the case with the main journals in the field: The majority of the articles use statistical analysis. Yet this sort of data analysis is what's missing from Comte's methods—and this wasn't an oversight.

Though trained in mathematics, Comte (1830–1842/1975b) didn't think it was appropriate for sociology. In fact, he thought it was dangerous: "mathematical analysis itself may betray us into substituting signs for ideas, and . . . it conceals inanity of conception under an imposing verbiage" (p. 250). Statistical analysis looks like it gives us hard data similar to the laboratory sciences, but Comte claimed that such analysis is empty—inane—and has the effect of shifting our focus away from theoretical ideas. In turn, this shift threatens to plunge sociology back "into the metaphysical domain by transferring to abstractions what exclusively belongs to observation" (Comte, 1822/1975a, p. 59).

The shift away from Comte's positivism began in the early decades of the twentieth century, as a result of the Vienna Circle, a group of philosophers, scientists, and mathematicians who gathered to talk about scientific methods. A number of things came out of these meetings, the most notable for us being quantification and unitary science. The Circle argued that science is defined through methodology and that therefore no differences exist among the sciences, be they natural or social, biological or physical. Comte would have so far agreed. It's the emphasis on quantification, however, that changed things. This emphasis made its way into sociology principally through the work of two men: Franklin Giddings and George A. Lundberg. Giddings's (1922) book *Studies in the Theory of Human Society* made the argument that sociology is "a science statistical in method" (p. 252). Lundberg's influence, however, was more pervasive than that of Giddings.

George A. Lundberg was the 33rd president of the American Sociological Society (later renamed the American Sociological Association). His most influential work was *Can Science Save Us?* The book was first published in 1947, just a few years after the close of World War II. It was a timely book. Lundberg argued that the scientific method could be applied to social issues and problems, things people were concerned about, having lived through the horrors of Nazism and seen the effects of nuclear weapons. Lundberg's (1961) recommended methods included "sampling, correlation, probability computation, and logical inference" (p. 19). During the latter part of his career, Lundberg served as chair of the Department of Sociology at the University of Washington. Under Lundberg's influence, the department claims it became one of the first in the United States to develop scientific sociology. "As a consequence, the graduate curriculum placed particular emphasis on general theory and quantitative methodology" (University of Washington, Department of Sociology, n.d.). Notice the link between scientific sociology and quantitative methods.

My desire isn't to weed out statistics from sociology, but I want you to see that the use of statistics as scientific methodology in sociology is historically specific and wasn't part of Comte's positivism. Comte's hesitation about quantification also has merit. An emphasis on numbers can move us away from what is fundamentally important—the enlightenment of the mind to see society and to use sociological ideas and theories to guide it. C. W. Mills is a well-known American sociologist from the mid-twentieth century, and he shared Comte's concern. In talking about students who had been trained in quantitative methods, Mills (1959) said, "I have never seen any passionate curiosity about a great problem, the sort of curiosity that compels the mind to travel anywhere and by any means, to re-make itself if necessary, in order *to find out*" (p. 105).

This enlightenment of the mind was extremely important to Comte. We've already seen that Comte wanted to use sociology to guide the rebuilding of society. He saw the regeneration of education as key, for "a mind suitably trained becomes able by exercise to convert almost all impressions from the events of life into sociological indications" (Comte, 1830–1842/1975b, p. 243). This is what Mills (1959) later referred to as "the sociological imagination" (p. 5). Like Mills, Comte saw that this state of mind would involve the totality of one's being, including emotion. In fact, Comte argued that positivism created a new feeling in people, different from those generated by fiction, religion, or human life in general. This "new form of social sentiment . . . is deeper, because in some sort personal; and more reflective, because it results from scientific conviction" (p. 249). That new feeling is the hope that empirically based knowledge can bring.

Practicing Theory—A Summary

We've come a long way in this chapter, and I appreciate your sticking with me. There are few things about theory that I'd like to pull out of the chapter and state more concisely for you. Hopefully, you'll be able to use this section as you go through the rest of the book. It'll help you to read and understand what the theorists say, and, it will help you do well on your tests and papers because the concepts will be more succinct in your head.

First, theoretical thinking is a way of seeing and being aware of the world. In his classic text, *Invitation to Sociology,* Peter Berger (1963) has an entire chapter dedicated to "Sociology as a Form of Consciousness." Practicing sociology, he says, "presupposes a certain awareness that human events have different levels of meaning, some of which are hidden from the consciousness of everyday life" (p. 29). This way of seeing is undoubtedly part of the reason you're studying sociology. Sociologists are fascinated with the "what else" of human action and interaction. We're all a bit like Toto in *The Wizard of Oz*—we want to know what's going on behind the curtain, what's really in back of the wizard. Yet the theoretical mind has an additional bent. Practicing theory means not only being curious about what something *is,* but it's also being curious about *the way something works.*

Definition

Theory is a logical explanation of how some empirical phenomenon works in general terms. Theory is built up from concepts, definitions, and relationships. The goals of scientific theory are to explain, predict, and control empirical phenomena.

The italicized part of that last sentence is incredibly important, yet it's also something that's difficult for many people to grasp. Borrowing from John W. Carroll, physicists Stephen Hawking and Leonard Mlodinow (2010, p. 28) give us an example using gold and uranium-235 that might help to clarify this concept. It is empirically true that all spheres of gold and uranium-235 are less than a mile in diameter—no human has ever seen one that big. In terms of simple observation, we could make a general statement and say that we will never see a ball of gold or uranium-235 larger than a mile in diameter—it's a fact. However, that's not quite true, at least not in terms of *scientific* facts. You see, there's nothing about the intrinsic properties of gold that make a mile-wide ball of gold impossible; we've just never seen one. However, a sphere of uranium-235 a mile in diameter is theoretically impossible because a ball of uranium-235 wider than 6 inches will create a nuclear explosion. That knowledge—that fact—is based on a theoretical explanation of how the substance works. Because physicists understand how uranium-235 works, they can *make predictions about its behavior.* No such prediction can be made about the sphere of gold.

There are a few things implicit in this idea of how things work that we need to make explicit. That is, the kind of theory that explains how things work is built out of abstract or general concepts, technical definitions, and causal-like relationships. *Concepts* or ideas can be more or less theoretical. So, what makes one concept theoretical and another not? There are at least three qualities that distinguish them: insight, explanatory power, and definition. Theoretical concepts inspire us to see the world with new eyes. I'm sure part of your education in sociology has involved such insights. Many people discover the concepts of race, class, and gender after taking a sociology class. Even more insightful are ideas like collective consciousness, rationalization, anomie, alienation, and commodity fetish. After you understand commodity fetish, you'll never see the shopping frenzy of Black Friday in quite the same way.

Concepts also determine how much we see and can explain, and those issues are determined by the concept's level of abstraction. Theorists usually call this facet of a concept its *explanatory power*. Generally speaking, the more abstract a concept is, the greater will be its explanatory power. For example, here is a quote from Marx that is tied strongly to a specific situation and is thus not very abstract: "Modern industry has established the world market, for which the discovery of America paved the way" (Marx & Engels, 1848/1978, p. 475). Marx explains a lot in that statement, but it's bound to a specific time period. However, if we can capture a more general social process in "the discovery of America," then we might have a statement that isn't so historically limited and would qualify, at least in a limited way, as a law-like principle. One way of making the concept expressed in Marx's quote more abstract is to change "discovery of America" to "geographic expansion." We can then make a more powerful theoretical statement: The greater the level of geographic expansion, the greater will be the level of market development. That proposition may sound dry, but it's extremely powerful.

The theoretical value of a concept is also determined by the kind of *definition* we give it. We use abstract concepts all the time, and if pressed, most of us could give some sort of definition for the concepts we use. We can think of these as dictionary definitions: They describe what is generally meant by a concept or idea by most people. In this way, dictionary definitions are not unique; they tell us how anyone would use a concept in a sentence. Theoretical definitions have to go beyond dictionary definitions in at least two ways. First, theoretical definitions are stipulative, which means they are specifically unique. Let's use religion as an illustration. Merriam-Webster (2002) gives us the following dictionary definition of *religion*: "the personal commitment to and serving of God . . . one of the systems of faith and worship" (n.p.). And that's how most of us think of it and would use it in a sentence, but it has little if any theoretical value.

However, Karl Marx (1932/1978d) defines religion as "the sigh of the oppressed creature, the sentiment of a heartless world, and the soul of soulless conditions. It is the *opium* of the people" (p. 54). And Émile Durkheim (1912/1995) defines religion as "a unified system of beliefs and practices relative to sacred things, that is to say, things set apart and forbidden—beliefs and practices which unite into one single moral community called a Church, all those who adhere to them" (p. 44). The first thing you probably noticed is that these two definitions are different; they are different from the dictionary definition and they are different from one another. That's precisely what we mean when we say that theoretical definitions are stipulative: *they stipulate the necessary characteristics to be such a thing in a specific case.* Another way to put this is that theoretical definitions are contextual; they are always understood within the broader context in which they appear. Marx's specific take on religion fits within the broader context of his theory of species-being and false consciousness, and Durkheim's within his theory of social solidarity and collective consciousness.

One way theory is built is through *synthesis*, taking elements from different theories in order to create a more robust explanation. Theory synthesis begins through comparison and contrast. For example, Marx and Durkheim's definitions seem worlds apart. However, they connect on the issue of consciousness, which would be a good place to begin thinking about how Marx's and Durkheim's theories of religion could be brought together.

Hint

The second way theoretical definitions go beyond dictionary ones is that they explain how a concept works. Theories are active things, concerned with processes and factors that make other things happen. Look carefully at the definitions that Marx and Durkheim give us, and you'll see what I'm talking about. They both have this kind of statement: "Religion is _____, and religion does _____." For Marx, religion is the "sigh of the oppressed," but it works like a narcotic: It not only dulls the pain of oppression, but it also stupefies the user and renders him or her unable to change society. For Durkheim, religion is a set of practices and beliefs that work to create a collective awareness and social solidarity.

The third issue we need to make explicit about theoretical explanations is that they propose causal-like *relationships* among and between concepts. The last quality of theoretical definitions actually begins to take us down this road. Explicating relationships just takes us one step further: This is what it is, this is what it does, and this is how it does it. While it's just one step, it's the most important one because it holds the heart of theory—it's also the biggest and most difficult step.

Let's use Durkheim's theory of religion for our example here. Making Durkheim's statement a bit more abstract might help us see what we need to do: **Practices + beliefs → collective consciousness + social solidarity = religion.** The part we need to explain is found in the arrow. What do practices and beliefs do that create collective consciousness and social solidarity? How does this relationship work? The really short answer is this: *Interactions characterized by high levels of copresence, a common focus of attention, and common emotional mood create high levels of emotional energy that is in turn invested in a set of symbols particular to the group. Because these emotionally infused symbols feel sacred and are held in common by everyone in the collective, they form a group-specific, shared by all, awareness of the world and produce a strong sense of unity, which together form a group's religion.* Right now, understanding the meaning of what I just said isn't as important as seeing what I did. I simply expanded and connected each part of the statement.

I'd like you to do me a favor right now. Please take the italicized statement apart using the statement in bold. I want you to see which part of the bold statement fits with what part of the italicized. The obvious reason I'm asking you to do this is that I want you to see how to build theoretical statements. That's why understanding what I said isn't important right now—we'll wait for our chapter on Durkheim for that. The important thing here is that you understand *the form of theoretical statements.* Theory isn't easy, but it is straightforward. Plus, if you understand and keep in mind the form theory takes, it'll be easier to understand the theory itself. You won't feel as overwhelmed by all the information.

There's also another reason I would like for you to think through the two statements. Theory isn't like most other subjects. Understanding theory requires you to think through the material. With most other subjects, just having a sense that you understand the material is enough—but that usually isn't enough with theory. The reason for this is that theory is found in the relationships. They are the core of every theory; they contain the theory's logic. It is possible to understand every word in the book and still miss the theory. To understand theory, your mind has to travel through that arrow in the bold statement (an arrow that is always there, whether I write it out or not). Theory isn't simply knowing the material; theory is a specific way of thinking—and to learn it, you have to practice it, by engaging your mind in the process of thinking through the relationships.

BUILDING YOUR THEORY TOOLBOX

At the conclusion of every chapter, there will be a box like this with review questions, suggestions for further study, ways of applying the theory to your world, and theory-building exercises. I'm keeping this one fairly brief because I'd like you to concentrate on specific issues that will carry through the rest of the book. As I said, I've told you a specific story about modernity—it's all true, but it's not the only story that can be told. This story includes certain ideas that we will come back to many times in our time together. Answering the following questions should help keep these important ideas clear in your mind.

- *Modernity:* Explain the Enlightenment and its place in modernity. How did society change as a result of modernity? What are the specific institutional arrangements in modernity? Explain the modern view of the person.

- *Comte:* Explain Comte's sociological imagination. In other words, what specific historical, social, and intellectual forces impacted Comte's thinking? How did these social factors influence his understanding of sociology and theory?

- *Sociology:* Explain how sociology is the "queen of the sciences." What purpose did Comte see for sociology? How do you see your involvement with sociology differently now that you've read the chapter? What are the advantages and goals of positivism, especially as it relates to sociological theory? What are the appropriate methods for sociology? Why does Comte think that statistics are inappropriate for sociology?

- *Theory:* Write a summation of theory that you can use throughout the book and your class. When your professor gives you a test or paper topic, he or she is expecting to get theoretical ideas in return. Writing theory is different from any other sort of writing, and to do well on your tests and papers, you'll need to know how to write theoretically. That starts by understanding what theory is, what it does, and how it's constructed. In your summation, be sure to include the purpose of theory as well as its building blocks.

CHAPTER 2

Seeing Society for the First Time

Herbert Spencer

t's amazing to me that there was a time when people didn't think much about society. As I pointed out in Chapter 1, even the word *society* didn't mean the same thing it does now before the middle of the eighteenth century. Society simply meant a group of people, like a society of friends or masons; it certainly didn't mean a large-scale system of interrelated parts. What's even more amazing is that many people today still don't have a clear idea of society. For example, when Margaret Thatcher (1987) was British Prime Minister of Great Britain she said, "There is no such thing [as society]. There are individual men and women and there are families" (n.p.). I see this difficulty each time I teach Introduction to Sociology, when I note the struggle that many students have in coming to grips with this idea.

One of the reasons I have chosen Herbert Spencer as our first theorist is that he was really the first to see and explain society in a sociological way. He was the first to clearly see society as a large-scale system of interrelated institutions and structures. He saw society as something *other*, something more than simple associations of people. In my mind, his explanation is where every student of society should begin—he tells us what modern society is, how it works, and how it changes. Knowing these things gives us a firm base to develop our own thinking about society.

More than that, Spencer's understanding of society gives us a clear point of contrast to understand other kinds of theory. In many ways, Spencer is the founder of *structural-functionalism*, one of the three main sociological perspectives I'm sure you learned about in your introduction to sociology class. Knowing Spencer's sociology gives us a foundation from which we can understand the many other perspectives, through comparison. Spencer also gives us a foundation for understanding such current issues as globalization, because by definition, globalization is the process of transcending the boundaries of modern society, which is precisely what Spencer helps us see. His ideas about how modern societies are structured also put us in a position to talk about the possibility that contemporary societies may have postmodern elements. We'll begin our conversation about that possibility at the end of this chapter.

THEORIST'S DIGEST

Central Sociological Questions and Issues

Spencer is primarily concerned with how societies change and function. He wants to understand what the basic parts of society do, how they relate to one another, and what force pushes societies to change. Remember, for most of human history, social groups were very slow to change. Modernity, however, is synonymous with progress and change. Thus, Spencer is centrally concerned with explaining modern societies and change in the most general terms.

Simply Stated

Spencer argues that all systems—whether organic or social—need to draw resources from their environment (operative function), distribute those resources through the system (distributive function), and be well-organized (regulatory function). As the population of a society grows, these three system needs are divided up among different structures. Society thus becomes come complex and better able to adapt and survive. However, the danger is that society will become so complex, it will be unable to function well because its organization isn't strong enough. The social system thus pushes for increased regulation, but there's a danger of too much regulation, which can stagnate society. In the long run, then, complex societies tend to cycle through periods of growth and stagnation.

Key Ideas

Jeremy Bentham, utilitarianism, laissez-faire government, Jean-Baptiste Lamarck, Charles Darwin, Thomas Robert Malthus, social evolution, organismic analogy, social structures, requisite needs, regulatory function, operative function, distributive function, differentiation, specialization, structural differentiation, compounding, problems of coordination and control, militaristic and industrial societies, domestic institutions, ceremonial institutions, political institutions, ecclesiastical institutions.

The Sociological Imagination of Herbert Spencer

I'll begin each of the chapters in this section of the book with the theorist's sociological imagination. The idea comes from C. Wright Mills (1959), and you probably heard about it in your introduction to sociology class. This imagination is the capacity to see how history, social structure, and a person's individual biography come together to shape his or her life. In our book, I'm using the idea to give you a sense of how and why these theorists came up with the ideas they did. We saw in Chapter 1 that theory itself is historically grounded: It came out of the Enlightenment and a positivistic way of understanding the world around us. In the same way, each of the specific theories we'll talk about came out of the theorist's personal experiences within a specific structural and historical context.

In this section, you'll learn how the social factors of industrialization, sustained economic growth, and the expansion of the British Empire all influenced Spencer's sociological imagination. You'll also see how Spencer's peculiar childhood and education shaped the way in which he theorized and the goals he had for his work. You'll find that Spencer was one of the most widely read and influential philosopher-scientists of his time. Today, his work continues to influence us through the general perspective he first developed, structural-functionalism. Spencer's sociological imagination is also valuable to us because he teaches us how to think in terms of the entire social system.

Spencer's Life

Spencer's personal life and the intellectual climate of his time had noticeable impacts on his sociological imagination. He was born April 27, 1820, in the English industrial city of Derby. Spencer was the oldest of nine children and the only one of his siblings to live through childhood. Though he survived, Spencer was sickly for most of his life and was thus homeschooled, first by his father and then by his uncle. Both his father and uncle were followers of Jeremy Bentham's (1748–1832) utilitarian philosophy. Bentham argued that human happiness and unhappiness are based on the two sovereign masters of human nature: pleasure and pain. The "utility" in utilitarianism refers to those things that are useful for bringing pleasure and thus happiness. Bentham developed the *felicific* or *utility calculus,* a way of calculating the amount of happiness that any specific action is likely to bring. The calculus had seven variables: intensity, duration, certainty, propinquity, fecundity, purity, and extent. Society's job was thus to create and guard structures that encourage each individual's pursuit of happiness.

In addition to this background in utilitarianism, Spencer's upbringing and physical condition set the course for his manner of study. Spencer never had formal training; after his father and uncle, Spencer's knowledge grew through associations in scholastic clubs and personal networking. He was a member of both the X Club and the Athenaeum. The X Club was founded in 1864 by Thomas Henry Huxley and eight other men. They met together monthly for scientific discourse and, more important, to influence the development of scientific knowledge. The X Club was disbanded in 1893, but in its time, the club powerfully influenced the general development of science and specifically the Royal Society, the oldest scientific society still in existence today.

The Athenaeum Club had a broader social base and many more members than the X Club. It was founded in 1824 and continues to this day in London. The purpose of the club was similar to the X Club: to bring together and support men of science and the arts (women were not admitted until 2002). James Collier (1904) tells us that Spencer "at no time received systematic instruction in any branch of science. What is more surprising, it may be doubted if he ever read a book on science from end to end" (p. 206). Where, then, did Spencer get his ideas and facts?

He *picked up* most of his facts. Spending a good part of every afternoon at the Athenaeum Club he ran through most of the periodicals . . . [and] he habitually met with all the leading savants, many of whom were his intimates. From these, by a happy mixture of suggestion and questioning, he extracted all that they knew. (pp. 208–209)

This kind of personal background and these practices, coupled with the scientific spirit of the age, made Spencer rather unique in the way he saw the world. Because he was not trained in any one discipline, he wasn't restricted by any disciplinary parameters. Spencer wrote freely in biology, philosophy, psychology, sociology, and anthropology. This broad base influenced Spencer to think that everything in the universe could be explained using one theory. While this sounds grandiose, many physicists today are working toward a theory of everything. Though it won't come out clearly in this chapter, Spencer based his thinking in the physics of his time. For example, he thought of population dynamics in the same way as force and matter in physics.

Many of his books were the standard texts of their time—such as *First Principles, The Principles of Biology, The Principles of Psychology,* and *Principles of Sociology.* And Spencer wrote extensively: Some estimate that Spencer's books sold over a million copies in his day, which, if true, would make him one of the best-selling philosophers of all time. Spencer's influence in his time matched his publication record. The American philosopher William James (1904) said,

> Spencer is thus the philosopher of vastness. Misprised by many specialists . . . he has nevertheless enlarged the imagination, and set free the speculative mind of countless doctors, engineers, and lawyers, of many physicists and chemists, and of thoughtful laymen generally. (p. 104)

While Spencer usually didn't cite other authors and thinkers in his work, we know that he was strongly influenced by such people as Jean-Baptiste Lamarck (1744–1829), Charles Darwin (1809–1882), and Thomas Robert Malthus (1766–1834). Malthus is most noted for his theory of population growth. He argued that populations tend to grow in size until checked by the four horsemen of the apocalypse: war, pestilence, famine, and disease. Spencer took Malthus's emphasis on population growth and his use of competition and struggle, but he argued that these dynamics result in social *evolution,* not tragedy. Lamarck and Darwin fuelled Spencer's interest in evolutionary theory. Interestingly, it was Spencer and not Darwin who first used the phrase "survival of the fittest." While Darwin gave credit to Spencer in the fifth edition of his *The Origin of Species,* a good part of the scientific community has forgotten Spencer's contributions. However, the important thing for us to see here isn't who should get credit, but rather, that Spencer was an important person in establishing the place evolutionary theory holds today, which, according to a recent issue of *Discover* magazine, has left "no sphere of human thought or activity . . . untouched" (Wright, 2009, p. 36).

Spencer's Social World

Herbert Spencer was born into a society already in the grips of the Industrial Revolution. The Industrial Age began in England toward the end of the eighteenth century. We can point to a number of beginning points, but among the most important was the invention of the steam engine, perfected by James Watt around 1775. This invention provided the pivotal shift in energy from wind and water power to fossil fuels (refined coal) that was needed for sustained industrial growth. Watt's work, along with improvements in the quality and production of iron, also paved the way for one of the most important developments in transportation technology since the wheel: railroads. The first intercity railroad—the Liverpool and Manchester Railway (L&MR)—was completed in 1830. The L&MR rail system was built to supply raw material and finished products from port to city. National markets, a necessary factor in the growth of capitalism, began to emerge as such railroads expanded across England and Europe.

The use of machines in the production of textiles also fueled the English Industrial Revolution. Several inventions spanning at least 70 years moved the industrialization of British textiles along, including John Kay's invention of the flying shuttle for weaving cloth, James Hargreaves' spinning jenny, and Edmund Cartwright's power loom. Prior to such inventions, textiles were by and large produced in homes with all family members participating. The mechanization of textiles irresistibly moved people from small, kin-based rural communities to large cities—a process called *urbanization*. During the nineteenth century, the population of London grew by almost 700% to become the world's largest city at that time. London was also the site where British industrial power was given center stage at the 1851 "Great Exhibition of the Works of Industry of All Nations." The massive Crystal Palace of the exhibition held some 13,000 to 14,000 exhibits and entertained over 6 million guests. The real income of people in England grew remarkably as well, about doubling from 1760 to 1860.

Britain's span of political control also grew tremendously during Spencer's lifetime. This growth began to firmly take hold as a result of the Congress of Vienna, 1814–1815. The Congress consisted of delegates from major European powers who met to forge new national alignments after the fall of the Holy Roman Empire and the French Revolutions and Napoleonic Wars. The purpose was to create a balance of power that would assure European peace. A key ingredient in this peace, which became known as *Pax Britannica* ("British peace"), was the supremacy of the British Royal Navy. At its height, the British Empire was the world's largest formal empire in history, covering approximately one quarter of the world's surface and population. The sun truly never set on the British Empire.

Spencer thus lived in an exceptional time. The positive effects of modern life were readily apparent and appeared to be on a decidedly upward path. In order to illustrate how this historical moment influenced Spencer's theorizing, let's do a bit of comparison with Auguste Comte (Chapter 1). While the lives of Spencer (1820–1903) and Comte (1798–1857) overlapped chronologically, in some important ways they lived worlds apart. Comte's world was filled with political upheaval and violence. He was born scant years after the beheading of the French monarchs and Robespierre's Reign of Terror, and in his life he saw numerous wars and regime changes: the First and Second French Republics, two Napoleonic Empires, and two restorations of the Bourbon monarchy. Spencer, on the other hand, lived in England during a time of unprecedented political and economic growth. These differences in experience gave the two men different sociological perspectives, though they would both be considered functionalists.

One central distinction concerns how each saw the relationship between the individual and government. For Comte, individuals are naturally inclined toward building society. This inclination to sociability is based in humanity's unique emotions. Our social connections have a great deal more emotion infused in them than those of any other animal. This emotional connection is very important to Comte (1830–1842/1975b) because he saw people as intellectually lazy, or as having very little motivation for intellectual pursuits: "The intellectual faculties being naturally the least energetic, their activity, if ever so little protracted beyond a certain degree, occasions in most men a fatigue that soon becomes utterly insupportable" (p. 264). Thus, social emotions keep us energized to do the work of human survival: tool building, cooperative labor, passing on culture, and so forth. Government is thus natural—"society without government is no less impossible than a government without society" (Comte, 1851–1854/1975c, p. 421)—and the extension of government is under most circumstances for the good of all, because natural cooperation becomes systematized. Comte (1822/1975a) further argued that while in most cases it may appear that an individual makes a difference, in actuality, "forces external to him act in his favor according to laws over which he has no control" (p. 43).

Spencer, on the other hand, argued that people were rational, not emotional, in their social connections, and that we are evolving toward less state infringement on individual actions. The best society is the one "in which *government* will be reduced to the smallest amount possible, and *freedom* increased to the greatest amount possible" (Spencer 1864/1968, p. 17). Comte experienced what government based on religious beliefs and individualism can create, and he wanted none of it. He assumed, then, that human nature is intrinsically social and is most strongly expressed in the human propensity for emotional connections. This nature is perhaps why people can be led astray—religious beliefs are intensely emotional. What is needed, Comte argued, is government based on scientific research, an objective sociology that could provide the basis of agreement in governing and for peace. As reasoned government expands, Comte argued, deity-centered religion would fade away.

Spencer, however, had experienced society's first blush of economic and political success. Capitalism and industrialization were successful precisely because individuals were free to pursue new ways of thinking and doing without fear of government reprisals. He envisioned, then, a kind of rugged individualism with simple rational exchanges and individual benefits forming the core function of society. "The society exists for the benefit of its members, not its members for the benefit of society" (Spencer, 1876, p. 479). Further, in Spencer's time, government had achieved a working and expanding constitutional monarchy, and, to his British eyes, *Pax Britannica* had achieved global peace. Spencer was also completely comfortable with religion as a way of understanding "the unknowable" and saw it evolving along with society in general; religion is thus seen as functional.

Spencer's life in England let him see society as gradually progressing through evolution and being functional, or exactly the way it should be. Most social scientists of the eighteenth and nineteenth centuries either assumed an evolutionary point of view or explicated one. For instance, both Marx and Durkheim, the theorists we'll discuss in Chapters 3 and 4, assumed an evolutionary-like position. Marx assumed that societies are changing through the material dialectic, though the "evolutionary" change was not seen as progressive until the very last movements he predicted, when history would move from capitalism to socialism and, finally, to communism. Durkheim argued for a moral–cultural evolution. This evolution moves from particularized to generalized culture and morality. More complex societies demand a more generalized culture and value system. For example, a social group that only has one kind of person in it can afford to have a very specific culture (in fact, it *must* have such a culture). Urban gangs are a good example. But a larger social group that embraces a number of different types of people must have a more general identity and value system. Durkheim argued that religion would facilitate this evolution, particularly in its final phase from nation-state to universal humanity—this kind of religion would be very humanistic and accepting of diversity and thus functional for modern society.

Yet both Marx and Durkheim saw problems associated with the evolution of society. Herbert Spencer, on the other hand, saw it more in terms of positive progress coming out of social evolution in general and the Industrial Revolution in particular. Spencer saw evolution as a movement from simple forms to complex forms with greater adaptability. Spencer was born and lived his life in England, which, in contrast to France (where Comte and Durkheim grew up), had experienced slow and steady growth, politically and economically. There was thus a tendency to see the world in terms of gradual and peaceful change. Spencer, like most British people at the time, saw the Empire as the pinnacle of social evolution. Capitalism and the Industrial Revolution were seen as expressions of this superiority. Spencer therefore saw laissez-faire capitalism, the division of labor, free markets, and social competition as part of the survival of the fittest.

Spencer's Sociological Imagination: Functionalism

Functionalism is a theoretical perspective that began with the work of Herbert Spencer in the nineteenth century and was systematized by Talcott Parsons in the twentieth (see Chapter 9). Functionalism is built around the organismic analogy and has at least five defining features: concern for requisite needs, structural differentiation, specialization, integration, and system equilibrium. Functionalism came under a great deal of criticism in the latter half of the twentieth century. Some of this criticism was no doubt due to functionalism's apparent inability to explain the social upheavals of the 1960s. But part was also due to the place that Talcott Parsons held in mid-century sociology. For much of the twentieth century, Parsons was "the major theoretical figure in English-speaking sociology, if not in world sociology" (Marshall, 1998, p. 480). As Victor Lidz (2000) notes, "Talcott Parsons . . . was, and remains, the pre-eminent American sociologist" (p. 388). A good portion of contemporary theories were penned in reaction to Parsons and were thought of as alternative approaches to his vision of social order.

But functionalism is still part of contemporary theory. *Neo-functionalism* reconceptualizes (see Alexander, 1985) some of these issues and is concerned with the integrity and interrelationship of social structures, and social change as generated by structural differentiation and by the tension produced by the relationships among various social structures. *Equilibrium* in neo-functionalism is seen as simply a reference point for studying an empirical system. Perhaps more important has been functionalism's continuing influence on our thinking as sociologists. The functionalists taught us to think in terms of social systems and the effects that different structures or social units can have on one another and the system as a whole.

This way of thinking is at the core of our ideas of globalization. Another specific idea that has currency in contemporary theory is what Spencer called the problem of coordination and control, but is more currently termed the problem of *steering*. This issue is at the heart of modernity, for society must be able to be controlled or guided in order to meet the goals of the social project. One theorist who looks at this issue is Anthony Giddens, who argues that late modern societies can no longer be steered or guided. Giddens (1990) characterizes late modern society as a juggernaut, "which threatens to rush out of our control and which could rend itself asunder" (p. 139).

Concepts and Theory: Social Evolution

In sociology, there are *different levels of analysis,* and there are different theories for these areas of analysis. Earlier, I said that some physicists are searching for the theory of everything, and that's true. However, not all physicists think such a theory is possible, and in practical terms, it may not be needed at all. Physicists Hawking and Mlodinow (2010) put it this way: "So though the components of everyday objects obey quantum physics, Newton's laws form an effective theory that describes very accurately how the composite structures that form our everyday world behave" (p. 67). One set of theories explains and predicts behavior at the subatomic level, and another works to explain larger phenomena. The interesting thing is that these theories don't connect to one another—even though all larger bodies are made up of subatomic particles, they obey a different set of laws.

Hint

The idea of *levels of analysis* is an important theoretical tool. The idea refers to the fact that large phenomena work according to different processes from those of small ones. At the macro level, the unit of analysis may be a social institution, relationships among social institutions, or an entire social system (including global systems). Meso-level analysis concerns formal and informal organizations; the unit of analysis at the micro level is the social encounter; and at the individual level, the focus is on the inner workings of the person.

It's like that in sociology as well. Even though large social structures are made up of small, individual particles (people), their workings are explained using different theories. In sociology, we refer to these as levels of analysis. Most sociologists talk about three levels, but I think that we actually have four. *Macro-level* analysis looks at widespread sociohistorical processes. These sorts of theories are good for explaining what goes on in society as a whole or the dynamics among groups of societies. This is the kind of sociology that Spencer represents. He gives us theories of how entire social systems work and how they have changed over millennia. Among the theorists in this book working at this level are Harriet Martineau, Karl Marx, Émile Durkheim, and Charlotte Perkins Gilman.

Meso-level analysis in sociology is cast at the level of organizations and groups. The university where I teach is an organization that is embedded in the social institution of education, which, in turn, is part of the entire social system. Within the organization, there are different social groups, such as the faculty senate. Compared to the macro level, this level is concerned with the relations among smaller social actors for shorter periods of time. The sociology of organizations is a large field now, but in terms of classical theorists, Max Weber is the strongest representative. He gave us amazing insights into how bureaucracies work and the effects they have on society as a whole and on the life of the individual. As you'll see in Chapter 5, Weber also works at the macro level.

The *micro level* is the level of "face-to-face" social encounters. The social actors at this level are people, but the focus is generally on the relationships *between* people rather than the individual. Most often, we're concerned with things like the interaction, the exchange network, or the ritual order. Sometimes these concerns move into the fourth level: analysis at the *individual level*. But it's fruitful to keep in mind that some micro-level theories, such as exchange theory, aren't concerned with what goes on inside the person at all, and thus in that sense don't qualify as social psychology. On the other hand, there are theories that concentrate on what goes on inside the person as a result of social factors, and these do qualify as social psychology. In this book, we have George Herbert Mead, Georg Simmel, and W. E. B. Du Bois who consider the inner life of the person in response to external forces. Mead and Simmel also give us theories about the micro-level social encounter.

The Social System

Spencer works primarily at the macro level and sees society as a system. In order to begin thinking about society, Spencer used the *organismic analogy*—this eventually became the cornerstone of functionalism. The organismic analogy is a way of looking at society that understands the form of society and the way society changes as if it were an organism. For example, in order for you to survive, you need oxygen, and you get your oxygen from air. Because of that need and because of that source, your body has a specific structure built inside it—your lungs. It also has other structures and systems (such as the circulatory system) that aid the lungs in fulfilling this need. Fish don't have lungs, and neither do plants. They have other structures and systems that are designed to meet their specific needs within their environment.

Definition

Analogies are quite often used in sociology as ways of understanding how society works. With an analogy, there is resemblance in some particulars between things that are otherwise unlike—and analogies are used to explain something unfamiliar using something that is well-known. The *organismic analogy* is specific to functionalist theorizing and implies that society works like a biological organism in that it has survival needs and evolves to greater complexity. The analogy also implies that society, just like complex organisms, operates like a system of interrelated parts that tend toward stasis or balance, and any derivation from that life-balance is seen as illness or pathology.

There are three specific theoretical ideas that come from looking at society as an organism: the relationship between needs and functions, systems thinking, and systemic equilibrium. Societies have specific needs, just like your body, called *requisite needs.* The term underscores the idea that these needs are required for survival. Societies develop certain structures or institutions that meet those needs. Just as your lungs are built differently from your stomach because they fulfill a different function, so, too, different social institutions are built differently from one another because they function to fulfill diverse social needs. Another similarity with your body is that institutions are taken for granted. That is, they work without our usually being aware of them. In fact, the more we are aware of social institutions, the less power they have over our lives.

Thinking in terms of society as an organism also implies that society works like a system. *Systems* are defined in terms of relatively self-contained wholes that are made up of variously interdependent parts. That's a mouthful, but it really isn't as daunting as it might seem. Your body is a system, and it's made up of various parts and subsystems that are mutually dependent upon one another (your lungs would have a hard time getting along without your stomach), each part contributing to the good of the whole. Thinking about the body also lets us see that systems are *bounded.* That is, they exist separate from but dependent upon the environment. There's a definite place where your body ends and the world around you begins. In that sense, your body is relatively self-contained. We have to negotiate the boundary between ourselves and the physical environment, but they are separate. As we will see, that negotiation is part of the needs of the system. So, for Spencer, society acts like a system with mutually dependent parts that are separate from but interacting with the environment.

In addition, systems tend toward *equilibrium.* If you are reading this book, then you are alive. If you are alive, your body is regulating its temperature, among other things. The point is that it is probably either hotter or colder in the environment than your body wants or needs. Your body senses the difference between its goal (98.6 degrees Fahrenheit) and its environment and uses more or less energy in heating or cooling itself. Your body keeps itself in balance. Spencer posits that society does the same kind of thing: The social system has internal pressure mechanisms that work to keep society in balance. One of the things to notice as we move through Spencer's theory is that change and integration are social responses to system pressures, rather than individual people making decisions. According to Spencer, the mechanisms for integration and change are in the system itself, not in the people who live in the system.

System Needs

Just as every living organism has certain needs that must be met for it to survive, society, too, has specific needs. For example, every society needs some standardized method of providing food, shelter, and clothing to its members. Every society also needs some agreed-upon method of communication, some way of passing down its culture to succeeding generations, some way of achieving a general but morally binding identity and value system, and so on. Each functionalist has his or her own proposed list of needs. As expected, these lists are pretty similar, but they are usually pretty abstract as well, particularly in the case of a grand theorist such as Spencer.

Definition

The concept of *requisite needs* is an essential assumption of the functionalist perspective. Requisite needs are requirements that every system must meet in order to survive. Because every system has the same needs, functionalists argue that all systems can be understood using the same set of ideas.

Spencer argues that because all systems basically work in the same manner, all systems should have the same requisite needs. For Spencer, there are three requisite needs or functions: regulatory, operative, and distributive. The *regulatory function* involves those structures that stabilize the relationships between the system and its external environment and between the different internal elements. In short, it regulates boundaries. An example of a regulatory social structure is government, or polity. Governments, both local and national, set and enforce laws that control relationships, whether between internal units (such as individuals or corporations) or between our society as a whole and other external systems (such as a foreign country). The printing and regulation of money is another example of how governments manage relationships. When exchanges occur between individuals or companies or nations, it is money and its determined value that standardize those exchanges.

The *operative function* concerns those structures that meet the internal needs of the system. For a society, these needs could be cultural or material. The cultural needs of a society are met through institutions such as education, and the material needs are met through the economy. The operative system also includes such things as values (we can either value material gain or spiritual enlightenment) and communities (the social networks that meet our emotional needs). The *distributive system* involves those structures that carry needed information and substances. It is the transportation (roads, railways, airlines) and communication (telephone, mail, and the Internet) networks that move goods and information through society.

These three different functions are not simply a way to categorize and understand what is going on in society, though they certainly do that. Spencer also says that there are at least two kinds of issues associated with these functions. First, Spencer notes that each subsystem will display the same needs. What that means is that every subsystem acts just like a system and can be understood as having the same needs. So, the distributive system, for example, has needs for distributive, operative, and regulatory functions. Spencer doesn't make subsystem needs and analysis a central feature of his theory, but, as we will see, Talcott Parsons (Chapter 9) does. A second dynamic that Spencer points out is that in social evolution, there is a tendency for the regulatory function to differentiate first. For example, as we move from simple collectives to more complex societies, the first structure to differentiate and specialize is government.

Differentiation and Specialization

Generally speaking, evolution is a scientific theory of the process of continuous change. The process is marked by transformation from simpler, less adaptable forms, to the more complex. Complexity is usually understood in terms of the number, types, and interrelationships among biological structures within an organism. The basic tenets are the same for social evolution: Societies change gradually over time into structurally more complex forms. More complex societies are judged superior by virtual of increased adaptation and survivability. Complexity comes about through the differentiation, specialization, and integration of structures.

Definition   *Social structure* is a central idea for sociology. The notion of structure is an answer to the problem of patterned human behavior. People tend to act in predictable ways across time and space, even though humans are not directed by instinct and are assumed to have free will. Thus, there are two fundamental characteristics of structures: Structures are made up of connections, and structures create and sustain predictable patterns and shapes. Generally speaking, social structures are built from status positions, roles, norms, values, and resources. Structures are usually understood as existing outside the individual (objective) and having the force necessary to conform the person's behaviors to social expectations. Structures are thus responsible for the general patterns we see in society.

We can think of *social structures* as made up of status positions, roles, norms, values, and resources. When human groups first emerged, the primary way they organized was through kinship relations, and kinship supplied all the operative, distributive, and regulatory functions. As societies grew through population growth and conquest, the single structure couldn't address all the needs; kinship was becoming inefficient. Soon, different social structures emerged and took some of the workload off the family. This process is called *structural differentiation.* Structural differentiation in society, then, is the process through which social networks break off from one another and become functionally *specialized.* That is, the network of status positions, rules, and resources becomes peculiar to a specific function. Social evolution has involved a movement from simple social forms (such as hunter-gatherer societies) to more complex forms (such as postindustrial societies). I've pictured this progression in Figure 2.1.

Figure 2.1 is not meant to be a dynamic model, nor does it attempt to map out all the steps. It simply presents a picture of the general idea of Spencerian social evolution. Notice that in the first circle, which would represent hunter-gatherer societies, there is only one structure: kinship. What that indicates is that all the needs of society are being met through one social institution. So, for example, the religious needs of the group are met through the same role and status structure as family. That is, the head of the family is also the "priest."

The second and third circles indicate that there is some differentiation, but the different institutions are still very closely linked. For example, the first-born son would be expected to go into government and the second-born son into religion. The final picture shows a society that is structurally differentiated and specialized,

Figure 2.1 Social Evolution—Simple to Complex

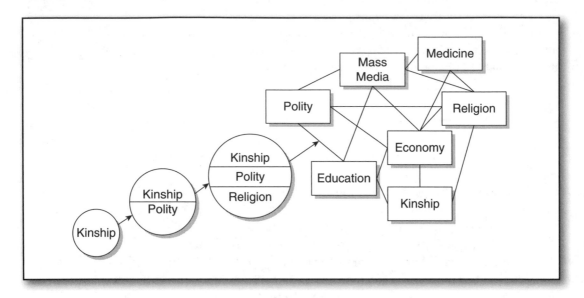

much like our own. In our society, your family role is generally not associated with the role you will play in religion, school, government, or the economy (to the degree that it is, there is another structure at work—a structure of inequality).

Once an organism has multiple structures performing specialized tasks, *integration* becomes a need. Think of single-celled amoebas. Because of their nature, integration isn't an issue: If there is only one cell, there is nothing with which that cell can be integrated. Differentiated and specialized structures by definition perform dissimilar tasks and will tend to move in different directions. Thus, multi-celled animals must create structural solutions to the problem of integration. The human body, for instance, uses the central nervous system to integrate all its different structures and subsystems.

As I implied previously, the force that drives social evolution is population. As populations grow, they need to differentiate their structural system in order to meet the needs of the collective. Population growth happens in two ways. First, there is the growth that comes from a higher birthrate than death rate. Since this is usually rather slow growth, society's evolution tends toward slow change and equilibrium. The second manner in which societies can experience population growth is through *compounding:* the influx of large populations through either military or political conquest. As populations grow, the social system will start differentiating. The system's initial response to increasing social differentiation is regulatory because of the need to integrate diverse structures.

Types of Society

Spencer gives us two related typologies of society. *Typologies* are used to categorize and understand some phenomenon, and several of our theorists use them. We think in terms of typologies all the time. For example, we use music typologies: Beethoven falls under classical music, Pete Seeger under folk, and Eminem is considered hip-hop. One of the big differences between our everyday typologies and those used in sociology is that the ones in sociology are more rigorously constructed and defined. We will see that a number of theorists use typologies to understand various aspects of society. Spencer's first societal typology is defined around the processes of compounding (population growth) that we have been talking about.

Regulatory Complexity

There may be simple, compound, and doubly compound societies. A *simple* society is "one which forms a single working whole unsubjected to any other, and of which the parts co-operate, with or without a regulating centre, for certain public ends" (Spencer, 1876–1896/1975a, p. 539). In this definition of simple societies, we see that the central feature is that the whole must not be subject to political rule from more than one group. I mentioned earlier that the first structure to differentiate is government. The levels of government that form in response to population compounding are, then, the basis of Spencer's first typology. In simple societies, like hunter-gatherer groups, the vertical dimension of government is flat. In other words, if the group has a chief, there is only one chief with no one above him. *Compounded* societies, on the other hand, not only have larger populations, but they also have several governing heads subordinated to a general head. In *doubly compounded* societies, there are additional layers of governing bodies. Obviously, we are talking about increasing complexity in social structures; this typology notes the important role the political structure plays.

When all the functions were carried out by one social structure, coordinating the way it fulfilled those functions was easy. However, as structures differentiate, they also become segmented and specialized. Each structure develops its own set of status positions, norms, goals, methods of organizing, values, and so on. This process of differentiation continues as each of the subsystems becomes further differentiated through the principle of segmentation and multiplication of effects. For example, airlines, financial markets, automobiles, the Internet, and newspapers are all part of the distributive system, but each has its own language and values. Spencer identifies these issues as *problems of coordination and control:* As societies become more differentiated, it becomes increasingly difficult to coordinate the activities of different institutions and to control their internal and inter-institutional relations. As a result of these problems, pressures arise to centralize the regulatory function (government). Notice the way that is phrased: It is not the individuals or political parties that decide there needs to be increased governmental control; it is the system itself that creates these pressures.

Spencer's basic model of evolution is outlined in Figure 2.2. For societies, population growth is the basic driving force; the rapidity and diversity of the population growth is what creates different sorts of societies. Society responds to influxes of population by structurally differentiating. Each of the differentiated structures is functionally specialized; each fulfills one rather than several functions. As a result,

Figure 2.2   Structural Differentiation and Integration

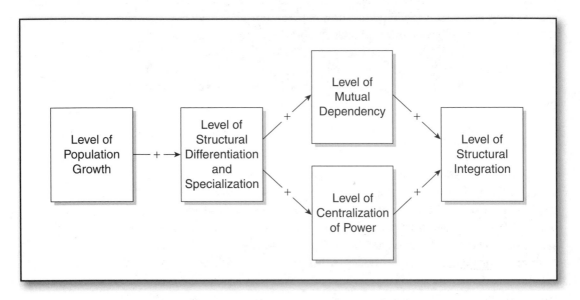

structures become mutually dependent. So, if the economic system only produces goods and services necessary for collective and organic survival, then it must depend upon another structure like the family for socialization. Because diverse structures require organization, the regulatory sector tends to become more powerful and exert greater influence on social units (both organizations and people). Together, mutual dependency and a strong regulatory subsystem facilitate structural integration. The positive feedback arrow from integration to population growth indicates that a well-integrated social system makes possible further population growth and aggregation (gathering together diverse populations through processes like migration).

Industrial and Militaristic

Increasing regulation thus solves the problems of coordination and control, but it in turn can create productive stagnation and resentments over excessive control. Increased authority over people's behavior makes them less likely to innovate. It also hampers the exchange of information, the flow of markets, and so forth. Freedom brings innovation; thus, stagnation in the long run creates pressures for deregulation. As you might already be able to tell, this is a kind of cycle that societies go through, but there are a variety of complications in the process.

This cycle of increased and decreased regulation brings us to Spencer's second societal typology. In some ways, we can think of his first typology as a rough guide to understanding the structural aspect of social evolution. It focuses on the kind of polity, or government, a society has defined in terms of its structure. That is, the structure surrounding governmental actions can be measured along a continuum from simple, with one level of governing structure, to complex, with multiple levels of governing structures. Understanding evolution in terms of the regulation function makes a certain amount of intuitive sense. In terms of evolutionary survival, the most complex and adaptive species is the human being, and it is mainly our brain and mind that give us the evolutionary advantage. Evolution, then, can be seen as the progress toward more and more complex regulatory systems.

Spencer's second typology also focuses on the regulatory system. While with the first he is interested in understanding the structural differentiation of polity, in this one he is looking more at the content of government and its relationship to society—in particular, the economic system. There are two types in this scheme—militaristic and industrial—that are defined around the relative importance of the regulatory and operative functions. The fundamental difference between them is that the militaristic type operates according to the principle of compulsory cooperation and the industrial according to voluntary cooperation.

I've listed the defining features of each type in Table 2.1. In *militaristic* societies, the state is dominant and controls the other sectors in society. The religious,

Table 2.1 Militaristic and Industrial Typologies

	Militaristic	*Industrial*
Dominant Institution	The state: the nation (including all social institutions and actions) seen as synonymous with the army; power is centralized	The economy: freedom of economic exchange and association; profit motivation; regulatory function is diffused
System Equilibrium	Compulsory cooperation	Functional equilibrium maintained through individual choices
Religious Function	Oriented toward legitimating government—direct regulatory function: • Government identified as God-ordained • Homogeneous religious types • Legitimating political myths tend to contain ultimate values • Social deviation is defined as sin	Oriented toward the individual and indirect regulatory function: • Greater separation of church and state • Religious pluralism • Focus on individual salvation and happiness
Status Structure	Greater status differentiation with distinct categories and clear ceremonial practice	Overlapping and vague status positions with unclear and infrequently practiced ceremonial distinctions
Educational Function	Directly controlled by state with strong ideological socialization	Indirectly monitored by state with emphasis on general and scientific knowledge
Mass Media	Information tightly controlled by state	Information freely flows from the bottom up
Conception of Individual	Individual exists for the benefit of the whole; individual behaviors, beliefs, and sentiments are of interest to the state	State exists to protect rights of the individual; duty of individual to resist government intrusion; belief in minority rights

economic, and educational structures are all used to further the state's interest. The economy's first mission is to provide for military needs rather than consumer products. In militaristic societies, religion plays a key role in legitimating state activities. The religious structure itself is rather homogeneous, as religion needs to create a singular focus for society. Religion also provides sets of ultimate values and beliefs that reinforce the legitimacy and actions of the state. Information is tightly controlled, and the government exercises strong control over the media. Public meetings and individual behaviors are controlled as well. Religion in particular provides ultimate sanctions (death and hell) for social deviance. The status structure also assures individual compliance. The status hierarchy in militaristic societies is much more pronounced, and the rituals surrounding status (like bowing or using status titles such as *madam, lord, mister,* and so forth) are clear, practiced, and enforced.

As you can see from Table 2.1, *industrial* societies are less controlled by the state and are oriented toward economic freedom and innovation. Rather than the collective being supreme, the individual is perceived as the focus of rights, privileges, consumer goods, and so forth. Religion is more diverse, and there is a greater emphasis on individualistic reasons for practicing religion than on collective ones. Individual happiness and peace are associated with salvation and religiosity. Education shifts to scientific and inclusive knowledge rather than increasing patriotism through selective histories and ideological practices (like the Pledge of Allegiance). Information is freed and flows from the bottom up.

Both the typology of compounding and the centralization of power are meant as ways of classifying and understanding social evolution. According to Spencer, the typology of compounding is clear: Societies have a tendency to become structurally more complex over time, especially with regards to the regulatory function. This progression is analogously seen in organisms. There has been an overall inclination toward more complex entities because they have a greater chance to survive.

Much less clear is the evolutionary path between militaristic and industrial societies. It seems that there is a general evolutionary trend toward industrial societies. Spencer characterizes the movement from industrial to militaristic as regressive. All societies begin as militaristic. Evolutionary routes are characterized by competition and struggles of life and death. Aggressive and protective behaviors became pronounced for humans once we began to use agriculture for survival; it was also at this point that people became aware of themselves as a society and not simply as a family. Land was paramount for survival and something that could be owned and thus taken or protected. The regulatory function, then, grew to be most important and differentiated from the rest. Early society developed through warfare, and eventually standing armies and taxation were created. Once established, it is then functional for society in the long run to move toward the industrial type. However, society can "regress." Having set the general tendency, Spencer delineates the reasons why there is likely to be a "revival of the predatory spirit" and the various complications involved in such a regression.

There are three main reasons why a society would, once it has become industrialized, regress back to the militaristic type. First, there is pressure in the system toward militaristic society originating from the military complex itself. A standing army and the parts of the economy that are oriented toward military production form a military complex. It's a fact of evolution: Once an organism comes into existence, it will fight for its own survival; this is no less the case for social entities than for natural organisms. Thus, a military complex by its very existence is not only available for protection; it is in its best interests to instigate aggression: "Of the traits accompanying this reversion towards the militant type, we have first to note the revival of predatory activities. Always a structure assumed for defensive action, available also for offensive action, tends to initiate it" (Spencer, 1876–1896/1975a, p. 569). This is not to deny the existence of legitimate threat, which is the second reason for regression to militarism. Spencer argues clearly that there are times when a system is threatened by other societies, internal pressures, or the natural environment. Under such threat, the system will revert back to militaristic configurations for its own survival. However, a constant pressure toward militaristic society is exerted by the military complex.

The final reason a society might revert is the presence of territorial subjects. A *territorial subject* is a social group that lives outside the normal geographic limits of the state, yet is still subject to state oversight. A current example of this kind of relationship for the United States is Puerto Rico. Puerto Rico is a commonwealth of the United States with autonomy in internal affairs, yet its chief officer is the U.S. president and its official currency is the U.S. dollar. The United States not only has a vested interest in Puerto Rico, but it is also organizationally and politically involved. As such, the United States is ultimately responsible for Puerto Rico's internal stability and external safety. Issues of investments and responsibilities provide opportunity for military action and the creation of a militaristic society.

However, there are complications on the road to regression. The first complexity is *social diversity.* When Spencer talks about social diversity and societal types, he generally has race in mind. He also, however, acknowledges that what he has in mind is best "understood as referring to the relative homogeneity or heterogeneity of the units constituting the social aggregate" (1876–1896/1975a, pp. 558–559). Race, in this case, is the most stable social difference because it has the clearest visible cues associated with it. Spencer argues that groups have natural tendencies to band together and to create their own social structures. If such different groups do not affiliate with one another in larger collectives, then society will not be internally integrated.

Spencer terms this the "law of incompleteness" and says that it represents a risk to society if an institutional shock were to occur. In other words, segregated societies are unstable because functional unity is difficult to achieve. This situation represents a potential threat: It indicates that the society may not be able to respond quickly enough to changes in the environment (societal or physical). Because of this potential threat, societies with diverse groups that do not associate with one another will tend to be militaristic. However, if the groups *do* affiliate with one another and the differences are perceived to be small, then the society is "relatively well fitted for progress." Such societies tend to produce heterogeneous groups that are functionally linked together, and are thus more flexible and gravitate toward the industrial type.

Another complication of regression is the effects of different institutions and their previous growth. For a society to regress to militarism, it must first have been industrial. Under the conditions of the industrial type, the economy grows and its power can rival or even exceed that of the state; religion diversifies and becomes more focused on the individual; and the culture of individualism grows. Returning to a militaristic posture, then, must counter all of these institutional developments.

I've diagrammed these factors and their relationships in Figure 2.3. As I mentioned before, there is a back-and-forth movement from centralized to decentralized state control. The general evolutionary movement is toward industrial forms, but there are also pressures to revert back to militarism that are themselves countered by other social pressures and complications. In the diagram, both the regressive factors (military complex, threat, and territorial subjects) and impediments (social diversity and institutional configurations) are pictured as mitigating forces. In other words, a society that has moved from militaristic to industrial will tend to stay industrial unless the regressive factors mitigate the general evolutionary trend. Moreover, once a revival of the predatory spirit begins, it will continue to move society to militarism unless the impediments mitigate the effects. Thus, modern society is generally seen as moving along the continuum from military to industrial forms in response to population pressures, system stagnation, internal or external threats (other nations or changes in physical resources), previous institutional

Figure 2.3 Militaristic-Industrial Cycle

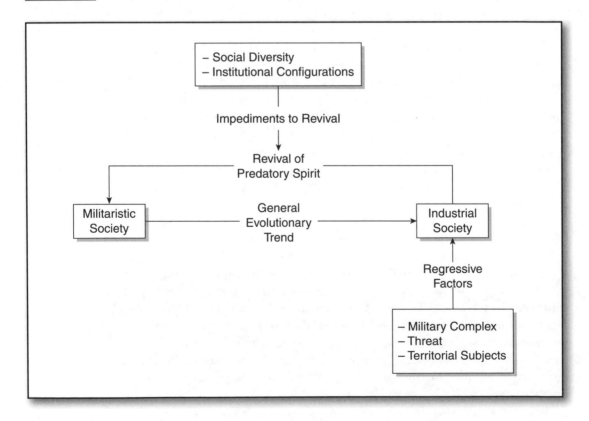

arrangements (such as the military complex or diversified religion), and the level of social integration.

Spencer does allow himself a speculative moment about future types. He says that the change from militaristic to industrial and back again involves a change in belief. In militaristic societies, people believe that the individual exists for the state and common good. Industrial societies represent a reverse of that belief: The state exists for the individual, to protect his or her rights and freedoms. Spencer says that the next type will also require an inverting of belief. In industrial societies, people believe that life is for work. As we'll see when we get to Weber, this is the work ethic of capitalism. The next step in social evolution inverts the belief from life is for work, to work is for life. As an illustration, Spencer (1876–1896/1975a) points to "the multiplication of institutions and appliances for intellectual and aesthetic culture and for kindred activities not of a directly life-sustaining kind, but of a kind having gratification for their immediate purpose"; unfortunately, he also says, "I can here say no more" (p. 563).

Concepts and Theory: Social Institutions

Spencer's ideas and theories have become part of the normal way sociologists talk and see the world, though we don't credit him as we do Marx and others. These ideas of Spencer's, such as social institutions, are part of our general life as sociologists because they are fundamental. In fact, I'd go so far as to say, unless someone understands the basic gist of what Spencer talks about in this section, he or she probably doesn't understand the sociological perspective. Don't get me wrong: There are many other sociological perspectives, but this is basic—it's foundational. In this section, you'll learn that society is made up of interrelated institutions; you'll learn what social institutions are, and you'll see how some specific institutions work and a bit about how they've changed historically.

Spencer uses the term *institution* but never provides a clear definition. I'm going to give you a definition that is in keeping with Spencer, but it probably goes a bit beyond him as well. *Social institutions* have four interrelated elements. First, institutions have functions: They are collective solutions to survival needs that provide predetermined meanings, legitimations, and normative scripts for behavior. This notion of function is probably the most important defining feature of an institution. As we've seen, functionalists assume that societies, like every other system, have requisite needs. Institutions are perceived to meet those needs, so they tend to be quite meaningful and people tend to believe in them. That's what we mean by legitimation—*legitimations* are stories we tell others and ourselves that give social relationships a moral basis. And because institutions are seen as necessary for social existence, the scripts for behavior (roles) are normative. *Norms* are expected behaviors that have sanctions attached to them. Sanctions can be positive or negative, reward or punishment. A good example of all this is family, or what Spencer calls domestic institutions.

Theory is as much about the process of thinking as it is about knowledge. So, be sure to think through the concepts, definitions, and relationships when given. Here, use the concept of family to look at examples of these elements of an institution: meanings, legitimations, and norms.

Hint

The second defining feature of institutions is that they are not reducible to individual actions or agency. Institutional behaviors aren't something that one person does. Institutions are part of what everybody knows; there is a taken-for-grantedness about them. Further, institutions resist modification by the individual or even groups of individuals. In addition to not being reducible to individual agency and resisting modification, the third defining feature of institutions is that they are not subjectively available. Let's take marriage as an example. There isn't one of us living that was around when marriage was created. We don't have a subjective memory of it or personal orientation toward it. Everything we know about the reasons why marriage was instituted and the functions it fulfills is contained in stories that we tell each other, and these legitimations give us a justifiable basis for believing in them.

The fourth characteristic of a social institution is that it tends to be wrapped in morality—institutions are moral phenomena. Here's an important point: Sociologists don't use "moral" to mean something intrinsically good. One reason for this is that we are only concerned with empirical phenomena, and empirically, nothing is intrinsically good or bad. Morality simply means something a society has infused with a great deal of emotion and meaning. An idea or belief can be moral in one society and not another. For example, the institution of slavery was based upon strong beliefs about racial purity and destiny, many of which were legitimated through Christianity (the "mark" God gave to punish Cain was believed to be black skin). Today most of us don't believe in slavery, but three hundred years ago, it was a morally infused belief system.

Social institution is a key concept in many macro-level theories of society. For most sociologists, social institutions are collective moral sets of predetermined meanings, values, legitimations, and scripts for behavior that resist individual agency and are perceived to meet the survival needs of a society. The main institutions studied by sociologists are family, education, religion, law, government, and the economy.

Definition

Domestic Institutions

Kinship or family is one of the most basic of all social institutions. Its essential function is to facilitate biological reproduction, without which any species will die.

In the long run, family also functions to provide for patterned relationships between men and women, emotional and physical support, lines of inheritance, care and socialization for the young, and—in traditional societies—care for the elderly.

In the absence of alternatives, kinship is the chief organizing principle of a collective. Before there were state bureaucracies or economic organizations, people were organized by kinship. The roles and status positions found in the kinship system informed people regarding how to act toward one another and what their obligations and rights were. All social functions were met through the status positions and roles of family. Spencer argues that there is a clear association between the structure of family and the regulation of social action and relationships generally. This idea makes sense when we consider it in the evolutionary model. Humans depend more upon social networks for survival than any other creature. However, these networks aren't created through instincts or sensory experiences (like smell) as they are with most other social animals; they are produced through culture. In fact, it is the inability of humans to generate extensive social networks and behaviors through instinct or the senses that pushed us to move toward monogamy.

In the beginning of human history, Spencer argues, there was almost complete and universal promiscuity. However, under general promiscuity, social relationships are weak and can only extend as far as the mother and child. Children are obviously linked to their mothers. However, it wasn't until just recently, with DNA testing, that we could be certain that a child was related to his or her father. This implies that fatherhood was the first true social relationship. Under general promiscuity, then, political subordination and control were limited to brute force. Only the strongest led, and leadership was based on contests of strength and was thus continually changing.

The development of religion was also hampered under promiscuity. Spencer argues that early forms of religion involved ancestor worship. When human beings first began to see past the objective line of this life, they saw it in terms of social relationships. The thing that connected us to the next life was kinship. We saw ancestors who had gone before as paving the way for us. It was through them that communication with the spiritual world could occur. Further, some of our first inklings of the idea of the soul came through the idea of ancestral reincarnation. Thus, when kinship only extended as far as mother to child, the ability of people to see past this life was limited.

Over time, monogamy was naturally selected as the primary form of kinship. It's important for you to see that it wasn't any religious or political group that chose monogamy; it was *naturally selected through survival of the fittest* and then *legitimated* by religious and political groups. Monogamy was selected for a number of evolutionary reasons. First, it added to the political stability of the group. Leadership could be established through family lineage rather than physical prowess. Spencer also notes that using inheritance to establish a family line is conducive to stability because it secures the supremacy of the elder; the use of elders for ruling tends to create what Max Weber would later call traditional authority. Further, the practice of monogamy was an important component in the evolution of religion, as it favors a single line of ancestral worship. The establishment of religion, of course, also helped to systematize socialization, which led to further social stability. In addition, monogamy tends to increase the overall birthrate and decrease the childhood morbidity rate. Interestingly, a feminist, Charlotte Perkins Gilman, gives us this same evolutionary theory about monogamy. However, she also argues that monogamy is now dysfunctional. We'll talk about her ideas in Chapter 7.

Because of the relationship between family and gender, I want to pause to note that Spencer felt that men and women should have equal rights. He saw them as performing different functions, but the functions themselves are equally needed by society. I must also note that one of the criticisms of Spencer is that he was a social Darwinist, seeing certain societies and races as more advanced—but in his view of equal rights for women, he was in many ways ahead of his time, as the following quotes illustrate:

> Equity knows no difference of sex. In its vocabulary the word man must be understood in a generic and not in a specific sense. The law of equal freedom manifestly applies to the whole race—female as well as male. (1882/1954, p. 138)

> Perhaps in no way is the moral progress of mankind more clearly shown, than by contrasting the position of women among savages with their position among the most advanced of the civilized world. (1876–1896/1975a, p. 713)

Definition

Initially, all requisite needs—regulatory, operative, and distributive— were met through family. As social groups began to differentiate structurally, the functions of family became more specialized and monogamy was selected to provide for political succession and clear lines of ancestral worship. *Domestic institutions* currently function to provide regulated ways of satisfying sexual and emotional needs, biological reproduction and support, socialization, and social placement.

Ceremonial Institutions

Spencer argues that there are three major forms of social control: ceremonial, ecclesiastical, and political. Most sociology students are familiar with the latter two. Religious, or ecclesiastical, institutions bring about social control internally. Religion functions to provide sets of beliefs about right and wrong behaviors. These moral standards exert their force from inside of the individual. Political entities govern human behavior externally, rather than internally, through laws, coercive force, and authority. One of the defining characteristics of nation-states is the monopolization of legitimate coercive force. This monopoly serves not only to protect the nation's external boundaries, but also to maintain internal peace.

Spencer gives us another, more basic type of social control in the form of ceremonial institutions. These are made up of formal or informal acts or series of acts that function to link people together hierarchically. Ceremonies may be simple, as in an act of politeness, or quite elaborate, as in prescribed rituals. Obviously, there are certain protocols that apply when a head of state visits another country. In the protocol, the relationship between the ruler and the ruled is played out in behaviors. Ceremonies also indicate the hierarchical relationship among nations: Some visiting dignitaries are afforded more elaborate ceremonies than others. Ceremonies can also be conventional and practiced every day. In this case, Spencer has in mind forms of address, titles and emblems of honor, forms of dress, and the like. For example, when we use titles like "Professor" or "Doctor" or "Miss," we are reproducing a hierarchical system of authority and control. This is a deep insight because it shows us how power is exercised insidiously in society.

Definition

Ceremonial institutions are ritualized behaviors that carry meaning and display social respect. In Spencer's evolutionary theory, these rituals create hierarchical social networks and are the most general and earliest form of social control. As societies develop economically and politically, and social groups become reflexive, rituals and ceremonial institutions become more symbolic, complex, and internalized through socialization, thus creating both formal and informal rituals and social control.

Ecclesiastical Institutions

There is an intrinsic and complex relationship that religion has with the rest of society. Émile Durkheim argues that society began in religion. It was in religion that humans first sensed something outside and greater than individual consciousness. It's the same sort of feeling that people can have about society—something larger and better than each individual, something worth "the ultimate sacrifice." Max Weber also argues that there has never been society without a god.

Religion functions on a number of levels. For one, it acts as an agent of social control. If social structures are seen as an extension of a god's spiritual order and behaviors are seen as holy directives, then questioning social arrangements becomes a matter of questioning the god, and inappropriate behavior becomes sin. This control isn't usually exercised outwardly, for religious people will conform to these norms, values, and beliefs in the name of their god. Religion is also the strongest legitimating force we know; it functions to reinforce and justify existing social structures, particularly those built around inequality. Religion also functions to alleviate personal anxieties and troubles—it provides answers to existential questions (such as, what happens after we die?) and explanations for loss.

Regarding social institutions, Spencer is primarily concerned with tracing their evolutionary development. Any student of religion is faced with one inarguable fact: Through the course of human history, religion has changed. Our initial forms of religion were not monotheistic or ethical. They were, in fact, marked by the presence of many gods displaying quite human behaviors of greed, lust, and so on.

The question with which we are faced is, why did our conceptions of deity change over time? Did God reveal her- or himself gradually, or was the idea of God linked to developments in society and human personality? Spencer (1876–1896/1975b) argues for the second alternative: "Among social phenomena, those presented by Ecclesiastical Institutions illustrate very clearly the general law of evolution" (p. 150).

Spencer argues that all religions share a common genesis. Religion began, according to Spencer, as "ghost-propitiation." Spencer argues that human beings became aware of themselves as double. We have the waking, walking self, and we also have a self that can be absent from the body. We became aware of this other-self through dreams and visions. During dreams, we have out-of-body experiences. We travel to far-off places and engage in some pretty amazing behaviors. In thinking about these experiences, primitive humans concluded that we are double. Part of us is tied to the earth and time, but there is also part of us that can transcend both.

Moving from conceptualizing the other-self from dream states to thinking that the other-self survives death is a short and quite intuitive step. Death seems a lot like permanent sleep; it's a sleep from which we don't awake. If we soul-travel when we are temporarily asleep, then we must also travel when we are permanently asleep. This idea led early humans to minister to the dead, especially dead relatives. These dead relatives could cause mischief if not taken care of, but they could also prove to be valuable allies in times of need. Further, these disembodied spirits also came to be seen as the source for life on this earth through reincarnation. Early humans were quite in tune with the cycles of nature. Trees that appear to die in winter came back to life in the spring. Thus, it was not a very big step, once souls were seen as surviving death, to see them also as the source for the other-self in life. In this way, early humans began their ancestor worship.

Obviously, ancestor worship is linked to family, and the head of the family was also the head of religious observance. However, two factors that came from military conquest brought about significant social and structural changes. First, the religious function differentiated from family. As human beings moved from nomadic existence to settled farming, war became more and more common. People were tied to the land in a way never before realized, and land became a limited resource that had to be protected from other groups. The increased involvement in war by the clan chief meant that he had to shift his priestly responsibilities to another family member. Eventually, these religious functions were separated structurally as well.

The other factor that influenced the development of religion, which was also the result of military conquest, was the incorporation of more and more conquered people into the collective. Each of these groups of people had its own family god and ancestral worship. *Polytheism* (worshipping more than one god) was therefore the natural result of war and conquest. Further, the presence of this competitive chaos of gods created pressures for there to be a professional class of priests. The priests began to impose order on this array of deities, which of course reflected the social order of society: The gods that became most important were the gods of the most important families.

As societies continued to evolve to more complex forms of organization, so did the structure of religion. Organized pantheons developed as the political structure became more centralized and class inequalities increased. The hierarchy of higher and lower gods thus reflected the organized divisions of society. Because polity relies a great deal on religion for legitimation, the more warlike, and thus more successful, societies developed images of vengeful and jealous gods. And as the political structure became more centralized and bureaucratized, the religious structure did as well. When governments rely on religion for legitimation, it is in the best interest of the professional priestly class to organize and unite. As religious leaders organize and unite, they have to simplify the religious belief system, which in turn leads to monotheism.

Definition

Ecclesiastical institutions (religion) function to legitimate other social institutions and to provide motivation for and assurance of social conformity. Religion also functions to relieve individual anxieties and troubles. Religion initially came to be as people became aware of an existence apart from fully awake life, through dreams and visions. These alternative states helped create a belief in life after death, out of which developed ancestral worship and propitiation. Religion evolved as ecclesiastical functions differentiated from kinship and as religion became more organized and symbolic in response to the political need to organize increasingly diverse populations. As an effect of the professionalization of a priestly class, ecclesiastical institutions became more complex, in terms of beliefs and practices, and generally evolved toward monotheism.

Thinking About Modernity and Postmodernity

In the first chapter, I made the case for understanding our time period as the modern age. There are certain qualities and social arrangements that are specific to modernity, and Spencer's theory explains a good part of those institutional and structural contours. Spencer's theory explains the evolution of society from primitive, simple structures to modern, complex structures. His argument is that as populations grow and compound, structures differentiate. Differentiated structures are held together within a system by mutual need and a powerful regulatory system.

You might remember from Chapter 1 that there are those who argue we are no longer modern. People talk about these changes in various ways, but one of the most common is postmodernity. Obviously, postmodernity is what comes after modernity. But what precisely does that mean? Our theorists of modernity give us different keys to understanding what's changed.

Defining Postmodernity

Spencer holds an important key because of his ideas of structural differentiation occurring within a cohesive, interdependent system. A postmodern society is one wherein social institutions are not simply differentiated but, rather, they are fragmented. There are at least five important reasons for this fragmentation and deinstitutionalization: the level of structural differentiation, the level of transportation and communication technologies and infrastructures, the level of market velocity and expansiveness, the level and rate of commodification, and the level of institutional doubt (Allan & Turner, 2000). I don't expect you to completely understand each of these variables right now (we will be looking at them in subsequent chapters), but I do want you to have a basic idea of what we mean by modernity and postmodernity.

Modernity and *postmodernity* are terms that are used in a number of disciplines. In each, the meaning is somewhat different, but the basic elements are the same: Modernity is characterized by unity, and postmodernity is distinguished by disunity. In modern literature, for example, the unity of narrative is important. Novels move along according to the plot, and while there might be twists and turns, readers can feel fairly confident in how time is moving and where the story is going. Postmodern literature, on the other hand, doesn't move in predictable patterns of plot and character. Stories will typically jump around and are filled with indirect and reflexive references to past styles of writing or stories. The goal of a modern novel is to convey a sense of continuity in the story; the goal of a postmodern work is to create a feeling of disorientation and ironic humor.

In the social world, modernity is associated with the Enlightenment and is defined by progress; by grand narratives and beliefs; and through the structures of capitalism, science, technology, and the nation-state. To state the obvious, postmodernism is the opposite or the critique of all that. According to most postmodern thought, things have changed. Society is no longer marked by a sense of hope in progress. People seem more discouraged than encouraged, more filled with a blasé attitude than with optimism. We'll explore some of the specifics of the postmodern critique of grand narratives, science and technology, capitalism, and the nation-state as we work our way through this book.

However, right now I want to give you a couple of handles on postmodernism. First, let me say that any attempt to capture postmodern ideas is doomed to fail on one level or another. Postmodern theory is based on the idea of the inability of categories to express human reality. So, if I say that I'm going to reduce postmodernism to a few concepts or categories, I'm already outside the postmodern project. Keep in mind that I can't capture what is meant by postmodernity—postmodernism is an open-ended discussion.

The various postmodern theories seem to boil down to two main issues or effects: culture and the individual. Because of certain social changes, culture has become more important in our lives than social structure, and we are more individually oriented than group oriented. Thus, culture and the individual are more important now than they ever have been. Nevertheless, both culture and the individual are less effectual than ever before as well. Culture is important, but it has become fragmented and unable to aid in the production of a unified society. Further, a decisive break between reality and culture has occurred, such that all we have are texts and subjective meanings. The individual is more important, too, but has become decentered and unable to emotionally connect in any stable or real fashion.

Religion: A Postmodern Case in Point

We will explore these themes as we move through the theorists, but let me give you an example of what we're talking about. Spencer argues that there is a symbiotic connection between religion and the state. Religion serves to legitimate the state, and the state in turn provides religion with structure and protection. In this relationship between the state and religion, Spencer also explores the evolution of religion from polytheism to monotheism. Let's ask ourselves a rather natural question: If religion is evolving, what's next? In his work, Spencer actually considers the future of religion, but he doesn't get very far. To see what might be going on with religion in a postmodern society, let's look at what Thomas Luckmann, a contemporary social theorist, has to say.

According to Luckmann (1991), society has always used religion to create ultimate meanings. Like Spencer, Luckmann acknowledges that these meanings are used to legitimate social institutions, as well as to answer impossible questions (such as, what is the meaning of life?). Religion also serves to stabilize our meaning systems. We can intentionally change our meanings—the changes we brought about in the meaning of race and gender in the United States are good examples. However, because meaning is socially constructed, it will tend to shift and change on its own. Let's look at it in terms of what we do to the natural environment. Most of us have lived in houses with yards. "Yards" don't naturally happen. We have to kill the weeds, rototill the soil, or bring topsoil in. We also have to plant all the trees, grass, and bushes in the exact places that we want them. I remember as a child, my parents had a plan drawn up to help them build the yard. Once we put in the yard, we had to work to maintain it. If we didn't, the yard would revert back to nature and be gone. The same is true with symbolic meanings. Once in place, they have to be kept up and stabilized in order to continue to exist. Religion does an amazing job of stabilizing our meaning by equating our constructed meanings with eternal truths and by equating deviance and disorder with sin.

However, according to Luckmann (1991), modern structural differentiation and specialization have made the great transcendences (the ultimate meanings of life) that religion provides structurally unstable. This structural instability has resulted in the "privatization of religion" (p. 176), a new form of religion.

Other parts of society have filled in the gap left by the loss of highly structured religion and turned it into profitable business. As a result, the individual is faced with a de-monopolized market created by mass media, churches and sects, residual nineteenth century secular ideologies, and substitute religious communities. The products of this market form a more or less systematically arranged meaning set that refers to minimal, intermediate, but rarely great transcendences. In other words, the meanings provided through mass media and commodities are small and fleeting. Under these conditions, a set of meanings may be taken up by an individual for a long or short period of time and may be combined with elements from other meaning sets.

What Luckmann is describing is an environment where meanings—particularly ultimate meanings—are held in doubt and where individual people can pick and choose which meanings they want to hold and which they don't. Most of us think that we ought to be free to choose what we believe and what we don't believe. I think this level of freedom is certainly preferable to the enforced meaning of more traditional societies (think of the Salem witch trials), but there are other effects of this level of freedom as well. If all people can pick and choose what meaning suits them, then meaning by definition is unstable and borders on meaninglessness.

Not too long ago, I met someone who exemplifies what Luckmann is talking about. When she introduced herself to me, she told me she was a "Christian-Pagan." That threw me for a loop. The word *pagan* originally meant "country dweller," but its current meaning of "heathen" or "irreligious hedonist and materialist" came from some Christians who used it to refer to nonbelievers and people who practiced magic. So, tell me, what is a Christian-nonbeliever? Or, better yet, what is a Christian-non-Christian? You might respond and say, "Well, that's her choice and she should be able to create her own meanings for her own life." I don't disagree, but this kind of situation begs a larger, more important question: How will this loss of predictable meaning influence society?

We will return to these issues surrounding modernity and postmodernity in most of the chapters that follow. Throughout the book, we are going to explore the main issues of postmodernism—culture and self—using some of the ideas that our theorists will provide. My hope for this exercise is to better attune you to the possibilities of the historical moment in which we live.

Summary

- Spencer's perspective is in many ways foundational for sociology. He explicates the basic premises with which most sociologists will either agree or disagree. Spencer's point of view is founded on positivism, the search for the invariant laws that govern the universe. He approaches society as if it were an object in the environment, and his purpose as a theorist is to simply describe how this object works. For Spencer, society works much like an organic system, with various parts functioning for the welfare of the whole. He extends the analogy to explicate the evolutionary changes in society, from simple to complex structural differentiation.

- All systems have universal needs: regulation, operation, and distribution. Every system, including society, meets these needs through various kinds of structures. The ways in which the needs are met produce the unique features of each system and society. Generally speaking, there is an evolutionary trend toward greater structural differentiation and specialization, with system needs being met through separate and distinct structures. For society, this trend is driven by increases in population size. As the population increases, especially through compounding, structures differentiate and specialize in function. This process produces not only inter-structural dependency, but also problems in coordination and control. In response, society tends to centralize the regulatory subsystem, which, along with dependency, facilitates system integration. However, too much regulation can cause stagnation and pressures for deregulation. In reaction to these system pressures, societies in the long run move back and forth on a continuum between militaristic and industrial.

- Social institutions are special kinds of structures. They are organized around societal needs, resist change, and are morally infused. All institutions, however, change slowly through evolutionary pressures, and because they are functionally related, change is mutual and generally moves in the same direction. Generally speaking, family has evolved from universal promiscuity to monogamy. (Monogamy is evolutionarily chosen because it provides more explicit and stable social relations.) Religion generally moves from polytheism to monotheism. (Monotheism has the evolutionary advantage of providing a single legitimating and enforcing mechanism for diverse populations and bureaucratic states.) The state generally moves from militaristic with simpler political structures to industrial with more complex structures. Ceremonial institutions generally recede in importance as religion, family, and polity become successful structures. In spite of this, the political and ceremonial structures tend to move back and forth on a continuum in response to vested interests, perceived threat, and the level of social inequality.

- Modernity for Spencer is characterized by high levels of structural differentiation. This increased complexity gives modern societies greater adaptability and increased chances of survival. Postmodernity is characterized by institutional and cultural fragmentation. The main difference between differentiation and fragmentation revolves around unity: In differentiated systems, the different elements are linked together, whereas in a fragmented system, the parts are not as organized and have greater freedom. The effects of postmodernity are contradictory: Culture and self (subjective experience) are simultaneously becoming more important and less real and meaningful. Religion is a case in point: Religion in postmodernity is less centrally organized and is thus open to increasingly idiosyncratic interpretations.

BUILDING YOUR THEORY TOOLBOX

Learning More—Primary and Secondary Sources

For primary readings, I suggest the following:

- Spencer, H. (1961). *The study of sociology.* Ann Arbor: University of Michigan Press;

And, if you're feeling particularly adventuresome,

- Spencer, H. (1975). *The principles of sociology.* Westport, CT: Greenwood Press.

(Continued)

(Continued)

The single-best secondary source for Spencer is

- Turner, J. H. (1985). *Herbert Spencer: A renewed appreciation.* Beverly Hills, CA: Sage. (Turner is probably the foremost expert on Spencer.)

For an overview of functional theorizing, including anthropology, see

- Turner, J. H., & Maryanski, A. (1979). *Functionalism.* New York: Benjamin-Cummings.

Seeing the Social World (knowing the theory)

- Write a 500-word synopsis of Spencer's sociological imagination, making certain to include not only how he sees the world (perspective) but also how that perspective came about (history, social structures, and biography).

- After reading and understanding this chapter, you should be able to define the following terms theoretically and explain their theoretical importance to Spencer's functionalism: evolution, segregation, organismic analogy, requisite needs, regulatory function, operative function, distributive function, differentiation, specialization, structural differentiation, compounding, problems of coordination and control, militaristic and industrial societies, domestic institutions, ceremonial institutions, political institutions, ecclesiastical institutions.

- After reading and understanding this chapter, you should be able to answer the following questions (remember to answer them theoretically):

  1. Explain how society becomes more complex.

  2. How are complex societies integrated?

  3. Explain how modern societies move between militaristic and industrial forms.

Engaging the Social World (using the theory)

- Spencer argues that monogamy developed because of its functional use in creating social relationships. Many people feel that monogamy is currently threatened. If monogamy is in danger, how do you think Spencer would theorize about it? (Remember, he would think about it in progressive, evolutionary terms.)

- Spencer argues that ceremonial institutions are at the heart of social control. Do you think the use of ceremonial institutions has gone up or down since Spencer's time? Theoretically, what do you think this change would indicate?

- Recalling our definitions of structures and institutions, do you think there are more social institutions today than 100 years ago? If so, what does this imply about society's survivability? What does it imply about the state?

- Using Google or your favorite search engine, do a search for "censored news stories." Read through a few of the sites. Based on your reading, do you think that the news is being censored in the United States? If so, or if not, how would Spencer's theory explain it?

Weaving the Threads (building theory)

There are central themes about which most of our theorists speak. These themes include modernity; social institutions such as the state, economy, and religion; culture; diversity, equality, and oppression; social cohesion and change; and empiricism. A good approach to theory is to pay attention to how these themes are developed. One of the ways we can build theory is through synthesis: Compare, contrast, and bring together elements from different theorists. The following questions are based on Spencer's theorizing and are meant to begin your thinking about these themes:

- What are the functions of kinship in society? How did the structure of kinship aid in the evolution of society? If the structure of kinship has changed since Spencer's time, what do you think that means (remember to think like Spencer)?

- What is religion's function in society? How has religion evolved over time? In other words, what are the social factors that contributed to changes in religion?

- What is the basis of political power in society? What factors bring about changes in the type of state and its structure?

- How do societies differentiate? Once societies differentiate, how do they integrate? Thinking like Spencer, would you conclude that the United States has higher or lower structural complexity and differentiation as compared to even 50 years ago?

CHAPTER 3

Class Inequality

Karl Marx

The idea of civil society has been integral to democracy since its modern beginnings. It's an organic idea, in the sense that there aren't clear boundaries indicating what's included or excluded and the emphasis changes according to whom you read. Nevertheless, there were few, if any, thinkers at the dawn of modern democracy that didn't see civil society as a necessary institutional sphere for the success of the democratic experiment that began most notably in France and the United States. In some ways, its organic nature is appropriate because democracy as originally conceived is organic as well, expanding as the ideas of freedom and equality move through the ever-changing landscape of society. However, don't let the indefiniteness of the idea lull you into thinking you can discount it: Without civil society, democracy would surely die. This institutional sphere was so important to the people who framed democracy in the United States that it finds its place in the First Amendment of the U.S. Constitution: The freedom of speech, press, and assembly are vital to civil society, as are the freedom from state-sponsored religion and the right to redress the government.

One of the people whose work is still seen as essential for any discussion of civil society is Alexis de Tocqueville and his study of early American democracy (1835–1840/2002). Jeffrey Alexander (2006), a contemporary sociologist, summarizes Tocqueville's understanding: "It included *the capitalist market and its institutions,* but it also denoted what Tocqueville called voluntary religion . . . private and public associations and organizations, and virtually every form of cooperative social relationship that created bonds of trust" (p. 24, emphasis added). I want to call your attention to the part in italics—originally, capitalism was seen as an important element in civil society; it was supposed to be the way through which men and women could achieve their own social standing. The interesting thing about this is that, generally speaking, capitalism is no longer seen as part of civil society. In fact, it is often specifically excluded "for the reason that free markets, by reducing all decisions to the calculation of self-interest, weaken the bonds of loyalty, friendship, and trust upon which civil society depends" (Eberly, 2000, p. 8).

Beginning a chapter about Karl Marx with a discussion of democracy may seem a bit odd, but I hope that you're beginning to see why I've done it. Marx's systematic analysis marked the separation of capitalism from civil society. The power of Marx's critique was less about the economic system and more about the effects of capitalism on people's thinking and awareness of the world. Democracy requires critical thinking and insightful awareness—Marx wants us to consider the possibility that capitalism may in fact dull those attributes. There is no doubt that capitalism has brought tremendous benefits. More people are materially better off than ever before in human history, and we've benefited from the improvements in technology and medical care spurred by capitalist motivation. Yet, capitalism has a dark side as well, one that may reach down into the very soul of humankind, at least from Marx's perspective.

THEORIST'S DIGEST

Central Sociological Questions

Marx's interests were broad, including such philosophic concerns as human nature, epistemology (the study of knowledge), and consciousness, along with more social science concerns with politics and capitalism. Specifically, Marx was concerned with the unanticipated effects of capitalism. Capitalism was to be an important mechanism through which social equality would come about in modernity. Instead, capitalism tended to produced high levels of inequality and abuse. Marx wanted to understand how these undesirable effects came about; and he wanted to explain how history would move from capitalism to another, more humane economic system.

Simply Stated

Marx argues that capitalism is based on certain structural factors that create tensions in capitalism—most notably exploitation and overproduction. These tensions invariably create the recurring cycles of economic recession to which capitalism is susceptible. These cycles generate a small class of wealthy capitalists and a large class of dependent and deprived workers. Each economic crisis is deeper than the previous, which in turn concentrates wealth and power in fewer and fewer hands, with the working class experiencing greater levels of economic, physical, and psychological suffering. In the long run, these cycles will cause capitalism to implode, which, in turn, brings economic and social change.

Key Ideas

German Enlightenment, sense-certainty, consciousness, reification, self-consciousness, species-being, material dialectic, means and relations of production, class bipolarization, exploitation, surplus value, commodification, overproduction, alienation, private property, commodity fetish, ideology, false consciousness, class consciousness.

The Sociological Imagination of Karl Marx

The political, economic, and intellectual contexts of the world into which Marx was born were decidedly different from those of Herbert Spencer. You remember that Spencer's world was one that appeared to be set on a glorious upward path of social evolution. England was in her political heyday, the boundaries of the empire stretched around the globe, and *Pax Britannica* ruled. Economically, Britain was at the forefront of the Industrial Revolution, and the intellectual community in London was pushing the boundaries of the known sciences. During Spencer's lifetime, Great Britain was leading Western Europe in the modernization of the world. Marx's Germany, however, was quite a bit different.

Marx's Life

Karl Marx was born to a German Jewish family on May 5, 1818, in Trier. Karl's father, Heinrich Marx, came from a family where the men usually became rabbis. In fact, Heinrich was the first to receive a secular education and espouse Christianity, though the conversion was socially motivated because of the oppression of Jews. Marx's father practiced law and was an avid reader of Leibniz and Kant, which is undoubtedly where the young Marx was first exposed to these philosophers.

The Marx family lived next to the home of Baron Johann Ludwig von Westphalen. Westphalen was a government official and Prussian aristocrat who befriended the family and specifically took young Marx under his wing. Undoubtedly, Westphalen had a profound influence on Marx—Karl eventually married his daughter, Baroness Johanna Bertha Julie "Jenny" von Westphalen—but perhaps the most important effect was that Westphalen introduced Marx to the ideas of Henri de Saint-Simon (1760–1825), whom we first met in Chapter 1.

Marx undoubtedly began thinking about socialism in response to Saint-Simon's writings. Marx's work is a kind of economic determinism that gives central place to the economy. Saint-Simon makes the same sort of argument: "Society . . . is based entirely on industry. Industry is the sole guarantee of its existence" (as cited in Normano, 1932, p. 8). Thus, under socialism, the economy is publicly controlled to bring the greatest good to the greatest number of people, and the state shifts from controlling people to directing things. Marx's collaborator, Friedrich Engels (1880/1978a), tells us that this Marxian idea "that economic conditions are the basis of political institutions" was based on Saint-Simon's work (p. 689). Engels intimates that the idea of class antagonisms also stems from Saint-Simon, though he spoke of "idlers" and "workers" and missed the true meaning of the proletariat under capitalism (p. 685). In addition, Marx's view of history as being driven by class conflict was also inspired by Saint-Simon who "wanted to found a science of progress which by studying the past would give men a sure guide for working out the future" (MacIver, 1922, p. 239).

Marx continued his education first at the University of Bonn, then at the University of Berlin, finally completing his PhD in philosophy at the University of Jena in 1841. His Berlin years were in some ways his most formative. There he came in contact with the ideas of Georg Wilhelm Friedrich Hegel (1770–1831) and the New Hegelians. In a letter to his father, dated November 10, 1837, Marx (1978c) talked about his interest in Hegel and said that during a recent illness, "I got to know Hegel from beginning to end, together with most of his disciples" (p. 8). We'll hear more about Hegel in a bit.

While in Berlin, Marx also became a member of a "Doctor's Club." These clubs were made up of professors and intellectuals who would endlessly debate philosophical points well into the night over mugs of beer (they met in Berlin's beer cellars). Marx also sought academic employment but had very little luck. He began working for a radical newspaper and within a short period of time, became its editor. The articles that Marx wrote gained quite a bit of attention, especially from the political elite, and he thus had to contend with censorship on a regular basis. After writing a particularly unflattering piece about the Russian emperor, the Prussian government shut the paper down. As Lewis Coser (2003) notes, Marx was "an amateur philosopher who had made a marked impression in advanced salons and bohemian gatherings, but had otherwise no prospects for a career" (p. 60). So Marx moved to Paris.

In Paris, Marx became acquainted with the works of reformist thinkers who had been suppressed in Germany and began his close friendship and collaboration with Friedrich Engels (1820–1895). Engels was the son of a wealthy German businessman and was significant in Marx's life for a few reasons. First, it was through Engels that Marx learned firsthand of the abuses of industrialized capitalism, most notably child labor and the impoverished working class. As the son of a well-to-do capitalist, and someone who later managed several of his family's businesses, Engels knew the economic system from the inside out. Note that Marx didn't actually learn of the abuses of capitalism from his German experience; most of his concerns originated with Engels. Engels sent Marx various articles depicting the deplorable conditions, and in 1844, Engels published his first book: *Conditions of the Working Class in England* (1844/2009). Engels also provided and arranged for a good deal of financial support for Marx, especially during his London years. Engels and Marx shared an interest in Hegel and association with the Young Hegelians, though the two didn't meet in Germany. Marx and Engels worked together for nearly 40 years and coauthored *The Communist Manifesto* in 1847 (1848/1978). After Marx's death, Engels finished the last two volumes of *Das Kapital—Capital: A Critique of Political Economy* (Marx, 1867/1977)—a massive work analyzing the historical development, political-economic roots, and practices of capitalism.

Engels was a significant scholar in his own right and published several pieces. One of the most interesting is *The Origin of the Family, Private Property, and the State* (1884/1978a). In it, he analyzed the historical development of monogamy. Engels argued that monogamous marriage originated "to bequeath this wealth to this man's children and to no one else's" (p. 745). In monogamy, Engels saw the beginnings of class oppression. Because it was based upon the unequal control of material wealth, monogamous marriage contained within it the "embryo" form of "all the antagonisms which later develop on a wide scale within society and its state" (p. 737). True to this understanding of marriage, Engels never married his lifelong partner, Mary Burns, who was instrumental in introducing Engels to the working class neighborhoods in London.

Marx was kicked out of Paris. In fact, he was banned from France by the prime minister, Francois Guizot, in 1845, and spent the next few years in Brussels. It was there that Marx first became associated with the Communist League, who, in 1847, commissioned Marx and Engels to write *The Communist Manifesto*. The publication of the *Manifesto* in 1848 was timely, as a wave of revolutions swept around the globe that year. This wave began in January with the Italian states. In February, a much bloodier and more significant revolution hit France; Germany was engulfed in March; and the wave soon hit most of Europe and extended as far as Brazil.

Surprisingly, neither Britain nor the United States was much affected. There were several causes for this wave of revolution: Europe had suffered through a number of famines in the preceding years; the Industrial Revolution created both increasing prosperity and misery; and there were a number of ideological belief systems—such as democracy, nationalism, liberalism, and socialism—that had gained widespread audiences through the popular press. As you can well imagine, Marx was ecstatic. He returned to Germany, took the editorship of a radical paper, and attempted to trigger a tsunami of social change.

However, it was not to be. The wave of revolutions was effectively put down within a year, and few if any reforms resulted. Marx and Engels went into exile in London. Both thought the exile would be short, fully expecting the revolutions to build again; both were wrong. Marx labored in London for the next 13 years, spending most of his days researching in the British Museum from 10 AM to 7 PM. Most of that work formed the basis of *Das Kapital*.

Things began to pick up again for Marx in 1863. That was the year that a group of French labor leaders came to London to discuss the possibility of forming an international worker's union. The next year, the men met again, this time to hammer out the particulars of the organization, which became known as the International Workingmen's Association, or simply The First International. Soon after its inception, Marx took control of the organization and delivered the inaugural address. In it, he recalled the wait since the 1848 revolutions and connected it to the basic antagonism of capitalism: capitalists prosper—workers suffer: "It is a great fact that the misery of the working masses has not diminished from 1848 to 1864, and yet this period is unrivalled for the development of its industry and growth of its commerce" (Marx, 1864/1978f, p. 512). His conclusion echoes both the 1848 *Communist Manifesto* and Marx's expectation for the immediate future: "Proletarians of all countries, Unite!" (p. 539). Marx's hope lasted 7 years.

The beginning of the death knell came from France. After France's defeat in the Franco–Prussian War (1870–1871), the working class took control of Paris. They formed the Paris Commune, the first government by and for the working class. This new government lasted 3 short months, from March to May of 1871. After 8 days of fighting, the Army of Versailles brought down the Commune. As a result of the battle and its aftermath, 30,000 Commune members were slaughtered and 38,000 arrested. In an address given just days after the fall, Marx (1891/1978a) said the Commune "will be forever celebrated as the glorious harbinger of a new society. Its martyrs are enshrined in the great heart of the working class" (p. 652). In an attempt to preserve the First International, Marx moved its headquarters to the United States. The move, however, was unsuccessful, and the International disbanded in 1876.

Marx never recovered his revolutionary flame and suffered from illness for the rest of his life. His wife died in 1881, and one of his remaining daughters (the Marx family lost four children in infancy) a year later. Marx died in 1883. At his graveside, Engels (1883/1978c) said, "On the 14th of March, at a quarter to three in the afternoon, the greatest living thinker ceased to think. He had been left alone for scarcely two minutes, and when we came back we found him in his armchair, peacefully gone to sleep—but forever" (p. 681).

Marx's Social World

Marx's birthplace, Trier, is one of the oldest cities in Germany. From 883 CE until the French occupation in 1807, Trier was the seat of the Archbishop and Elector of the Holy Roman Empire. Napoleon controlled Germany from 1807 to 1813, during which time feudalism was officially brought to a close, the government opened to common citizens, and Jews were given political standing. These reforms were short lived, however. After Napoleon's defeat, the Vienna Congress repartitioned Germany into a loose affiliation 39 states. "Prejudice and discrimination [of Jews] remained firmly entrenched in German society and intensified in the postwar years" (Kitchen, 2006, p. 47).

While serfdom and feudalism were officially ended in the early 1800s, it took until 1823 to go into effect. This is one of the reasons why the Industrial Revolution didn't have much of an impact in Germany until decades after Britain. In addition, each territory and state practiced protectionism with high tariffs and duties, which slowed the development of a market economy. Similarly, Germany lacked the transportation infrastructure needed to facilitate trade, markets, and demand, which generally push for industrialization. This slowdown of industrialization affected two key indicators: textiles (by 1846, only 2.2 percent of the textile looms were run by machine) and the production of iron (by 1837 less than 10 percent came from coke-fired furnaces). Thus, for much of the nineteenth century, the German economy was largely agricultural, and its people suffered through at least three major famines.

The Enlightenment, too, was different in Germany from how it was in Britain or France. While it's difficult to set a definite date, we won't be far off by saying that generally the Enlightenment as a whole began with the work of British philosophers Thomas Hobbes (1588–1679), who first posed the problem of social order and argued for individual natural rights and equality, and John Locke (1632–1704), who is considered the father of liberalism. France had its own outstanding figures, such as Voltaire (1694–1778) and Montesquieu (1689–1755); more importantly, it was in France where the ideas of enlightenment were collectively pulled together in the *Systematic Dictionary of the Sciences, Arts, and Crafts,* an encyclopedia of the Enlightenment literature published in France between 1751 and 1772. Generally speaking, the British and French Enlightenments argued for the supremacy of human reason in the search for timeless truths based in empiricism.

The German Enlightenment can be dated to the work of Gottfried Leibniz (1646–1716), who published his *Discourse on Metaphysics* in 1686 and *Monadology* in 1714 (both included in Leibniz, 1714/1992). Leibniz was a mathematician and philosopher whose philosophy moved completely away from empiricism. He argued that the universe is made up of "monads," which are in some sense like spiritual atoms. Monads are solitary bits of existence that do not bind together and that contain preprogrammed instructions that create the best of all possible universes. Monads aren't necessarily small like atoms; both God and the individual human are complete monads. The important thing to see here is that Leibniz's philosophy is metaphysical; he argued that reason should be applied to understanding what is in back of empirical phenomena.

Immanuel Kant (1724–1804) is probably the best known and most influential of German Enlightenment thinkers. Like Leibniz, Kant argued for a more metaphysical understanding of the universe. One of the problems that English and French thinkers faced was that the mechanistic, law-bound approach left little room for the human soul or spirit. If all empirical phenomena can be reduced to natural law, then human consciousness is nothing more than chemical reactions among bits of brain matter. To solve this issue, Kant argued that the mind has intrinsic categories and ideas that are used to order our experience of the empirical world. In other words, the mind isn't a blank slate upon which experience writes itself; rather, humans are born with mental structures already in place. Though Kant argued that there is an objective, empirical world "out there," he also asserted that raw empiricism isn't possible because what we see is prefigured by our minds. Humans therefore can never grasp the thing-in-itself because the mind imposes its ideas upon material reality. The German Enlightenment, then, had a strong base in mysticism and was concerned with preserving the metaphysical existence of human nature and reason.

Marx was structurally an outsider. He was born a Jew in a nation that would become known for its persecution of Jews. This outsider status haunted Marx throughout most of his life and certainly during his formative years. This position was exacerbated by the fact that during the early part of Marx's life, Germany was fragmented into various political and economic territories. Political tensions thus ran high. While these tensions undoubtedly inspired Marx, they also created a situation ripe for his estrangement from his homeland. Marx saw and experienced firsthand the ways in which elite groups could use governments for their own ends. His voice was silenced, which undoubtedly sensitized him to the political use of ideology. Marx also saw where Germany was heading, thanks to his friend Engels. Engels knew from experience the abuses of industrial capitalism; it was from Engels that Marx learned what an entrenched capitalist system could do.

Marx was strongly influenced by the metaphysical or spiritual bent of the German Enlightenment. It is likely this intellectual context affected Marx's sociological imagination more than any other. The *zeitgeist*, the spirit of the age, into which Marx was born was decidedly different from that of either Comte or Spencer. He thus had a different sense about him, a distinct way of seeing the world. Rather than the raw empiricism and brute-force hope found in Comte and Spencer, German intellectuals were concerned with preserving the metaphysical existence of humanity while building in the use of reason to undergird faith. Where English and French intellectuals were busy separating church and state, German philosophers were finding ways of melding modern government with traditional faith. Marx came out of this environment with powerful ideas about human nature, ideology, and a concern for ideology's expression in religion.

Marx's Intellectual World

We've already spent a good deal of time considering the sources of Marx's sociological imagination. But to truly understand this "greatest living thinker," we have to delve deeper into the intellectual world in which Marx lived. Unlike Spencer, Marx was a true intellectual in the sense that he spent years of his life immersed in books and philosophical arguments. Marx's theory is thus deeply embedded in philosophical concerns—most of them, as you would guess from the unique qualities of the German Enlightenment, revolve around metaphysics.

The simplest (and most simplistic) way of understanding *metaphysics* is to take the word literally. The suffix "meta" means with or after. Metaphysics thus refers to that which is alongside of or after the physical. So, imagine the universe ends in flames, burning away all planets, stars, and matter itself; whatever is left over would be metaphysical. It should be clear that the only animal that could even think of such a thing is the human being. We're the only ones that ask the question, "What is the meaning of life?" It appears that all other animals take life as it is; humans, on the other hand, need to have things mean something. We humans have long been convinced that we have something other than a purely physical existence. You could call it the spirit, soul, or mind, but for most of our history, we have thought there's something other than or added to pure physicality for us. That idea, of course, is in back of every religion; and it also forms the beginning point for philosophy.

The important thing I want you to see here is that almost everything that Marx talks about is based upon or comes back to these sorts of transcendent issues. He is very concerned about human nature, for example. Marx didn't just say that capitalism is unfair to the working class; he argued that capitalism perverts or destroys human nature. To say something like that, Marx had to have a clear idea of what human nature is in the first place. I'm sure you've heard the terms *false consciousness* and *ideology.* The only way to have false consciousness is to first assume a true consciousness. The word *ideology* simply means the study of ideas, but in Marx's hands, it becomes the study of *false* ideas, and to talk about that, Marx must first assume what real ideas are. All of these foundational ideas are deeply philosophical, and all speak of something other than simple physical existence.

That doesn't mean that Marx isn't concerned about physical existence; he is. In fact, it is his primary concern, but he's concerned about material existence *because of* his metaphysical base. The way Marx connects our material existence to such things as consciousness is utterly unique and stunning in its elegance. In order to see such things in Marx, we'll have to look at the ideas of Hegel, whom I briefly mentioned previously, as well as those of Ludwig Andreas von Feuerbach (1804–1872) and Adam Smith (1723–1790). Please take the time to work through the ideas that follow. If you do, the rest of Marx will be a breeze, comparatively speaking, anyway.

Georg Wilhelm Friedrich Hegel

Like Kant, Hegel was an idealist, but he criticized Kant's approach. Kant left the human being divided in two ways. First, because the mind has and uses built-in categories to understand the world, we can never truly know the thing-in-itself. We can never directly know the world of physical objects because our mind gets in the way. Second, Kant set reason and passion against each other. While reason is good and possible, the "disease of the soul," human passions, stood in the way. Thus, while Kant elevated the mind above pure materialism, he left us isolated and in struggle.

Enduring Issues

There are several intertwined philosophical issues that have concerned sociology since its inception, most notably epistemology (the study of knowledge) and ontology (the study of existence). Materialism and idealism are two philosophical schools that are concerned with these questions. *Materialism* makes the argument that the only thing that exists is physical matter. Our knowledge, then, is based on intrinsic characteristics of material objects and the way they impact the brain. As you would expect, *idealism* argues that ideas exist independently of material objects. This was basically Kant's argument, and Hegel is seen as an absolute idealist. This split between idealism and materialism becomes really interesting when sociology gets involved. Marx and Durkheim both speak directly to the issue of existence and knowledge. Both argue for a material base to reality, but the "matter" in this case is society, not physical matter. In simple terms, Marx's materialism is based in economic production, and Durkheim's is founded in social morphology. More recently, sociologists Peter Berger and Thomas Luckmann (*The Social Construction of Reality,* 1966) argue for another approach: a phenomenological sociology of knowledge, which simply looks at how what we take to be "common knowledge" creates our sense of reality. This approach has become significant in recent sociological literature, with multiple works looking at the social construction of race, gender, and sexuality. The issues of epistemology and ontology continue in the sociology of science, postmodernism, poststructuralism, and so on.

Hegel began his argument with sense-certainty. *Sense-certainty* is pure experience apart from any language. Such a thing is difficult for us to imagine because we casually assume the reality of our experiences. Let's take something mundane for an example, like an apple. Imagine you have an apple in front of you. What do you see? If you say you see an "apple," you're not experiencing the thing-in-itself because "apple" is part of language. In fact, you would have to be able to get rid of all linguistic ideas associated with apple, such as "fruit," "red," "hard," "crunchy," ad infinitum in order to experience it with sense-certainty.

There's an intrinsic problem with sense-certainty for humans, and we just experienced it: Sense-certainty has no meaning, nor can it be communicated. It is, in fact, not consciousness. This part is a little tricky, but it is imperative for understanding Marx. *Consciousness* is always awareness *of* something—it isn't simple sense-certainty. With sense-certainty, an organism is *in* the experience; there's no secondary awareness of the experience. The story of human history, then, is one where the mind evolved toward consciousness.

Defining *consciousness* is difficult, even though philosophers have been studying it since Aristotle. The most common definition is that "an organism is conscious if there is something it is like to be that organism" (Chalmers, 1995, p. 200). In question form, is there a subject (*you*) of the experience? If there is, then what is that subject? Notice that being conscious of an experience is *something other than* the pure experience itself. The question is, what is that other? Many disciplines have weighed in on this issue, most recently neuroscience and quantum mechanics. A goodly number of sociologists have theories of consciousness as well. Marx is one we look in this book, as are Durkheim and Mead. When considering sociological theories, focus on this one question: What is the source of the other—the *subject* of experience?

The easiest way for me to help you see what Hegel is talking about is to use the example of language. Imagine going to the zoo with your dog (if you don't have one, imagine a dog as well). Now imagine that you and your dog are standing in front of the gorilla cage. How will your experience be different from your dog's? Chances are very good that your dog will have sense-certainty about the other animal—she'll probably bark her head off. You, however, will understand and make meaningful that imposing biological entity by recognizing it as a caged-in-the-zoo gorilla, and if you've taken particular classes, you may recognize it as a primate. But "primate" and "gorilla" don't exist in the physical world; they are, as Kant would point out, categories of the mind.

However, *there's an intrinsic problem with consciousness*. The problem is that once these categories are used, we take them as physical reality, but they aren't. Here's an easy example: Up until 2006, all astronomy textbooks taught that Pluto was a planet. So when you were in grade school, you and your teachers took Pluto's planet status as physical reality. Of course, we now "know" that Pluto isn't a planet; the problem, again, is that most will take that new category as physical reality. Humans make up categories and explanations, impose them on the material world, and then see their own creations as objective, material reality. This, in Hegel's eyes, is considered *alienation*.

This problem of *reification* (making something real that isn't) has been around since we started using language, but it became acute because of science. "Classic science is based on the belief that *there exists a real external world* whose properties are definite and independent of the observer who perceives them" (Hawking & Mlodinow, 2010, p. 43, emphasis added). This is where Kant comes in. Remember that Kant argued that categories and theories originate in the human mind, and we can thus never get to the thing-in-itself. Kant also argued that scientific laws, such as Newton's law of gravity, likewise originate in the mind. Kant's importance is that he systematically exposed the mind's role in not only the perception but also the creation of material reality. Hegel sees Kant's work as the beginning of self-consciousness, as we will see, which is the next step in his evolution of consciousness.

There are a couple of things to notice. First, Hegel is giving us a history of the mind or consciousness. The mind has historically evolved from sense-certainty to consciousness, and from consciousness to self-consciousness. For Hegel, history is nothing more and nothing less than the process of the mind and ultimately the universe becoming conscious. The second thing to notice is that Hegel isn't simply mapping history; he is also giving us *the dynamic in back of history*—the movement principle, the reason things change.

The dynamic in back of the evolution of consciousness is dialectic in character. The word *dialectic* comes from the Greek word *dialektikos,* meaning discourse or discussion. Dialectic is different from debate because in a debate, each side is committed and seeks to win over the other; in a dialectic, there are opposing elements, but the goal is to reach resolution, not to win. The key to the dialectic is that the beginning state generates its own contradiction. Science might be a good example to consider in isolation. Notice that the preceding quote from physicists Stephen Hawking and Leonard Mlodinow talks about "classic science." That implies that there is a new kind of science. Hawking and Mlodinow (2010) call this new science "model-dependent realism." The only thing we need to know about model-dependent realism right now is that it rejects the fundamental assumption of empiricism upon which classic science was based. This means that classic and contemporary science contradict one another. How did science go from assuming empirical reality to rejecting it? The short answer is that science got there by working according to the scientific method. For example, modern physics started with Newton and then over time got to Einstein's relativity by using scientific methods. It then moved from Einstein to quantum physics, and quantum physics to string theory in the same way. Science, then, produced its own contradictions that changed the very nature of science.

Definition

A *dialectic* is a theoretical concept that describes the intrinsic dynamic relations within a phenomenon. Dialectics contain different elements that are naturally antagonistic to or in tension with one another—this antagonism is what energizes and brings about change. Dialectics are cyclical in nature, with each new cycle bringing a different and generally unpredictable resolution. The resolution contains its own antagonistic elements, and the cycle continues (though Marx's theory does resolve because of his assumptions about human nature). A goodly number of contemporary theorists continue to use this concept.

For Hegel, the dialectic is intrinsic to consciousness. Hegel assumed that the universe is meant to be conscious. As such, sense-certainty is a contradiction; it knows only the material world. That tension pushed for consciousness to evolve, and we became conscious through language, religion, and so forth. But this form of consciousness—awareness of being in the world—isn't perfect consciousness; it perceives a separation between the observer and observed. That separation is basic to classic science: The world is *empirical,* and humans can *discover* the empirical world. We can't discover something if we are part of it; discovery implies separation. So the history of consciousness reached critical mass with classic science. Kant took the next step by telling us that the categories and explanations of the material world that we use exist first in the mind.

This step of Kant's was the beginning of *self-consciousness* (awareness becoming aware of itself), and Hegel saw his work as perfecting the beginning that Kant instituted. Hegel is an absolute idealist; that is, he argued that self-consciousness is Absolute Truth. It's the point to which the universe has been evolving since its inception. In self-consciousness, Hegel saw humanity moving toward the realization that there is nothing beyond consciousness, least of all the thing-in-itself. The "outer" material world is simply objectified spirit, not something fundamentally different from consciousness. Self-consciousness is becoming aware that consciousness is the final and ultimate reality. *Self-consciousness is, in fact, the mind of God.* It is vital that you don't read "God" here as an entity separate from humanity and that you don't posit God as existing before all things. Both those suppositions create separation, and with true self-consciousness there is no separation. All things exist within the mind of God, and individual consciousnesses are nothing more than the particularization of that universal mind. "The essence of spirit, then, is self-consciousness" (Hegel, 1830/1975, p. 51).

Definition   Hegel's definition of *self-consciousness* is tricky for us because we normally assume that "self" refers to the individual person; but that's not at all what Hegel has in mind. This "self" is the Universal Self, or more specifically, it's *the universe becoming aware of itself.* Rather than seeing God and creation as separate, Hegel sees the universe as developing its own God-consciousness, with humanity as the instrument of this development. It's important to see that at the "moment of creation," consciousness and matter were both present; one cannot exist without the other, according to Hegel. Yet both were also undeveloped and so evolved. The Absolute Truth of the universe is that matter and consciousness are locked together dialectically, and history will cease to exist once this Truth is embraced.

Marx built his theory on these Hegelian ideas: human nature, reality, dialectical history, and consciousness. I'll explain these in greater detail later, but let me give you a hint about the direction Marx will take using his own words. In explaining the sources of his theory, he credits Hegel with "a guiding thread for my studies. . . . It is not the consciousness of men that determines their being, but, on the contrary, their social being that determines their consciousness" (Marx, 1859/1978g, p. 4).

Ludwig Andreas von Feuerbach

Feuerbach was a member of the Young Hegelians, a group of young men who wanted to take Hegel's ideas further. Hegel basically argued that his period of time was the dawn of a new age of spiritual enlightenment. In addition to Kant's insights, Hegel also felt that the Lutheran Reformation was key to the evolution of consciousness. Luther proclaimed that each individual believer was responsible for her or his relationship with God. No outside authority (the Catholic Church) or intermediary (the priesthood) was necessary; an individual's conscience in harmony with the Spirit of God is all that was needed. The Young Hegelians criticized Hegel for leaving even this brand of religion untouched.

Feuerbach argued that religion and God are nothing more than projections of human nature. This is where, Feuerbach contended, true alienation exists. Mankind created God in his own image and then worshiped God as if He were Other. Feuerbach also criticized Hegel for seeing the dialectic as spiritual. Feuerbach argued that material objects are not simply mental projections (as in Kant), and neither are they held within the mind of God (Hegel) and thus fundamentally spiritual. Rather, reality is material, and our consciousness of it is based on contemplation of the objective world.

Marx found much to agree with in Feuerbach. He agreed that the world is material and concurred with the alienating work of religion. Yet Marx takes Feuerbach further by arguing first that Feuerbach didn't understand the true material basis of human existence. Humans aren't generally concerned with purely material objects. Rather, the objective, material world in which humans live is a produced one, created through economic production. In this, Marx claimed, Feuerbach missed true human nature; he, like Hegel, was too focused on humans as individual entities. Marx contends that human nature is social: "the human essence is no abstraction inherent in each single individual. In its reality it is the ensemble of social relations" (Marx, 1888/1978h, p. 145). In addition, Feuerbach fell into the trap of philosophy, believing that abstract thought is an end in itself. Marx argued that after philosophizing, the "chief thing" still remains: "The philosophers have only *interpreted* the world, in various ways; the point, however, is to *change* it" (p. 145). This is the source for the Marxian idea of *praxis*, practical engagement with the world.

Adam Smith

Hegel and Feuerbach informed the metaphysical background of Marx's theory. Yet Marx is primarily concerned with material existence and for that portion of his theory, he drew on Adam Smith. Smith was a philosopher of the Scottish Enlightenment and is considered the father of classic economics. His most important economic work was *An Inquiry Into the Nature and Causes of the Wealth of Nations*, which was the first systematic defense and explanation of free market capitalism. The first edition of the book sold out within 6 months of publication, and Smith's theories began informing debates and policy decisions in Britain and the United States as early as a year after its publication.

Enduring
Issues

The question of government involvement in capitalist economy is a politically hot topic and a central issue in the social sciences. Adam Smith formed the base for what's known as "laissez-faire" (let do) economics or *free market capitalism*. This doctrine of minimal state involvement is based on the assumption that free markets have inherent mechanisms that keep it functional—this is an evolutionary idea much like that of *natural selection*. Marx sees something quite the opposite happening. He argues that the state actively supports capitalism in its exploitation of the working class. The question of the relationship between capitalism and government continues today. For example, Jürgen Habermas argues that increasing state involvement in capitalism has had disastrous effects on democracy. Another contemporary theorist, Immanuel Wallerstein, theorizes that government subsidizes capitalist enterprise in core nations and is thus a central dynamic in globalizing capitalist exploitation, which will, in the long run, lead to "the end of the world as we know it."

Smith didn't actually use the word *capitalism* in the book. Rather, he called it the system of perfect or natural liberty. Just as Hegel did with consciousness, Smith saw capitalism as the natural result of social evolution; its evolutionary purpose was to bring individual freedom. This economic system of perfect liberty was seen as part of the process of doing away with traditional systems of stratification, where one's social position was determined by birth, and creating a system of merit (meritocracy) in which natural selection could work through competition in the marketplace. This relationship between capitalism and freedom not only set a goal for Smith's work, but it motivated Marx's writing as well. "The conception of capitalism as the period in history when freedom is finally established sets an intellectual agenda for Smith, just as the conception of capitalism as a way station toward freedom sets a different one for Marx" (Heilbroner, 1986, p. 9).

Smith argued that national mercantilism's protective tariffs and gold-based economies were hindrances to the evolutionary move to capitalism. Rather than regulating imports and exports, the national economy will prosper as states remove market controls. Specifically, Smith argued that humans have a natural inclination to barter, to realize profit from the exchange of goods and services. This motivation is based on the "natural effort of every individual to better his own condition, when suffered to exert itself with freedom and security" (Smith, 1776/1937, p. 508). In bettering her or his own condition, the individual unintentionally helps everyone. The individual "neither intends to promote the public interest, nor knows how much he is promoting it. . . . [H]e intends only his own gain, and he is in this . . . led by an invisible hand to promote an end which was no part of his intention" (p. 421).

Marx argued that Smith's idea of the *natural individual* was a myth. Generally speaking, the literature by philosophers and political economists of Smith's time was filled with this notion of human nature as being innately individualistic. However, these writers missed the fact that the idea is historically specific: The individual "appears as an ideal, whose existence they project into the past" (Marx, 1939–1941/1978e, p. 222). Marx contended that if we simply let history tell its tale, we'll find that mankind is *an intrinsically social animal.* Economic production is thus socially based, not individually as Smith saw it. According to Marx, seeing isolated, individual productivity as leading society "is as much an absurdity as is the development of language without individuals living *together* and talking to each other" (p. 223).

Smith also had a theory of value that informed Marx's theorizing. Smith argued that the natural value of a commodity is determined by human labor, in terms of costs to both produce and acquire goods. To increase productivity and profit, labor must be divided, or the *division of labor* must increase. The division of labor is the process through which the work done to create a good is broken down into small increments. Smith uses the example of making a wooden pin that's used to secure two pieces of wood together. A man working the trade by himself may make a pin a day. Smith observed that a small manufacturing firm that used 10 men to produce pins was able to produce 12 pounds of pins. Each person, then, "might be considered as making four thousand eight hundred pins in a day" (Smith, 1776/1937, p. 5).

Enduring Issues

At first glance, the *division of labor* doesn't seem like an important question. Yet, in our book, Marx, Durkheim, and Simmel directly address this issue, and it is a key issue in understanding race and gender. In addition, the division of labor is an important concept in studying globalization. The division of labor is simply the way in which work is divided in any economy. It can vary from everyone doing similar tasks to each person having a specialized job. In previous epochs, labor was more holistic in the sense that the worker was invested in a product from beginning to end. One of the distinctive traits of modernity is the use of scientific management, or Fordism, to divide work up into the smallest manageable parts. Today the issue of the division of labor is global. The global division of labor impacts political relations (e.g., the United States is utterly dependent upon China for its steel) and individual job opportunities (e.g., moving the textile industry to developing nations took jobs away from thousands of people in North Carolina).

Marx pushed Smith's idea further, first by including industrialization, which increased production even more dramatically, and by arguing that both the division of labor and industrialization increased workers' alienation. This is a clear instance where we can see how Marx's humanism determined his perspective. While capitalism may have economic benefits in terms of increased production and profit, it also takes a toll on the human spirit. Later on, we'll also see how Marx extends our understanding of the division of labor to include the division between mental and material labor. That division, Marx argued, dramatically influences the kind of knowledge we create.

Smith also contrasted the natural value of a commodity with its market value. The law of supply and demand in the short run determines the market value of a product. If the market value is lower than the natural value, the market value will increase. On the other hand, if the market value is higher than the natural, other capitalists will begin producing the product to take advantage of the higher profits. This in turn drives prices down to their natural level. There is thus an "invisible hand" at work in price control as well. If left free, the market will always produce the best product for the least cost, according to Smith.

Marx used these ideas to make distinctions among exchange value, use value, and surplus value. *Surplus value*, the difference between a commodity's use and exchange values, is the source of capitalist profit. Capitalists, then, are driven to increase surplus value at any cost. This motivation in the long run creates business cycles of overproduction, which for Marx is one of the primary dialectical contradictions of capitalism that will in the end generate socialism. Further, surplus value itself contains a contradiction: It is both the source of capitalist profit and workers' negotiating power. For Marx, these contradictions are the invisible hand of the marketplace. Like Smith's idea, these contradictions work as unseen mechanisms. Unlike Smith, however, Marx saw the collapse of capitalism as the outcome of this kind of invisible force. We'll talk more about these issues in a bit. Now, I want to bring all these ideas together and give you a more cohesive description of Marx's perspective.

Marx's Sociological Imagination: Critical Conflict Theory

The above section title implies that there is a *non*critical conflict theory, and that's true, though we usually refer to the two perspectives simply as conflict theory and critical theory. Marx is the foundational thinker in both those approaches. Conflict theory generally seeks to explain the social factors and processes that lead to conflict in society, as well as the results of conflict. Theorists and researchers in this camp usually take the classic science model of discovering the universal laws of human conflict. From this conflict perspective, the social factors that create inequality today are basically the same as the ones that stratified the Roman Empire. Moreover, this approach is nonevaluative; it isn't necessarily oriented toward changing society.

Critical theory, on the other hand, isn't as convinced of the existence of invariant laws, generally seeing social factors and processes as historically specific. So the power inequities in Roman society worked differently from those in today's society. Critical theorists are also generally interested in how inequality influences people and knowledge. Critical theorists are more humanistic in this way, and conflict theorists are more scientific. Critical theory is evaluative, ethical, and very much interested in changing society.

Please be aware that I just made some very clear distinctions between the two approaches, whereas real life is usually messy. There are obviously people who see themselves as practicing humanistic science, for example. In fact, as the founder of these ways of doing sociology, Marx is a prime example of someone who fits with both perspectives. So think of these as two poles on a continuum, with sociologists, political scientists, economists, and so forth doing work at different points along that line.

Human Nature

Marx argues that the unique thing about being human is that we create our world. All other animals live in a kind of symbiotic relationship with the physical environment that surrounds them. Zebras feed on the grass, and lions feed on the zebras; in the end, the grass feeds on both the zebras and the lions. The world of the lion, zebra, and grass is a naturally occurring world, but not so for the human world. Humans must create a world in which to live. They must in effect alter or destroy the natural setting and construct something new. The human survival mechanism is the ability to change the environment in a creative fashion in order to produce the necessities of life. Thus, when humans plow a field or build a sky-scraper, there is something new in the environment that in turn acts as a mirror through which humans can come to see their own nature. Self-consciousness—our awareness of being human—comes about as we see human nature reflected back in what we've created. There is, then, an intimate connection between pro-ducer and product: The very existence of the product defines the nature of the producer.

The idea that there is an intrinsic connection between humans and their cre-ations is sometimes difficult to grasp, so let me give you a small analogy. Imagine you are invited to a friend's house for a dinner party, and they ask you to bring a cake. So, you look through your grandmother's recipes and find the German chocolate cake she used to bake for family gatherings. You bake the cake using real German's Sweet Chocolate, just like granny did, and you add a couple of your own ideas: a bit more vanilla and more pecans. After dinner, the cake is served with fresh brewed coffee. One of the guest's exclaims, "This is the best German chocolate cake I have ever had!" A sense of pride and remembrance of your grandmother sweeps over you. You say "thank you" and pause for a moment to think about the last time your grandmother baked this cake for you, and perhaps you tell a story or two about her.

Now imagine a different scenario. You're still invited for dinner and asked to bring a cake, except this time you stop off at the local grocery store and buy the German chocolate cake. The same thing happens after dinner: A guest exclaims, "This is the best German chocolate cake I have ever had!" Now how do you feel? Rather, what *don't* you feel? You don't feel pride and you don't sense that intimate connection with your grandmother. At the first imagined dinner, the cake wasn't simply a commodity, a product of the local grocery designed to gain profit. No, you were personally and intimately invested in the first cake, and you sensed a deep connection to your grandmother as well. And, you know what? Almost anyone would feel the exact same thing. This personal and emotional connection isn't peculiar to you; it's something common to us as humans.

I think you can see what Marx is trying to tell us. It is our nature as human beings to create; when we do, we invest a portion of nature in the creation and we see ourselves in it. More than that, our handiwork carries deep, meaningful social connections. That first cake carried a bit of you and your relationship with your grandmother, and when you shared the cake with others, they experienced a connection with you and you with them. Your grandmother was there, too. Now, imagine that every single product that you and others own has such a character—every creation carrying the same personal and social investments. Your world would be aglow with this creative, aesthetic force. This is the meaning of *species-being*.

Definition

Species-being is one of Karl Marx's basic assumptions about human nature. The idea links the way humans as a species survive with human consciousness. According to Marx, every species is unique and defined by the way it exists as a biological organism. Humans exist and survive through creative production. Human consciousness, then, is created as people see the humanity in the world that has been economically produced. False consciousness and ideology increase as humans fail to perceive their intrinsic link to production

Marx implies that human beings in their natural state live in a kind of immediate consciousness. Initially, human beings created everything in their world by hand. There weren't supermarkets or malls. If they had a tool or a shirt, they had made it themselves, or they knew the person who did. Their entire world was intimately and socially connected. They saw themselves, their collective nature, purely in every product. Or, if they had bartered for something, then they saw an immediate social relationship with the person who had made the thing. When they looked into the world they had produced, they saw themselves; they saw a clear picture of themselves as being human (creative producers). They also saw intimate and immediate social relations with other people. The world that surrounded them was immediately and intimately human. They created, controlled, and understood themselves through the world that they had made.

Notice two very important implications of Marx's species-being: Human beings by their nature are social, and they are altruistic. Marx's vision of humanity is based on the importance of *society* in our species survival. We survive collectively and individually because of society. Through society, we create what is needed for survival; if it were not for society, the human animal would become extinct. We are not equipped to survive in any other manner. What this means, of course, is that *we have a social nature—we are not individuals by nature.*

Species-being also implies that we are altruistic. *Altruism* is defined as uncalculated commitment to others' interests. Not only are we not individualistic by nature, but we are also not naturally selfish. If human survival is based on collective cooperation, then it would stand to reason that our most natural inclination would be to serve the group and not the self. The idea of species-being is the reason why Marx believes in communism—it's the closest economic system to our natural state. Further, Marx would argue that under conditions of modern capitalism, these attributes are not apparent in humans because we exist under compromising structures. It is capitalism that teaches us to be self-centered and self-serving.

Hint

In Chapter 1, I mentioned that the assumptions theorists make carry implications, though they aren't always apparent. Here's one place we can see this important point: Marx assumes that people are by nature social and inclined to naturally cooperate. This implies that in most cases, social structures are detrimental to people. In this way, Marx is against big government! And this is why Marx argues for communism: It is most in keeping with his view of human nature. What I want you to see here is the importance of thinking through the implications of an idea or theory. The Latin root of *implication* literally means to be entangled. Theoretical and critical thinking demand that we think through everything that comes packaged (entangled) with our ideas, concepts, and theories.

Marx's theory of species-being also has implications for knowledge and consciousness. In the primitive society that we've been talking about, humans' knowledge about the world was objective and real; they held ideas that were in perfect harmony with their own nature. According to Marx, human ideas and thought come about in the moment of solving the problem of survival. Humans survive because we creatively produce, and our clearest and truest ideas are grounded in this creative act. In species-being, people become truly conscious of themselves and their ideas. Material production, then, is supposed to be the conduit through which human nature is expressed; the product should act as a mirror that reflects back our own nature.

Let's try an analogy to get at this extremely important issue. There are a limited number of ways you can know how you physically look (video, pictures, portraits, mirrors, and so forth). The function of each of these methods is to represent or reproduce our image with as little distortion as possible. But what if accurate representation was impossible? What if every medium changed your image in some way? We would have no true idea of how we physically look. All of our ideas would be false in some way. We would think we see ourselves, but we wouldn't. Marx is making this kind of argument, but not about our physical appearance. Marx is concerned with something much more important and fundamental—our nature as humans. We think we see it, but we don't.

We need to take this analogy one step further: Notice that with mirrors, pictures, and videos, there is a kind of correspondence between the representation and its reality. What I mean is that each of these media presents a visual image. In the case of our physical appearance, that's what we want. Imagine if you asked someone how you looked, and the person played a CD for you. That wouldn't make any sense, would it? There would be no correspondence between the mode of representation and the initial presentation. This, too, is what Marx is telling us. If we want to know something about our human nature, if we want to see it represented to us, where should we look? What kind of medium would correspond to our nature? Marx is arguing that every species is defined by its method of survival or existence. Why are whales, lions, and hummingbirds all different? They are different because they have different ways of existing in the world. What makes human beings different from whales, lions, and hummingbirds? Humans have a *different mode of existence*. We creatively produce what we need—we are the only species that makes products.

So, where should we look to understand our nature? What is the medium that corresponds to the question? If we want to know how we look physically, we look toward visual images. But if we want to know about our nature, we must look to production and everything associated with it. Thus, according to Marx, production is the vehicle through which we can know human nature. However, Marx says that there is something wrong with the medium. Under present conditions (capitalism), it gives a distorted picture of who and what we are.

If we understand this notion of species-being, then almost everything else Marx says falls into place. To understand species-being is to understand alienation (being cut off from our true nature), ideology (ideas not grounded in creative production), and false consciousness (self-awareness that is grounded in anything other than creative production). We can also understand why Marx places such emphasis on the need for class consciousness in social change. This understanding of human nature is also why Marx is considered to be an economic determinist. The economy is the foundation upon which all other structures (superstructure) of human existence come into being and have relevance.

History—The Material Dialectic

When we think of history, most of us think about what's written in books. That's our history; it tells the story of what we've been through. Marx sees history in that way as well, but the characters in the story are different. Most of our histories are of events and people. (In 1492, Columbus sailed the ocean blue.) Marx would see that sort of history as focusing on secondary effects rather than the important issues. Imagine a newspaper story that told you that visibility was horrible in San Francisco yesterday. It was difficult to see for a quarter mile, and even being able to breathe properly was a challenge. Now imagine going to your local bookstore for coffee that same day. On your way to the coffee counter, you see a different newspaper covering the same story, only this time the headline reads, "San Francisco Destroyed by Largest Quake in History—City Burns in Aftermath." You come to find out, the poor visibility and problems with breathing were due to the fire that resulted from the earthquake! That may sound like a silly example, because no newspaper would ever make that sort of mistake, but that is exactly the same sort of mistake we make in seeing history in terms of characters and events. When we focus on characters and events, we *misrecognize* what's really going on. Just like the poor visibility of our first imaginary newspaper, characters and events are simply the effects of social structures moving and shaking beneath the surface.

Thus, *accurate histories will chronicle the changing face of social structures.* Those are the important issues of history, not "men and their moments." The powerful story of history, for social historians and sociologists, concerns the structural forces that create those events and people. The fundamental question of such history is its driving force: What pushes for structural change? For Hegel, human history is driven by the tension that exists when the material world and consciousness are separate. That tension pushes movement forward to a resolution. For Marx, the defining feature of humanity is production, not ideas and concepts. In his view, Hegel's analytical lens was misplaced. The dialectic is oriented around material production and not ideas. This is the material dialectic. Thus, the dynamics of the historical dialectic are to be found in the economic system, with each economic system inherently containing antagonistic elements (see Figure 3.1). As the antagonistic elements work themselves out, they form a new economic system.

Figure 3.1 Marx's Material Dialectic

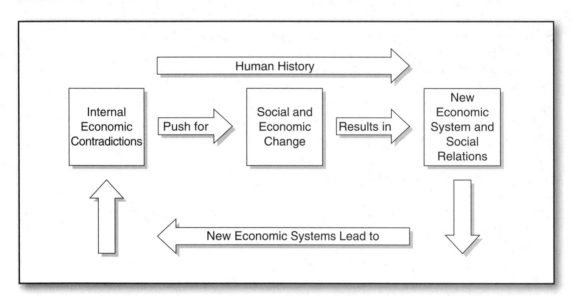

Notice that the engine of progress according to the dialectic is tension or conflict. This is an important insight because Marx is sometimes seen as a warmonger, pushing for revolution. It's true, he did support revolutionary efforts, but here we can see that it wasn't simply a personal issue: *Conflict is built into the dialectic—conflict is the driving force of history* (from a Marxian point of view). We can think of this way of seeing society as an upheaval model. According to this perspective, society is not like an organism that gradually and peacefully becomes more complex in order to increase its survival chances, as in functionalism. Rather, society is filled with human beings who exercise power to oppress and coerce others. Periods of apparent peace are simply times when the powerful are able to dominate the populace in an efficient manner. But, according to this model, the suppressed will become enabled and will eventually overthrow and change the system. Social change, then, occurs episodically and through social upheaval.

In mapping out the past of the historical dialectic, Marx categorizes five different economic systems (means of production along with their relations of production): pre-class societies, Asiatic societies, ancient societies, feudal societies, and capitalist societies. Pre-class societies are like hunter-gatherer groups. These were small groups of people with a minimal division of labor (one that Marx termed the *natural division of labor*) and communal ownership of property (termed *primitive communism*). Asiatic societies were a special form in that they had particular problems to overcome due to their large populations. To solve these problems, there was a tendency to form systems of government that gave emperors absolute power. Ancient societies developed around large urban centers, such as Rome. Private property and slave labor came into existence, as well as significant class inequality. Ancient societies were replaced by feudal systems wherein the primary economic form was serf labor tied to the land of the aristocracy. Feudal systems were replaced by capitalist systems. Eventually, the capitalist system would be replaced by socialism and that, in turn, by communism. The specific dynamics that Marx says cause these shifts in economic systems are not important in our consideration right now. What is important to note is that Marx argues that social change occurs because of inherent contradictions in the economic structure.

Hint

Marx is a structuralist (as are many sociologists). This means that change and stability happen because of social structures and *not individual people*. We can think of this using the analogy of the human body: Individuals are like biological cells that are relatively unimportant to the body as a whole. When working with structuralist theory, keep this in mind. In terms of causal force, the actions of social structures are important, not the actions of individuals. However, many sociologists are not structuralists. In this book, George Herbert Mead is a good example.

Concepts and Theory: The Contradictions of Capitalism

Taking his cue from Hegel, Marx gives us a theory of humankind. He obviously has a theory of capitalism and class relations, but those issues are only meaningful in the broader context of his theory of humanity. Interestingly, this is true of religion as well. All religions are based upon an assumed human nature. For example, Christianity holds that all people are born in sin as a result of Adam's original sin. The implication of that assumption is that humans need salvation from the effects of sin—death—which has been provided by Christ. Hinduism, on the other hand, assumes that true human nature is spiritual—there is no difference between *atman* (the soul) and *Brahman* (God or Universal Spirit). The problem here isn't sin; the problem is that people become attached to and desire the things of this world. The implication of this view of human nature is that people must detach from this physical realm and realize their true Godhood through meditation and reincarnation.

These are obvious simplifications, but I want you to see first the importance of making an assumption about human nature. Once made, the assumption influences everything else. I also want you to see that Marx's intention is to liberate us so that we can become fully human, just as it is the intent of religion. Marx's issue with religion isn't gratuitous. From his perspective, religion misunderstands human nature and in doing so, offers false hope. Most people don't understand this aspect of Marx and his theory and thus wrongly criticize the man. A well-known psychoanalyst of the twentieth century, Erich Fromm (1961), said that "Marx's philosophy constitutes a spiritual existentialism in secular language" (p. 5).

Hint

Here are two theoretical hints for you. First, one of the most important aspects of critical thinking is *implication*—what something implies. Ideas or concepts always come intertwined with other ideas or concepts. I encourage you to always ask, "What are the implications of this?" The second hint is that while I say that Marx's notion of human nature is an assumption, it's an assumption based on a logical argument, and this is quite often the case with theorists. Thus, if you find that you disagree with an assumption like Marx's, it isn't sufficient to simply say, "I think he's wrong." To legitimately disagree, you must address his argument and present a reasoned account of your conclusion.

Marx builds his theory on his view of human nature. Our nature, like every other animal on the planet, is tied up with the way the species survives. According to Marx, humans survive through economic production. Borrowing from Hegel, Marx sees the history of economic forms developing through dialectical contradictions. In the case of capitalism, the system contains inherent contradictions that will eventually lead to its demise. The most important contradiction is overproduction. In order for us to understand that contradiction, we have to first think through the concepts of value, labor, exploitation, commodification, and class.

Value and Exploitation

One of the important issues confronting early political economists concerned the problem of *value*. We may have a product, such as a car, and that product has value. But from where does its value come? More importantly, why would anyone pay more for the car than it is worth? Well, you say, only a sucker would pay more for a car than it is worth. Yet, as you'll see, there is a way in which we all pay more for every commodity or economic good than it is worth. That was what struck the early economists as a strange problem in need of a solution.

Drawing on Adam Smith's labor theory of value, Marx argues that every commodity has at least two different kinds of value: use-value and exchange-value. *Use-value* refers to the actual function that a product contains. This function gets used up as the product is used. Take a bottle of beer, for example. The use-value of a bottle of beer is its taste and alcoholic effect. As someone drinks the bottle, those functions are expended. Beer also has exchange-value that is distinct from use-value. *Exchange-value* refers to the rate of exchange one commodity bears when compared to other commodities. Let's say I make a pair of shoes. Those shoes could be exchanged for 1 leather-bound book or 5 pounds of fish or 10 pounds of potatoes or 1 cord of oak wood, and so on.

This notion of exchange-value poses a question for us: What do the shoes, books, fish, and potatoes have in common that allow them to be exchanged? I could exchange my pair of shoes for the leather-bound book and then exchange the book for a keg of beer. The keg of beer might have a use-value for me, where the leather-bound book does not; nonetheless, they both have exchange-value. This train of exchange could be extended indefinitely with my never extracting any use-value from the products at all, which implies that *exchange- and use-value are distinct*. So, what is the common denominator that allows these different items to be exchanged? What is the source of exchange-value?

Adam Smith (1776/1937) and Marx both argue that the substance of all value is human labor: *"Labour, therefore, is the real measure of the exchangeable value of all commodities"* (p. 30, emphasis added). There is labor involved in the book, the fish, the shoes, the potatoes, and in fact everything that people deem worthy of being exchanged. It is labor, then, that creates exchange-value. If we stop and think for a moment about Marx's idea of species-being, we can see why Smith's notion appeals to him: The value of a product is the "humanness" it contains. But there is something else going on as well.

Capitalists pay less for the labor than its actual worth. I may receive $75.00 per day to work (determined by the cost of bare sustenance for the worker plus any social amenities deemed necessary), but I will produce $200.00 worth of goods or services. The *necessary labor* in this case is $75.00; it's necessary because it provides a living for the worker. The amount of labor left over is what Marx calls *surplus labor* (in this case, $125.00). The difference between necessary labor and surplus labor is the level of *exploitation*. Different societies can have different levels of exploitation. For example, if we compare the situation of automobile workers in the United States with those in Mexico, we will see that the level of exploitation is higher in Mexico (which is why U.S. companies are moving so many jobs out of this country—labor is cheaper elsewhere). Surplus labor and exploitation are the places from which profit comes: "The rate of surplus-value is . . . an exact expression for the degree of exploitation . . . of the work by the capitalist" (Marx, 1867/1977, p. 326).

Exploitation is a central concept in Marx's theory of capitalism. It is the measurable difference between what a worker gets paid and the worth of the product produced—it is the source of capitalist profit. Exploitation also has the characteristic of giving workers leverage over capitalists. The dependency of capitalists upon exploitation for profit is what gives labor the power to negotiate and strike. Exploitation, then, is dialectical in nature.

Definition

Capitalists are in business for only one reason: profit. By definition, then, capitalists are pushed to increase their profit margin and thus the level of surplus labor and the rate or level of exploitation. There are two main ways in which this can be done: through absolute and relative surplus labor. The capitalist can directly increase the amount of time work is performed by lengthening the workday, say from 10 to 12 hours. He or she can also remove the barriers between "work" and "home," as is happening today as a result of increases in communication (computers) and transportation technologies. The product of this lengthening of work time is called *absolute surplus labor*. The other way a capitalist can increase the rate of exploitation is to reduce the amount of necessary labor time. The result of this move is called *relative surplus labor*. The most effective way this is done is through industrialization. With industrialization, the worker works the same number of hours, but his or her output is increased through the use of machinery. These two different kinds of surplus labor can get a bit confusing, so I've compared them in Figure 3.2.

Figure 3.2 Surplus Labor

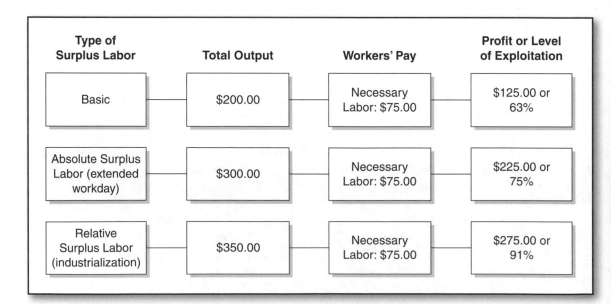

Type of Surplus Labor	Total Output	Workers' Pay	Profit or Level of Exploitation
Basic	$200.00	Necessary Labor: $75.00	$125.00 or 63%
Absolute Surplus Labor (extended workday)	$300.00	Necessary Labor: $75.00	$225.00 or 75%
Relative Surplus Labor (industrialization)	$350.00	Necessary Labor: $75.00	$275.00 or 91%

The figure starts off with the type of surplus labor employed. Since there is always exploitation (you can't have capitalism without it), I've included a "base rate" for the purpose of comparison. Under this scheme, the worker has a total output of $200.00; he or she gets paid $75.00 of the $200.00 produced, which leaves a rate of exploitation of about 63%, or $125.00. The simplest way to increase profit, or the rate of exploitation, is to make the worker work longer hours or take on added responsibilities without raising pay (this happens in downsizing, for example). In our hypothetical case, the wage of $75.00 remains, but the profit margin (rate of exploitation) goes up to 75%. By automating production, the capitalist is able to extract more work from the worker, thus increasing the total output and the level of exploitation (91%). If this seems legitimate to us ("capitalists have a right to make a profit"), Marx would say that it is because we have bought into the capitalist ideology. We should also keep in mind that in their search for maintaining or increasing the rate of exploitation, capitalists in industrialized nations export their exploitation—they move jobs to less-developed countries.

Industrialization, Markets, and Commodification

As we've seen, capitalists are motivated to increase the level of surplus labor and exploitation, and the most efficient way to accomplish these goals is through industrialization. *Industrialization* is the process through which work moves from being performed directly by human hands to having the intermediate force of a machine. Industrialization increases the level of production (by increasing the level of relative surplus labor), which in turn expands the use of markets, because the more products we have, the more points of purchase (markets) we need. The relationships among industrialization, production, and markets are reciprocal so that they are mutually reinforcing. If a capitalist comes up with a new "labor-saving" machine, it will increase production, and increased production pushes for new or expanded markets in which to sell the product. These expanding markets also tend to push for increased production and industrialization. Likewise, if a new market opens up through political negotiations (like with Mexico or China, for the United States) or the invention of a new product, there will be a corresponding push for increased production and the search for new machinery.

An important point to note here is that capitalism, in order to feed the need for continuous capital accumulations, requires expanding markets—which implies that markets and their effects may be seen as part of the dialectic of capitalism. To increase profits, capitalists can expand their markets horizontally and vertically, in addition to increasing the level of surplus labor. In fact, profit margins would slip if capitalists did not expand their markets. For example, one of the main reasons CD players were offered for sale is that the market for cassette tape players bottomed out (and now there is a push for MP3 and newer technologies). Most people who were going to buy a cassette player had already done so, and the only time another would be purchased is for replacement. So, capitalists invented something new for us to buy so that their profit margin would be maintained.

In general, *markets* refer to an arena in which commodities are exchanged between buyers and sellers. For example, we talk about the grocery market and the money market. These markets are defined by the products they offer and the social network involved. Markets in general have certain characteristics that have consequences for both commodities and people. They are inherently susceptible to expansion (particularly when driven by the capitalist need for profit), abstraction (so we can have markets on markets, like stock market futures or the buying and selling of home mortgage contracts), trade cycles (due to the previous two issues), and undesirable outputs (such as pollution). In addition, they are amoral (so they may be used to sell weapons or religion, or to grant access to health care).

The speed at which goods and services move through markets is largely dependent upon a *generalized medium of exchange,* something that can act as universal value. Barter is characterized by the exchange of products. The problem with bartering is that it slows down the exchange process because there is no general value system. For example, how much is a keg of beer worth in a barter system? We can't really answer that question because the answer depends on what it is being exchanged for, who is doing the exchanging, what their needs are, where the exchange takes place, and so on. Because of the slowness of bartering, markets tend to push for more generalized means of exchange—such as *money.* Using money, we can give an answer to the keg question, and having such an answer speeds up the exchange process quite a bit. Marx argues that as markets expand and become more important in a society, and the use of money for equivalency becomes more universal, money becomes more and more the common denominator of *all* human relations. As Marx (1932/1978d) says,

> By possessing the *property* of buying everything, by possessing the property of appropriating all objects, *money* is thus the *object* of eminent possession. The universality of its *property* is the omnipotence of its being. It therefore functions as the almighty being. Money is the *pimp* between man's need and the object, between his life and his means of life. But that which mediates *my* life for me, also *mediates* the existence of other people *for me.* For me it is the *other* person. (p. 102)

Expanding markets create more and more commodities. The simple definition of *commodity* is an economic good. However, it's vital to keep in mind that capitalist commodities are devoid of species-being. They reflect little if any humanity back to us. Whatever part of humanity commodities do contain is perverted—they contain the exploited, alienated nature of workers under capitalism. This perversion goes further. Remember the illustration of the cake? You had a personal investment in the cake. Thus, when your friends had some of the cake, they also experienced you and their friendship with you. In sharing the cake, you also shared your family life with them. True products—those created with species-being intact— thus contain not only human nature but also social relationships. Under capitalism, however, we don't see or experience those social relationships. Worse, we are blind to the exploitive relationships that are in the commodity.

The concept of *commodification* describes the process through which more and more of the human life-world is turned into commodities. Commodification is thus a variable. So, instead of creatively producing the world as in species-being, people increasingly buy (and sell) the world in which they live, and thus see the human world in terms of money rather than human nature and social relations. Because capitalists are driven to increase profits and thus markets, this process of commodification becomes more and more a feature of human life. Think of a farming family living in the United States around 1850 or so. They bought some of what they needed, and they bartered for other things, but the family produced much of their own necessities of life. Today, our world is composed of little else than capitalist commodities.

Definition

Commodification is a theoretical idea in Marxian theory that expresses the process through which material and nonmaterial goods are turned into products for sale. Such commodities do not reflect true human nature or social relations. Because commodities are created (nothing by its nature is a commodity), because humans have the ability to create subjectively felt needs, and because modern capitalism is defined by the endless pursuit of more capital, the process of commodification has no natural limit. Increasing commodification is a result of industrialization and market expansion and directly affects overproduction.

We have to add two more elements to this part of Marx's theory. First, modern capitalism is different from any previous type of capitalism. In terms of people making things to sell and get profit, capitalism has been around for ages. Traditional or old school capitalism was about getting money in order to live. Whatever profit was made went to support people's day-to-day existence. Modern capitalism is very different. Capitalists use some of the profit to support their lifestyle, but the majority of it is invested back into the business. The reason to invest the money (which is now capital) is to make more money, which, in turn, is reinvested with the purpose of making more money so that it can be invested as well. Thus, *modern capitalism is characterized by the endless pursuit of capital.*

The second factor we need to add is human need. All living organisms, including humans, have things they need to survive. We call these requisite needs. However, unlike other animals, humans can create secondary needs—things we want but in truth feel like needs. For example, the very first cell phone call was made on April 3, 1973. Obviously, most people didn't even know about the call, let alone "need" their own cell phone. It took 10 years to bring the cell phone to market. Very few people actually wanted one because they cost about $3,500. It wasn't until 1990, after the price had come down considerably, that there were 1 million cell phone users in the United States. Today, there are over 5 billion cell phones worldwide, with a global population of 6.9 billion. Having a cell phone is now a necessity for many—it's one that you probably feel personally. "I *need* a cell phone." Actually, *I need* a smart phone so I can answer e-mail and keep up with Facebook and Twitter; more than that, all my contacts are in there; my appointment calendar is in there;

and I need it for entertainment, songs, games, a camera—I'd be lost without it! This subjective need is so strong that parents now discipline their children by restricting cell phone use—children and teens *feel* punished if they can't have their cell phone.

Human beings can create subjectively felt needs well beyond requisite needs. The amazing thing is that there doesn't appear to be any boundary beyond which we can't imagine new things to build and new needs to create. Thus, the potential for commodification is endless. There is no natural limit to the ability of humans to create new needs, and there isn't anything in capitalism that will stem the endless pursuit of capital because that's its defining feature.

Putting all this together brings us to capitalism's first contradiction—*overproduction*. Here's how it works: Modern capitalism is defined by the endless accumulation of capital. This drive pushes capitalists to create more and different kinds of commodities for us to buy. Happily, for the capitalists, the capacity of humans to create new needs is boundless. These two open-ended dynamics are supercharged in today's society through added institutional support. Thus, "every person speculates on creating a *new* need in another, so as to drive him to a fresh sacrifice, to place him in a new dependence and to seduce him into a new mode of *gratification*" (Marx, 1932/1978d, p. 93).

Definition

In Marx's theory, *overproduction* is one of the central contradictions of capitalism. Modern capitalism is defined by the endless pursuit of capital accumulation. This essential characteristic of capitalism, coupled with the human capacity to create endless needs, drives ever-expanding markets, commodification, and the use of technology to increase productive output, all of which lead to an overproduction of goods compared to the current demand and, thus, an economic downturn. Every cycle of expansion and downturn is deeper and more widespread. And because there are no natural limits within capitalism to capital accumulation and no limits to the potential for new human needs, these cycles continue until capitalism fails.

Concepts and Theory: Class Revolution

Overproduction sets up a dynamic that will eventually lead to the destruction of capitalism. Remember, this is a dialectical issue: Contradictions intrinsic to capitalism (expanding markets, accumulation of capital, incessant commodification, expansive human needs) bring repeated economic crises. The United States, for instance, has gone through 33 economic cycles since 1854 (National Bureau of Economic Research, n.d.). Marx argues that each cycle will deepen and be more difficult to resolve. At some point, the cycling crises become so bad that capitalism simply can't sustain its business, and it collapses. As you would expect, there are other dynamics at work that go along with the rounds of economic downturns. The most important of those dynamics concerns class.

Class and Class Structure

For Marx, *class* involves two issues: the means and relations of production. The *means of production* refers to the methods and materials that we use to produce commodities. On a small scale, we might think of the air hammers, nails, wood, concrete mixers, and so forth that we use to produce a house. Inherent within any means of production are the *relations of production:* in this case, the contractor, subcontractor, carpenter, financier, buyer, and so on. In the U.S. economy, we organize the work of building a house through a contracting system. The person who wants the house built has to contract with a licensed builder who in turn hires different kinds of workers (day laborers, carpenters, foremen, etc.). The actual social connections that are created through particular methods of production are what Marx wants us to see in the concept of the relations of production.

Enduring
Issues

Marx uses the term *class* in the broadest or most basic sense. But to understand its significance, we should understand it historically. The word class first came into the English language in the seventeenth century (see Williams, 1983, pp. 60–69). At that time, it had reference mainly to education; our uses of *classic* and *classical* are examples. The modern use of the term class came into existence between 1770 and 1840 and was specifically linked to capitalism (originally termed *economic individualism*). Since Marx, the social disciplines have been concerned with how class structures inequality. The central question here is, how is class structured in such a way as to prevent upward mobility? One of the more interesting ideas in contemporary theory for the structuring of class is Pierre Bourdieu's notion of *habitus* (see *Distinction: A Social Critique of the Judgment of Taste,* 1979/1984). In habitus, class is structured physically in the body; class then is expressed and re-created in the tastes and practices that are part of the human body.

Of course, Marx has something much bigger in mind than our example. In classic feudalism, for instance, people formed communities around a designated piece of land and a central manor for provision and security. At the heart of this local arrangement was a noble who had been granted the land from the king in return for political support and military service. At the bottom of the community was the serf. The serf lived on and from the land and was granted protection by the noble in return for service. Feudalism was a political and economic system that centered on land—land ownership was the primary means of production. People were related to the land through oaths of homage and *fealty* (the fidelity of a feudal tenant to his lord). These relations functioned somewhat like family roles and spelled out normative obligations and rights. The point here, of course, is that the way people related to each other under feudalism was determined by that economic system and was quite different from the way we relate to one another under capitalism.

For Marx, human history is the history of class struggles. Marx identifies several different types of classes, such as the feudal nobility, the bourgeoisie, the petite bourgeoisie, the proletariat, the peasantry, the subproletariat, and so on. As long as these classes have existed, they have been antagonistic toward each other. Under capitalism, however, two factors create a unique class system.

First, capitalism lifts economic work out of all other institutional forms. Under capitalism, the relationships we have with people in the economy are seen as distinctly different from religious, family, or political relations. As we saw with Spencer, these relationships overlapped for most of our history. For example, in agriculturally based societies, family and work coincided. Fathers worked at home, and all family members contributed to the work that was done. Capitalism lifted this work away from the farm, where work and workers were embedded in family, and placed it in urban-based factories. Capitalist industrialization thus disembedded work from family and other social relations. Contemporary gender theorists point out that this movement created separate spheres of home and work, each controlled by a specific gender. As I noted in the introduction, Marx and Engels were among the first to theorize about gender inequality—we'll look more closely at their theory in the chapter dedicated to gender (Chapter 7).

The second unique feature of class under capitalism is that it tends to be structured around two positions—the bourgeoisie (owners) and the proletariat (workers). All other classes are pushed out of existence by the cycle of capitalist investment. Here's how it works: As capitalists acquire capital, the demand for labor goes up (as the labor theory of value would predict). The increased demand for labor causes the labor pool, the number of unemployed, to shrink. As with any commodity, when demand is greater than the supply, the price goes up—in this case, that means wages increase. The increase in wages causes profits to go down. As profits go down, capitalists cut back production, which precipitates a crisis in the economy. The crisis causes more workers to be laid off and small businesses (the petite bourgeoisie) to fold. The larger capitalists buy out these small businesses, and the once small-scale capitalists become part of the working class. The result of this process being repeated over time is that the class of dependent workers increases, and capital is centralized into fewer and fewer hands. Thus, the existing capitalists accumulate additional capital, and the entire cycle starts again. "But not only has the bourgeoisie forged the weapons that bring death to itself; it has also called into existence the men who are to wield those weapons—the modern working class— the proletariat" (Marx & Engels, 1848/1978, p. 478).

The *bipolarization of conflict* is a necessary step for Marx in the process of social change, and an important one for conflict theory as a whole. Overt and intense conflict are both dependent upon bipolarization. Crosscutting interests, that is, having more than one issue over which groups are in conflict, tend to pull resources (emotional and material) away from conflict. For example, during World War II, it was necessary for the various nations to coalesce into only two factions, the Axis and the Allied powers. So the United States became strange bedfellows with the Soviet Union in order to create large-scale, intense conflict. Note that each time the United States has engaged in a police action or war, attempts are made to align resources in as few camps as possible. This move isn't simply a matter of world opinion; rather, it is a structural necessity for violent and overt conflict. The lack of bipolarized conflict creates an arena of crosscutting interests that can drain resources and prevent the conflict from escalating and potentially resolving.

Overproduction

We're now in a position to bring all these factors together. I've diagrammed these relationships in Figure 3.3. Note that the plus (+) signs indicate a positive relationship (both variables move in the same direction, either more or less), and the minus (–) signs indicate negative relationships (the variables go in opposite directions). Starting at the far left, the defining feature of modern capitalism (incessant accumulation of capital) is the primary dynamic. As capitalists invest, markets, industrialization, and employment all go up. Notice that all the arrows are double-headed, which means these all mutually influence one another. Industrialization, markets, and employment all produce profit that is then reinvested in those same three factors. Also notice that the possible limit of markets is linked to our ability to create our own needs. Markets and human "need" feed off each other. It's also important to note that markets don't simply expand; they can also become more abstract. Such markets are the ones associated with the speculation that has been endemic in the world stock markets.

The effects of these first five factors—capital, employment, industrialization, market expansion, and human "need"—continue to feed one another, creating ever-increasing levels of commodity production until a threshold is reached—overproduction—and there's too much production for the current demand. When this happens, production is cut back, workers are laid off, and the economy slumps. Small businesses collapse, and capital is concentrated in fewer hands; capital is thus accumulated, and the failures of small businesses push toward the bipolarization of classes. With the renewed accumulation of capital, capitalists reinvest and the market picks back up, and the cycle starts again. However, each economic cycle is deeper than the previous, and the working class becomes larger and larger. In the end, Marx argues, this process of accumulation and overproduction will lead to the collapse of capitalism: "The development of Modern Industry, therefore, cuts from under its feet the very foundation on which the bourgeoisie produces and appropriates products. What the bourgeoisie, therefore, produces, above all, is its own gravediggers" (Marx & Engels, 1848/1978, p. 483).

Concepts and Theory: The Problem of Consciousness

We arrive at the result that man (the worker) feels himself to be freely active only in his animal functions—eating, drinking, and procreating, or at most also in his dwelling and in personal adornment—while in his human functions he is reduced to an animal. The animal becomes human and the human becomes animal.

—Karl Marx (1932/1995, p. 99)

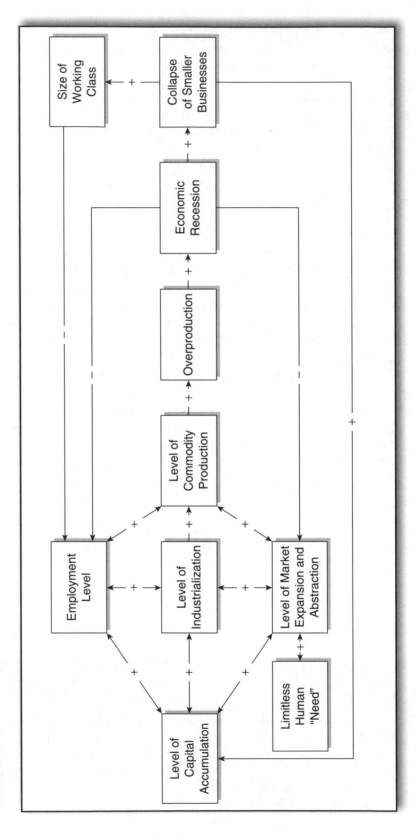

Figure 3.3 Overproduction Cycle

As I've said, Marx is a structuralist. However, we misread Marx if we think that his concern with the economic structure is paramount. True, that is the structure where the primary dynamics work, but his concern is with the human condition, how those structural issues impact people at their most basic level. *Human consciousness,* the awareness that makes us unique in the animal world, comes out of our innate ability to economically produce, because that is the way our species exists (species-being). Consciousness is fundamental, but there's more to our existence as humans than that. We know things. Humans are constantly trying to figure out the world. You're in school so that you can learn a small portion of what humanity knows. Knowledge seems straightforward; however, it is anything but that in Marx's hands. Marx tells us that what we know is a function of the society in which we live. Again, that seems straightforward: We know things today that the Athenians in ancient Greece didn't know. Marx pushes this further, too, and tells us that our knowledge, value system, and political interests are based upon our class position and are powerfully influenced by our level of involvement with religion.

Enduring
Issues

Knowledge is simply what we perceive to be factual. The *sociology of knowledge* is the study of how social factors and processes influence or determine what we think is true. There are two central questions. First, what is the relationship between social institutions and practices on the one hand and knowledge on the other? Second, how is knowledge legitimated or validated? Many of our classical theorists have something to say about this issue. Durkheim, for example, argues that the fundamental categories that people use to know things are determined by the way in which we organize society. Weber looks at how social diversity and specialization influence religious knowledge in particular. Marx is concerned with how capitalist elites create an ideology that legitimates their wealth and then present that ideology as universal. Two twentieth century thinkers stand out: Karl Mannheim and Robert Merton. Mannheim works from a Marxian position and argues that the concept of ideology is applicable to all knowledge, not just class-based knowledge (see Mannheim, *Ideology and Utopia: An Introduction to the Sociology of Knowledge,* 1936). Robert Merton comes at this issue from a functionalist perspective and is most interested in how scientific practices and norms form the basis of scientific knowledge (see Merton, *The Sociology of Science: Theoretical and Empirical Investigations,* 1973).

Alienation, Private Property, and Commodity Fetish

We often think of alienation as the subjective experience of a worker on an assembly line or at McDonald's. This perception is partly true. To get a little better handle on it, however, let's think about the differences between a gunsmith and a worker in a Remington plant. The gunsmith is a craftsperson and would make every part of the gun by hand—all the metal work, all the woodwork, everything.

If you knew guns, you could tell the craftsperson just by looking at the gun. It would bear the person's mark because part of the craftsperson was in the work. The gunsmith would take justifiable pride in the piece.

Compare that experience to the Remington plant worker. Perhaps she is the person who bolts the plastic end piece on the butt of the rifle, and she performs the task on rifle after rifle, day after day. It won't take long until the work becomes mind numbing. It is repetitive and noncreative. Chances are, this worker won't feel the pride of the craftsperson but will instead experience disassociation and depression. We can see in this example some of the problems associated with a severe division of labor and overcontrol of the worker. There's no ownership of the product or any pride as there is in craftsmanship. We can also see how the same issues would apply to the worker at McDonald's. All of it is mind-numbing, depressive work, and much of it is the result of the work of scientific management.

Frederick Taylor was the man who applied the scientific method to labor. He was interested in finding the most efficient way to do a job. *Efficiency* here is defined in terms of the least amount of work for the greatest amount of output. Under this system, the worker becomes an object that is directly manipulated for efficiency and profit. Taylor would go out into the field and find the best worker. He and his team would then time the worker (this is the origin of time-management studies) and break the job down into its smallest parts. In the end, Taylorism created a high division of labor, assembly lines, and extremely large factories.

The problem with understanding these Remington and McDonald's examples as alienation is that it focuses on the subjective experience of the worker. It implies that if we change the way we control workers—say, from Taylorism to Japanese management, as many American companies have—we've solved the problem of alienation. For Marx, that wouldn't be the case. While alienation implies the subjective experience of the worker (depression and disassociation), it is more accurate to think of it as an objective state. So, workers under the Japanese system are *not* structurally *less* alienated because they are more broadly trained, are able to rotate jobs, work holistically, and have creative input. According to Marx, workers *are* alienated under *all* forms of capitalism, whether they feel it or not. Alienation is a structural condition, not a personal one. Workers are of course more likely to revolt if they experience alienation, but that is a different issue.

Alienation always exists when someone other than the worker owns the means of production and the product being produced. Having said that, we can note that Marx actually talks about four different kinds of alienation. *Alienation* in its most basic sense is separation from one's own awareness of being human. We know we are human; Marx doesn't mean to imply that we don't have the *idea* of being human. What he means is that our idea is wrong or inaccurate. Recall Marx's argument of species-being: That which makes us distinctly human is creative production, and we become aware of our humanity as our nature is clearly reflected back to us by the mirror of the produced world. At best, the image is distorted if we look to see our nature in the things we own. It's the right place to look, but according to Marx, products should be the natural expression of species-being, and production should result in true consciousness. The commodity does not reflect humanity, and thus, everything about our commodities is wrong. At worst, the image is simply false. Marx argues that if we look elsewhere for our definition and knowledge of human nature (such as using language, having emotions, possessing a soul, religion, rationality, free will, and so on), it is not founded on the essential human characteristic—free and creative production. Thus, the human we see is rooted in false consciousness.

Definition

Alienation is a concept in Marx's theory of the effects of capitalism on consciousness and human nature and is based on the idea that there is an intrinsic connection between the producer and the product. The word *alienation* means to be separated from; it also implies that there is something that faces humans as an unknown or alien object. For Marx, there are four different kinds of alienation potentially affecting the worker: alienation from one's own species-being, alienation from other social beings, alienation from the work process, and alienation from the product. Alienation also forms the basis of private property.

From this basic alienation (from one's species-being), three other forms are born: alienation from the work process, alienation from the product, and alienation from other people. In species-being, there is not only the idea of the relationship between consciousness and creative production, but also the notion that human beings are related to one another directly and intimately. We don't create a human world individually; it is created collectively. Therefore, under conditions of species-being, humans are intimately and immediately connected to one another. The reflected world around them is the social, human world that they created. Imagine, if you can, a world of products that are all directly connected to human beings (not market forces, advertising, the drive for profit, and so forth). Either you made everything in that world, or you know who did. When you see a product, you see yourself or you see your neighbor or your neighbor's friend. Thus, when we are alienated from our own species-being, because someone else controls the means and ends of production, we are estranged from other humans as well.

Moreover, we are alienated from the process of work and from the product. There is something essential in the work process, according to Marx. The kind of work we perform and how we perform that work determines the kind of person we are. Obviously, doctors are different from garbage collectors, but that isn't what Marx has in mind. An individual's humanity is rooted in the work process. When humans are cut off from controlling the means of production (the way in which work is performed and for what reason), then the labor process becomes alienated. There are three reasons for this alienated labor, which end in alienation from the product. First, when someone else owns the means of production, the work is external to the worker; that is, it is not a direct expression of his or her nature. So, rather than being an extension of the person's inner being, work becomes something external and foreign. Second, work is forced. People don't work because they want to; they work because they have to—under capitalism, if you don't work, you die. Concerning labor under capitalism Marx (1932/1995) says, "Its alien character is clearly shown by the fact that as soon as there is no physical or other compulsion it is avoided like the plague" (pp. 98–99). And third, when we do perform the work, the thing that we produce is not our own; it belongs to another person.

Further, alienation, according to Marx, is the origin of private property. It exists solely because we are cut off from our species-being; someone else owns the means and ends of production. Underlying markets and commodification—and capitalism—is the institution of *private property*. Marx criticizes the political economists of his day because they assumed the fact of private property without offering any explanation for it. These economists believed that private property was simply a natural part of the economic process. But for Marx, the *source* of private property was the crux of the problem. Based on species-being, Marx argues that private property emerged out of the alienation of labor.

Definition

Private property is a distinctly modern, capitalist concept and a major idea in Marx's theory of capitalism. For Marx, the quality of "private" property can only be understood through the idea of alienation. Private property can only exist when the worker is first alienated from his or her humanity (species-being).

Marx also argues that there is a reciprocal influence of private property on the experience of alienation. As we have seen, Marx claims that private property is the result of alienated labor. Once private property exists, it can exert its own influence on the worker and it becomes "the realization of this alienation" (Marx, 1932/1995, p. 106). Workers then become controlled by private property. This notion is most clearly seen in what Marx describes as *commodity fetish:* Workers become infatuated with their own product as if it were an alien thing. It confronts them not as the work of their hands, but as a commodity—something alien to them that they must buy and appropriate.

Commodity fetish is a difficult notion, and it is hard to come up with an illustration. Workers create and produce the product, yet they don't recognize their work or themselves in the product. So, as workers, we see the product outside of ourselves and we fall in love with it. We have to possess it, not realizing that it is ours already by its very nature. We think *it,* the object, will satisfy our needs, when what we need is to find ourselves in creative production and a socially connected world. We go from sterile object to sterile object, seeking satisfaction, because they all leave us empty. To use a science fiction example, it's like a male scientist who creates a female robot, but then forgets he created it, falls in love with it, and tries to buy its affection through money. As I said, it is a difficult concept to illustrate because in this late stage of capitalism, *our entire world and way of living are the examples.* In commodity fetish, our perception of self-worth is linked with money and objects in a vicious cycle.

Definition

Commodity fetish—or, commodity fetishism—is a central concept in Marxian theory. In capitalism, products are owned and controlled by the owners of capital, and the product thus faces the worker as something alien that must be bought and appropriated. In misrecognizing their own nature in the product, workers also fail to see that there are sets of oppressive social relations behind both the perceived need and the simple exchange of money for a commodity. In addition, commodities substitute for real social relations: Under capitalism, people see their lives being defined through commodity acquisition rather than community relations.

Moreover, in commodity fetish we fail to recognize—or in Marxian terms, we *misrec-ognize*—that there are sets of oppressive social relations behind both the perceived need for and the simple exchange of money for a commodity. We think that the value of the commodity is simply its intrinsic worth—that's just how much a Calvin Klein jacket is worth—when, in fact, hidden labor relations and exploitation produce its value. We come to need these products in an alienated way: We think that owning the product will fulfill our needs. The need is produced through the commodification process, and in back of the exchange are relations of oppression. In this sense, the commodity becomes reified: It takes on a sense of reality that is not materially real at all.

Contemporary Marxists argue that the process of commodification affects every sphere of human existence and is the "central, structural problem of capitalist society in all its aspects" (Lukács, 1922/1971, p. 83). Commodification translates *all human activity and relations* into objects that can be bought or sold. In this process, value is determined not by any intrinsic feature of the activity or the relations, but by the impersonal forces of markets, over which individuals have no control. In this expanded view of commodification, the objects and relations that will truly gratify human needs are hidden, and the commodified object is internalized and accepted as reality. So, for example, a young college woman may internalize the commodified image of thinness and create an eating disorder such as anorexia nervosa that rules her life and becomes unquestionably real. Commodification, then, results in a consciousness based on reified, false objects. It is difficult to think outside this commodified box. There is, in fact, a tendency to justify and rationalize our commodified selves and behaviors.

False Consciousness and Religion

Alienation, false consciousness, and ideology go hand in hand. In place of a true awareness of species-being comes false consciousness, a consciousness built on any foundation other than free and creative production. Humans in false consciousness thus come to think of themselves as being defined through the ability to have ideas, concepts, and abstract thought, rather than through production. When these ideas are brought together in some kind of system, Marx considers them to be ideology.

Ideas function as *ideology* when they are perceived as independent entities that transcend historical, economic relations: Ideologies contain beliefs that we hold to be true and right, regardless of the time or place (like the value of hard work and just reward). In the strictest Marxian sense, ideologies legitimate capitalist relations and blind workers to the effects of class. Ideologies prevent class consciousness from forming and thus hinder social change. More generally, the concept of ideology helps us see that all ideas are linked to social relationships, that "the intellectual point of view held and the social position occupied" correspond (Mannheim, 1936, p. 78).

Definition

In Marxian theory, *ideology* is any system of ideas that blinds us to the truth of economic relations. In a more general sense, ideology is any set of cultural practices (norms, values, and beliefs) that undergird and legitimate relations of power and prevent workers from seeing their oppression, which in turn inhibits the possibility of social change.

Though Marx sometimes appears to use the terms interchangeably, it is important to keep the distinction between false consciousness and ideology clear. Ideologies can change and vary. For example, the ideology of consumerism is quite different from the previous ideologies of the work ethic and frugality, yet they are all capitalist ideologies. The ideologies behind feminism are different from the beliefs behind the racial equality movement, yet from Marx's position, both are ideologies that blind us to the true structure of inequality: class. However, false consciousness doesn't vary. It is a state of being, somewhat like alienation in this aspect. We are by definition in a state of false consciousness because we are living outside of species-being. The very way through which we are aware of ourselves and the world around us is false or dysfunctional. The very method of our consciousness is fictitious.

Generally speaking, false consciousness and ideology are structurally connected to two social factors: religion and the division of labor. For Marx, *religion* is the archetypal form of ideology. Religion is based on an abstract idea, like God, and religion takes this abstract idea to be the way through which humans can come to know their true nature. For example, in the evangelical Christian faith, believers are exhorted to repent from not only their sinful ways, but also their sinful nature, and to be born again with a new nature—the true nature of humankind. Christians are thus encouraged to become Christ-like because they have been created in the image of God. Religion, then, reifies thought, according to Marx. It takes an abstract (God), treating it as if it is materially real, and it then replaces species-being with nonmaterially based ideas (becoming Christ-like). It is, for Marx, a never-ending reflexive loop of abstraction, with no basis in material reality whatsoever. Religion, like the commodity fetish, erroneously attributes reality and causation. We pour ourselves out, this time into a religious idea, and we misrecognize our own nature as that of God or devil.

It is clear that the more the worker spends himself, the more powerful the alien objective world becomes which he creates over-against himself, the poorer he himself—his inner world—becomes, the less belongs to him as his own. The more man puts into God, the less he retains in himself. (Marx, 1932/1978, p. 72)

There is also a second sense in which religion functions as ideology. Marx uses the term ideology as *apologia,* or a defense of one's own ideas, opinions, or actions. In this kind of ideology, the orientation and beliefs of a single class, the elite, become generalized and seem to be applicable to all classes. Here, the issue is not so much reification as class consciousness. The problem in reification is that we accept something as real that isn't. With ideology, the problem is that we are blinded to the oppression of the class system. This is in part what Marx means when he claims that religion is "the opium of the people." As we have seen, Marx argues that because most people are cut off from the material means of production, they misrecognize their true class position and the actual class-based relationships, as well as the effects of class position. In the place of class consciousness, people accept an ideology. For Marx, religion is the handmaiden of the elite; it is a principal vehicle for transmitting and reproducing the capitalist ideology.

Thus, in the United States we tend to find a stronger belief in American ideological concepts (such as meritocracy, equal opportunity, work ethic, poverty as the result of laziness, free enterprise, and so on) among the religious (particularly among the traditional American denominations). We would also expect to see religious people being less concerned with the social foundations of inequality and more concerned with patience in this life and rewards in the next. These kinds of beliefs, according to Marx, dull the workers' ability to institute social change and bring about real equality.

It is important to note that this is a function of religion in general, not just American religion. The Hindu caste system in India is another good example. There are five different castes in the system: *Brahmin* (priests and teachers), *Kshatriya* (rulers), *Vaishya* (merchants and farmers), *Shudra* (laborers and servants), and *Harijans* (polluted laborers, the outcastes). Position in these different castes is a result of birth; birth position is based on *karma* (action); and karma is based on *dharma* (duty). There is virtually no social mobility among the castes. People are taught to accept their position in life and perform the duty (dharma) that their caste dictates so that their actions (karma) will be morally good. This ideological structure generally prevents social change, as does the Christian ideology of seeking rewards in heavenly places.

For Marx, then, religion simultaneously represents the farthest reach of humanity's misguided reification and functions to blind people to the underlying class conditions that produce their suffering. Yet Marx (1844/1978b) also recognizes that those sufferings are articulated in religion as the collective sigh of the oppressed:

Religious suffering is at the same time an *expression* of real suffering and a *protest* against real suffering. Religion is the sigh of the oppressed creature, the sentiment of a heartless world, and the soul of soulless conditions. It is the *opium* of the people. (p. 54)

Marx thus recognizes that religion also gives an outlet to suffering. He feels that religion places a "halo" around the "veil of tears" that is present in the human world. The tears are there because of the suffering that humans experience when they don't live communally and cooperatively. Marx's antagonism toward religion, then, is not directed at religion and God per se, but at "the *illusory* happiness of men" that religion promises (p. 54).

Most, if not all, of Marx's writings on religion were in response to already existing critiques (mostly from Georg Hegel and Ludwig Feuerbach). Marx (1844/1978b), then, takes the point of view that "the criticism of religion has largely been completed" (p. 53). What Marx is doing is critiquing the criticism. He is responding to already established ideas about religion and not necessarily to religion itself. What Marx wants to do is to push us to see that the real issue isn't religion; rather, it is the material, class-based life of human beings that leads to actual human suffering.

Marx also sees ideology and alienation as structurally facilitated by the *division of labor* (how the duties are assigned in any society). Marx talks about several different kinds of division of labor. The most primitive form of separation of work is the *natural division of labor*. The natural division was based upon the individual's natural abilities and desires. People did not work at something for which they were ill suited, nor did they have to work as individuals in order to survive. Within the natural division of labor, survival is a group matter, not an individual concern. Marx claims that the only time this ever existed was in pre-class societies. When individuals within a society began to accumulate goods and exercise power, the *forced division of labor* replaced the natural division. With the forced division, individual people must work in order to survive (sell their labor), and they are forced to work at jobs they neither enjoy nor have the natural gifts to perform. The forced division of labor and the commodification of labor characterize capitalism.

Definition

For Marx, there are two major ways in which labor is divided: natural or forced, and material or mental. Natural and forced both generally refer to material labor. A *natural division of labor* is one where each works the job she or he desires; under forced division, people must work to survive. The *forced division of labor* is a basic premise of capitalism. Marx is also critical of mental and material labor being divided: Knowledge divorced from practice can only be some form of ideology.

This primary division of labor historically becomes extended when mental labor (such as that performed by professors, priests, and philosophers) is divided from material labor (workers). When this happens, reification, ideology, and alienation reach new heights. As we've seen, Marx argues that people have true consciousness only under conditions of species-being. Anytime people are removed from controlling the product or the production process, there will be some level of false consciousness and ideology. Even so, workers who actually produce a material good are in some way connected to the production process. However, with the separation of mental from material labor, even this tenuous relationship to species-being is cut off. Thus, the knowledge created by those involved with mental labor is radically cut off from what makes us human (species-being). As a result, everything produced by professors, priests, philosophers, and so on has some reified ideological component and is generally controlled by the elite.

Class Consciousness

So far, we have seen that capitalism increases the levels of industrialization, exploitation, market-driven forces like commodification, false consciousness, ideology, and reification, and it tends to bifurcate the class structure. On the other hand, Marx also argues that these factors have dialectical effects and will thus push capitalism inexorably toward social change. Conflict and social change begin with a change in the way we are aware of our world. It begins with *class consciousness.*

Definition

Class consciousness is a concept from Marx's theory of capitalism. It is the subjective awareness that class determines life chances and group interests. With class consciousness, a class moves from a class *in* itself to a class *for* itself, a shift from a structured position to a political one. Class consciousness is one of the prerequisites to social change in Marxian theory, and it varies in direct accordance with the levels of worker communication, exploitation, and alienation.

Marx notes that classes exist objectively, whether we are aware of them or not. He refers to this as a "class in itself," that is, an aggregate of people who have a common relationship to the means of production. But classes can also exist subjectively, as a "class *for* itself." It is the latter that is produced through class consciousness. Class consciousness has two parts: (1) the subjective awareness that experiences of deprivation are determined by structured class relations and not individual talent, and (2) the effort and the group identity that come from such awareness.

Industrialization has two main lines of effects when it comes to class consciousness. First, it tends to increase exploitation and alienation. We've talked about both of these already, but remember that these are primarily objective states for Marx. In other words, these aren't necessarily subjectively felt—alienation isn't chiefly a feeling of being psychologically disenfranchised; it is the state of being cut off from species-being. Humans can be further alienated and exploited, and machines do a good job of that. What happens at this point in Marx's scheme is that these objective states can produce a sense (or feeling) of belonging to a group that is disenfranchised, that is, class consciousness.

As you can see from Figure 3.4, industrialization has positive relationships with both exploitation and alienation. As capitalists employ machinery to aid in labor, the objective levels of alienation and exploitation increase. As the objective levels increase, so does the probability that workers will subjectively experience them, thus aiding in the production of class consciousness. Keep in mind that industrialization is a variable, which means that it can increase (as in when robots do the work that humans once did on the assembly line—there is a human controlling that robot, but far, far removed from the labor of production) or decrease (as when "cottage industries" spring up in an economy).

The second area of effect is an increase in the level of worker communication. Worker communication is a positive function of education and ecological concentration. Using machines—and then more *complex* machines—requires increasing levels of technical knowledge. A crude but clear example is the different kinds of knowledge needed to use a horse and plow compared to a modern tractor. Increasing the use of technology in general requires an increase in the education level of the worker. (This relationship is clearly seen in today's computer-driven U.S. labor market.)

In addition, higher levels of industrialization generally increase the level of worker concentration. Moving workers from small guild shops to large-scale machine shops or assembly lines made interaction among these workers possible in a way never before achievable, particularly during break and lunch periods when

Figure 3.4 The Production of Class Consciousness

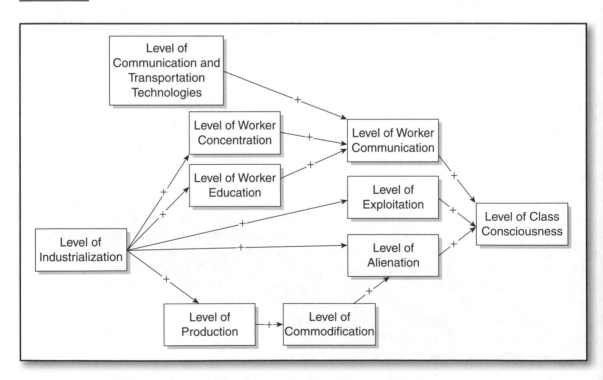

hundreds of workers can gather in a single room. Economies of scale tend to increase this concentration of the workforce as well. So, for a long time in the United States, we saw ever-bigger factories being built and larger and larger office buildings (like the Sears Tower and World Trade Center—and it is significant that terrorists saw the World Trade Center as representative of American society). These two processes, education and ecological concentration, work together to increase the level of communication among workers. These processes are supplemented through greater levels of communication and transportation technologies. Marx argues that communication and transportation would help class consciousness spread from city to city.

So, in general, class consciousness comes about as workers communicate with each other about the problems associated with being a member of the working class (like not being able to afford medication). The key to Marx's thinking here is to keep in mind that these things come about due to structural changes that result simply from the way capitalism works. Capitalists are driven to increase profits. As a result, they use industrialization, which sets in motion a whole series of processes that tend to increase the class consciousness of the workers. Class consciousness increases the probability of social change. As workers share their grievances with one another, they begin to doubt the legitimacy of the distribution of scarce resources, which in turn increases the level of overt conflict. As class inequality and the level of bipolarization increase, the violence of the conflict will tend to increase, which in turn brings about deeper levels of social change.

However, class consciousness has been difficult to achieve. There are a number of reasons given as to why this is true, but many Marxian approaches focus on the relationships among a triad of actors: the state, the elite, and workers. Marx argues that as a result of the rise of class consciousness, and other factors such as the business cycle, workers would unite and act through labor unions to bring about change. Some of the work of the unions would be violent, but it would eventually lead to a successful social movement. In the end, the labor movement would bring about socialism.

Marx sees two of the actors in the triad working in collusion. He argues that under capitalism, the state is basically an arm of the elite. It is controlled by capitalists and functions with capitalist interests in mind. Many of the top governing officials come from the same social background as the bourgeoisie. A good example of this in the United States is the Bush family. Of course, electing a member of the capitalist elite to high public office not only prejudices the state toward capitalist interests, but it is also an indicator of how ideologically bound a populace is.

In addition, as C. W. Mills (1956) argues, the elite tend to cross over, with military men (in 1956) serving on corporate boards and as high-placed political appointees, and CEOs functioning as political advisees and cabinet members, and so on. A notable example of these kinds of interconnections is Charles Erwin Wilson, president of General Motors from 1941 to 1953. He was appointed secretary of defense under President Dwight D. Eisenhower and famously said, "what was good for our country was good for General Motors, and vice versa" (*Time*, 1957).

In response to the demands of labor unions, certain concessions were ultimately granted to the working and middle classes. Work hours were reduced and health care provided, and so forth. Capitalists working together with a capitalist-privileging state granted those concessions. Thus, even when allowances are granted, they may function in the long run to keep the system intact by maintaining the capitalists' position, silencing the workers, and preventing class consciousness from adequately forming.

The state is also active in the production of ideology. Marx's idea of the state isn't clearly defined. In other words, where the state begins and ends under capitalist democracy is hard to say. The state functions through many other institutions, such as public schools, and its ideology is propagated through such institutions, not only through such direct means as the forced pledge of allegiance in the United States but also through indirect control measures like specific funding initiatives. A good deal of the state's ideology is, of course, capitalist ideology. As a result, the worker is faced with a fairly cohesive ideology coming from various sources. This dominant ideology creates a backdrop of taken-for-grantedness about the way the world works, against which it is difficult to create class consciousness.

The movement of capitalist exploitation across national boundaries, which we mentioned before, accentuates the "trickle-down" effect of capitalism in such countries as the United States. Because workers in other countries are being exploited, the workers in the United States can be paid an inflated wage (inflated from the capitalists' point of view). This is functional for capitalism in that it provides a collection of buyers for the world's goods and services. Moving work out from the United States and making it the world's marketplace also changes the kind of ideology or culture that is needed. There is a movement from worker identities to consumer identities in such economies.

In addition, capitalists use this world labor market to pit workers against one another. While wages are certainly higher due to exported exploitation, workers are also aware that their jobs are in jeopardy as work is moved out of the country. Capitalism always requires a certain level of unemployment. The business cycle teaches us this—that zero unemployment means higher wages and lower profits. The world labor market makes available an extremely large pool of unemployed workers. So, workers in an advanced industrialized economy see themselves in competition with much cheaper labor. This global competition also hinders class consciousness by shifting the worker's focus of attention away from the owners and onto the foreign labor market. In other words, a globalized division of labor pits worker against worker in competition for scarce jobs. This competition is particularly threatening for workers in advanced capitalist countries, like the United States, because the foreign workers' wage is so much lower. These threatened workers, then, will be inclined to see their economic problems in terms of global, political issues rather than as class issues.

Further, the workers divide themselves over issues other than class. We do not tend to see ourselves through class-based identities, which Marx would argue is the identity that determines our life chances; rather, we see ourselves through racial, ethnic, gender, and sexual preference identities. Marxists would argue that the culture of diversity and victimhood is part of the ideology that blinds our eyes to true social inequality, thus preventing class consciousness.

Thinking About Modernity and Postmodernity

Marx has a very clear notion of modernity, though he doesn't actually use this term. As we've seen, Marx argues that society in general evolves and changes due to variations in the economic system. The means of production determines the relations of production (or the social relations). As noted earlier, Marx named five different economic societies: pre-class societies, Asiatic societies, ancient societies, feudal societies, and capitalist societies. Societies that contain the capitalist dynamics that we have been talking about are, of course, modern. We would expect, then, for a Marxian view of postmodernity to argue that something has changed the basic configuration of capitalism. Interestingly, we will find as we move through each of our theorists that many of the postmodern thinkers we will consider have some relation to Marxian theory. For now, though, I want us to focus on the central issue for Marx: the means of production.

Machines of Production and Consciousness

Fredric Jameson (1984b) argues that modern capitalism has gone through three distinct phases, each linked to a particular kind of technology. Early market capitalism was distinguished by steam-driven machinery; steam and combustion engines characterized mid-monopoly capitalism; and late multinational capitalism is associated with nuclear power and electronic machines. The important Marxian issue here is the relationship between the mode of production and consciousness. As we have seen, Marx argues that consciousness is explicitly tied to production. Human production is supposed to act as a mirror that reflects our nature. Thus, the way in which production is performed is critically important.

This idea is not as extreme as it might appear. Think about digging in your garden. If you use a shovel and hoe to dig the earth, and you bend down and plant the seeds by hand, you will find that you'll notice things about the earth, yourself, and your work. You will smell the loam of the earth; you'll feel the consistency of the soil in your hands and on your knees as you kneel; you'll sense the soil, water, and seed as they mix together; and when the plants sprout, you will see your sweat and toil reflected in the ground and new life. But if you use a tractor, it won't be the same; further, if you use a computer to guide the tractor as it plows, plants, and harvests, you will be utterly removed from the smells, tastes, and feel of earth, sun, water, and the human body.

Let me give you another illustrative example. The aesthetics of airplane mechanics are generally different from those of symphony conductors. Part of the difference is undoubtedly due to psychological dispositions, but the majority of the differences are due to the kind of work they do. Constantly working on real machines, getting your hands dirty and banged up, and having to exert physical strength and mechanical ingenuity all day, gives a person a particular perspective and cultural disposition. Likewise, for a symphony conductor, interpreting a musical score and working to present that interpretation through an orchestra gives the conductor a certain perspective and disposition as well. We become what we do.

Consequently, there are significant social and individual effects when the general means of production change in a society.

Machines of Reproduction and Schizophrenic Culture

Jameson (1984b) argues that human reality and consciousness were non-problematically represented by the aesthetic of the machine in earlier phases of capitalism. In other words, workers could touch and see the machines that were used to create products. Although alienated, they were still connected to the production process and could, most importantly, also experience the alienating process through their senses. But in postindustrial societies, through multinational capitalism, workers are using machines less and less. Even if they are in a manufacturing sector, the machines are more and more controlled through computers. Those people not in manufacturing today, like the rising service class, use machines of reproduction (movie cameras, video, tape recorders, computers, and so forth) rather than production. Because the machines of late capitalism reproduce knowledge rather than produce it, and because reproduction is always focused more on the medium than the message, Jameson argues that the link from production to signification (culture and meaning) has broken down. Jameson characterizes this breakdown as the schizophrenia of culture and argues that our culture is filled with "free-floating signifiers."

Let's back up a minute and try and understand what Jameson is saying. What he and Marx are telling us is that there is, or was, a chain of real relationships from the material world to the human world of ideas. At one time, our ideas and the language we used were embedded in the physical world. Let's use our garden example again, but place it in a horticultural society (one that lives by planting and growing food). If you are working in that group, your knowledge of farming is firsthand and real. You know how to grow food, and the food is there for you when it is done growing. When you talk about growing, the language you use is firsthand, objective, and real. But if we put an owner between the product and the person, there is a break in that continuity. If we insert a tractor and even more equipment in the chain, there are still more breaks in the signification chain. Further, if instead of *production* equipment, we use machines of *reproduction,* then there isn't any objective reference for the signification chain at all. Machines of reproduction don't produce anything; they only reproduce images or text about something else.

Jameson (1984b) argues that the schizophrenic nature of culture and the immense size and complexity of the networks of power and control in multinational capitalism produce three distinct features of the postmodern era. One, the cultural system is now dominated by image rather than actual signed reality. These media images, or *simulacrum* (identical copies of something that never existed), have no depth of meaning; meaning is fleeting and fragmented. Two, there is a consequent weakening of any sense of either social or personal biographic history. In other words, it is becoming increasingly difficult for us to create and maintain consistent narratives about our national or personal identities. And three, postmodern culture is emotionally flat. The further culture is removed from actual social and physical reality, the harder it is to invest culture with strong emotion.

Besides offering us the provocative idea of postmodernity, these ventures into postmodern theory also give us an opportunity to see how a theorist's perspective can be used. In other words, it is very possible that a theory can be Marxian but not exactly what Marx himself might say. The theorist's perspective can be used apart from his or her explicit theory. That's what Jameson does with Marx. Jameson ends up saying that culture is set loose from its economic base and can have independent effects. This isn't something that Marx himself said. Nevertheless, Jameson's theory is Marxian because it originates in a Marxian perspective.

Summary

- Marx's perspective is created through two central ideas: species-being and the material dialectic. Species-being refers to the unique way in which humans survive as a species—we creatively produce all that we need. The material dialectic is the primary mechanism through which history progresses. There are internal contradictions within every economic system that push society to form new economic systems. The dialectic continues until communism is reached, because for Marx, communism is a system that is in harmony with species-being.
- The means and relations of production characterize every economic system. The means of production in capitalism is owned by the bourgeoisie and generally consists of commodification, industrial production, private property, markets, and money. One of the unique features of capitalism is that it will swallow up all other classes save two: the bourgeoisie and proletariat. This bifurcation of class structure will, in turn, set the stage for class consciousness and economic revolution.
- Capitalism affects every area of human existence. Through it, individuals are alienated from each aspect of species-being and creative production. The work process, the product, other people, and even their own inner being confront the worker as alien objects. As a result, humankind misrecognizes the truth and falls victim to commodity fetish, ideology, and false consciousness. However, because capitalism contains dialectical elements, it will also produce the necessary ingredient for economic revolution: class consciousness. Class consciousness is the result of workers becoming aware that their fate in life is determined primarily by class position. This awareness comes as alienation and exploitation reach high levels and as workers communicate with one another through increasing levels of education, worker concentration in the factory and city, and communication and transportation technologies.
- Jameson's argument concerning postmodernity is based on Marx's theory of consciousness: Human consciousness is materially based; that is, because of species-being, there is a direct relationship between the method of production and the ideas we have about ourselves and the way we perceive the world. In primitive communism, we had direct and objective knowledge of the world and ourselves. Under early capitalism, we suffered from false consciousness and ideology, but this knowledge was still materially grounded in that we were connected to machines of production. In that state, we could have come to a real sense of alienation and class consciousness. However, in

postmodernity, the economy is shifting from machines of production to machines of reproduction. With machines of production, there was a connection between culture and the material world. With machines of reproduction, that connection is broken. As a result, cultural signs and symbols are cut loose from their moorings and lose any grounded sense of meaning or reality. From a Marxian point of view, this shift from machines of production to reproduction, from materially grounded ideas to free-floating signifiers, contributes to the prevention of class consciousness.

BUILDING YOUR THEORY TOOLBOX

Learning More—Primary and Secondary Sources

For a primary source for Marx, it is best to start off with a reader. Here's my favorite:

- Tucker, R. C. (Ed.). (1978). *The Marx-Engels reader.* New York: Norton.

There are many good secondary sources for Marx. The following are a few of the good ones to start with:

- Appelbaum, R. P. (1988). *Karl Marx.* Newbury Park, CA: Sage. (Part of the Masters of Social Theory series; short, book-length introduction to Marx's life and work)

- Bottomore, T., Harris, L., & Miliband, R. (Eds.). (1992). *The dictionary of Marxist thought* (2nd ed.). Cambridge, MA: Blackwell. (Excellent dictionary reference to Marxist thought)

- Fromm, E. (1961). *Marx's concept of man.* New York: Continuum. (Insightful explanation of Marx's idea of human nature)

- McClellan, D. (1973). *Karl Marx: His life and thought.* New York: Harper & Row. (Good intellectual biography)

Seeing the Social World (knowing the theory)

- Write a 500-word synopsis of Marx's sociological imagination, making certain to include not only how he sees the world (perspective) but also how that perspective came about (history, social structures, and biography).

- After reading and understanding this chapter, you should be able to define the following terms theoretically and explain their theoretical importance to Marx's theory: species-being, idealism, materialism, dualisms, material dialectic, use-value, exchange-value, labor theory of value, industrialization, markets, commodification, the means and relations of production, class bipolarization, exploitation, surplus value, commodification, overproduction, alienation, private property, commodity fetish, ideology, false consciousness, religion, class consciousness.

- After reading and understanding this chapter, you should be able to answer the following questions (remember to answer them theoretically):

  1. Explain human nature through the idea of species-being and how it affects Marx's theory.

  2. Define dialectical processes, and analyze the structural dialectics of capitalism.

  3. Explain how overproduction occurs and its effects on capitalism.

  4. Explain Marx's theory of religion and ideology.

  5. Explain how class consciousness develops.

Engaging the Social World (using the theory)

- Apply the idea of exploitation to current economic relations. What has been happening to the U.S. industrial base? What do you suppose this implies about global capitalism?

- Explain how industrialization, markets, and commodification are intrinsically expansive in capitalism, and apply this dynamic to understanding contemporary capitalism and your or your family's buying patterns.

- Using Google or your favorite search engine, do a search for "job loss." Look for sites that give statistics for the number of jobs lost in the past 5 years or so (this can be a national or regional number). How would Marx's theory explain this?

- Get a sense about the frequency and uses of plastic surgery in the United States. You can do this by watching "makeover" programs on TV, or by doing web searches, or by reading through popular magazines, or in any number of ways. How can we understand the popularity and uses of plastic surgery using Marx's theory? (Hint: Think of commodification and commodity fetish.) Can you think of other areas of our life for which we can use the same analysis?

Weaving the Threads (building theory)

There are central themes in classical theory. As we proceed through our discussions of the different theorists, it is a good idea to pay attention to these themes and see how the theorists add to or modify them. In the end, we should have a clear and well-informed idea of the central dynamics in sociological theory. Thinking about Marx's theory, answer the following questions:

1. What is the fundamental structure of inequality in society? How is inequality perpetuated (in other words, how is it structured)?

2. How does social change occur? Under what conditions is conflict likely to take place?

3. What is religion? What function does religion have in society?

4. What is the purpose of the state? How is it related to other social structures? Where does power reside?

CHAPTER 4

Diversity and Social Solidarity

Émile Durkheim

Cultural diversity is a byword in contemporary society. It generally refers to racial or ethnic diversity. However, if we think about cultural diversity theoretically, the phrase "racial or ethnic diversity" begs the question, how is it that racial or ethnic groups come to have different cultures? Most people simply assume that different races and ethnic groups have diverse cultures. Yet there is no necessary relationship between what we think of as race and cultural diversity. Theoretically and sociologically, then, it is much better to ask how cultural diversity is created rather than simply assuming it exists. Besides, cultural diversity is much broader than merely race and ethnicity. For example, it is quite possible that the cultural differences between the elite and the poor are greater than the differences among racial groups within the same society.

So, how is cultural diversity created? More specifically, what are the general processes through which cultural differences are created, whether among racial, ethnic, class, or gender groups? Émile Durkheim provides us with answers to these kinds of questions.

Yet Durkheim is actually concerned with a more important issue, one that few people think about when considering cultural diversity. His concern is based on the insight that every society needs a certain level of cultural integration and social solidarity to exist and function. Durkheim's main concern is this: How much cultural diversity can a society have and still function? Think about an extreme situation as an example: Picture two people who speak totally different languages. How easy would it be for them to carry on a conversation? If it was necessary, they undoubtedly could find a way, but what they could talk about would be limited, and it would take a great deal of time to have even the simplest of conversations.

The same is true with cultural diversity. Cultural diversity includes language, but it also encompasses nonverbal cues, dialects, values, normative behaviors, beliefs and assumptions about the world, and so on. The more different people are from one another, the more difficult it will be for them to work together and communicate, which is the basis of any society. Durkheim, then, specifically asks, how can a diverse society create social solidarity and function?

THEORIST'S DIGEST

Central Sociological Questions

Durkheim is intensely concerned with understanding how social solidarity and integration could be preserved in modernity. He recognizes that society is built on a foundation of shared values and morals. Yet he also realizes that there are structural forces at work in modernity that relentlessly produce cultural diversity, something that could tear away this foundation of social solidarity. His project, then, is to discover and implement the necessary social processes that could create a new kind of unity in society, one that would allow the dynamics of modernity to function within a context of social integration.

Simply Stated

Durkheim sees individuals apart from society as concerned only with their own desires that, because of human nature, are insatiable. Thus, the one thing society needs above all else is a common, moral culture—a set of ideas, values, beliefs, norms, and practices that guide us to act collectively rather than individually. Given that moral culture is the basis of society, Durkheim argues that society first began in religion. Modernity, however, creates a problem that threatens to tear apart society's moral basis: social diversity created through structural differentiation and the division of labor. The main solution to this problem is an evolving, more general moral culture—able to embrace greater levels of social diversity—through the formation of intermediary groups, restitutive law, the centralization and rationalization of law, and social and structural interdependency.

Key Ideas

Social will, state of nature, social contract, primal society, four conditions for social science, social facts, society sui generis, collective consciousness, religion, sacred and profane, ritual, effervescence, social solidarity (mechanical and organic), punitive and restitutive law, the division of labor, social differentiation, cultural generalization, intermediary groups, social pathologies, anomie, suicide (altruistic, fatalistic, egoistic, and anomic), the cult of the individual.

The Sociological Imagination of Émile Durkheim

Durkheim was born in France when it was at the cusp of bringing to reality the republican vision of government and enlightenment that began in 1792. Its struggles with monarchs and emperors were coming to an end. The institution of education that had been the purview of the Catholic Church was moving toward scientific, positive instruction. During Durkheim's lifetime, France made significant strides forward in creating a transportation infrastructure, which gave a boost to economic development and personal income levels. It was a time of hope and forward vision, an environment in which Durkheim thrived.

Durkheim's Life

David Émile Durkheim was born in Epinal, France, on April 15, 1858. Durkheim's mother, Mélanie, was a merchant's daughter, and his father, Moïse, was a rabbi, descended from generations of rabbis. Durkheim studied the Hebrew language and Jewish law, no doubt in preparation for becoming a rabbi himself. However, shortly after his Bar Mitzvah at age 13, Durkheim considered Catholicism for a brief time. Durkheim didn't become Catholic, but neither did he return to his rabbinical studies. He continued on with his secular studies and was considered an outstanding student. With hopes of becoming a teacher, Durkheim moved to Paris to attend the *École Normale Supérieure,* the training ground for the new French intellectual elite. His first two attempts at admission failed, but he eventually earned a spot in 1879.

Durkheim's time at the École was formative. He met and became friends with a number of intellectually prominent men, such as Henri Bergson (1859–1941), best known for developing the concept of "duration," a philosophy of time and consciousness; the French historian Fustel de Coulanges (1830–1889); and Émile Doutroux (1845–1921), French philosopher of science and religion. From Coulanges, Durkheim learned his historical approach to collecting data. Doutroux was particularly influential, as he impressed on Durkheim the importance of different levels of analysis. Rather than seeing society as a continuous whole, Durkheim learned to see different dynamics working at different levels of society. (For a Theoretical Hint concerning levels of analysis, see Chapter 2, on Herbert Spencer.) It was also at the École that Durkheim studied the work of Montesquieu (1689–1755) at length. He is best known for the separation of powers doctrine that undergirds most democracies, especially that of the United States. Yet Durkheim focused on another issue in Montesquieu's work: the possibility of social science. We'll talk more about this in a bit.

There is, I think, another effect that the École had on Durkheim that is worth considering. While Durkheim no longer practiced religion, he remained convinced that every society needs an ethical, moral base in order to function. He was also convinced that the ethical base of society should be informed and guided by social science. Religion had disappointed, led astray, and was the core divisive issue in Durkheim's France. Yet Durkheim was also convinced that society would fail without a strong moral base. Both the disappointment and conviction undoubtedly came from Durkheim's early religious training. The only avenue open for Durkheim was to create a science of ethics, one that would base morality on empirical observations rather than imagined deities. There's little doubt that Durkheim put a lot of hope in his time at the École; he looked forward to systematic training in science, philosophy, and pedagogy. What he got was "dilettantism" and "elegant dabbling" (Coser, 2003, p. 144). Rather than discouraging Durkheim, his experience at the École made him even more adamant: "No doubt the day will come when the *science of ethics* will have advanced enough for theory to be able to regulate practice; but we are still far from that day" (Durkheim, 1887/1993, p. 134, emphasis added).

The first years after finishing school, Durkheim taught philosophy in Paris, but felt philosophy was a poor approach to solving the social ills that surrounded him. In response to educational reform (see "Ferry Laws" in subsequent section), the University of Bordeaux created France's first teacher training course, and in 1884, the state began to financially support a professorship in education. Initially, a man whom Durkheim admired was appointed to that position, but he soon moved on to become dean of the faculty. Durkheim's published articles had impressed the director of higher education, and in 1887, Durkheim was offered the position. To induce Durkheim to take the position, "social science" was added to the title. Durkheim was thus appointed as Charge d'un Cours de Science Sociale et de

Pedagogie at the University of Bordeaux, and Durkheim became the first teacher of sociology in the French system.

Durkheim's time at Bordeaux was scholastically productive. Within a space of 4 years, Durkheim published three of his most influential books: *The Division of Labor in Society, The Rules of Sociological Method,* and *Suicide.* In addition, in 1898 Durkheim created the first social science academic journal in France, *Année Sociologique.* The journal provided a yearly review of the best sociological work being done, and he published original research monographs. Durkheim drew together some of the best minds of the time and used the publication to place French sociology on a firm scientific base. To say the least, the journal did much to legitimate the new science of society; it continues to be published today.

In 1902, Durkheim took a post at the Sorbonne and by 1906, was appointed Professor of the Science of Education, a title later changed to Professor of Science of Education and Sociology. In December 1915, Durkheim received word that his son, Andre, had been declared missing in action (WWI). Andre had followed in his father's footsteps and gone to the École Normale; he was seen as an exceptionally promising social linguist. Durkheim had hoped his son would complete the research he had begun in linguistic classifications. The following April, Durkheim received official notification that his son was dead. Durkheim withdrew into a "ferocious silence." After only a few months following his son's death, Durkheim suffered a stroke; he died at the age of 59 on November 15, 1917.

Durkheim's Social World

Durkheim was born just 7 months after Comte's death. The France of Comte's life, you'll remember, was in constant turmoil and conflict. Comte lived through the First Republic, First Empire, the Bourbon Restoration, the July Monarchy, the Second Republic, and the Second Empire. By comparison, Durkheim's France was more settled. He was born shortly after Napoléon III's coup d'état and the beginning of the Second Empire. The Second Empire was no cakewalk for France, but the period following brought the long-sought stability. Beginning in 1870, "through the establishment of a strong centralised state, the Third Republic, [was] better able than its predecessors to influence and control the social groups and regional societies which divided the nation" (Price, 2005, p. 182).

Beginning in 1850, France experienced a period of accelerated industrialization and financial growth. A major impetus was the extension of the railroad system throughout the country. The first railroad was constructed in 1843; by 1847, there were just 1,830 kilometers (1,137 miles) of track, but just over 20 years later there were 20,000 kilometers (12,427 miles), and by 1914, France had 40,770 kilometers (25,333 miles) of rails (Price, 2005, pp. 169–172). This sudden burst of transportation technology had a dramatic impact on markets and production. The increased rail system meant that more areas of the country could be reached quickly and efficiently. This gave businesses a cost-effective way to ship goods to other cities and towns. New consumers increased the demand for goods, which, in turn, drove industrialization and innovation. Increasing industrialization, along with improved transportation systems, moved people from rural areas to cities (urbanization): The number of people living in French cities almost doubled between 1851 and 1911. In addition, the per capita income rose almost 60% between 1855 and 1913 (pp. 174–180).

In the light of these changes, historian Roger Price (2005) concludes, "The creation of larger and more competitive markets intensified *innovatory pressure* throughout the economy" (p. 167, emphasis added). Price's statement is extremely interesting because Durkheim said something similar. One of Durkheim's (1893/1984) primary concerns was the division of labor in society, which "increases both the productive capacity and skill of the workman" (p. 12). The cause of increasing division of labor is something he called dynamic or moral density, which is defined as the reduction of "real distance between individuals" (p. 201). Dynamic density is created through increasing transportation technologies and urbanization, which, in turn, creates innovation. Price's work as a historian lets us clearly see how the structural changes Durkheim lived through influenced his theory.

Another structural feature that dramatically impacted Durkheim was the changes in the French education system. "French educational history in the 19th century is essentially the story of the struggle for the freedom of education, of the introduction at the secondary level of the modern and scientific branches of learning" (Anweiler, et al., 2011). You may remember from Chapter 1 that Nicolas de Condorcet had begun the work of moving the French education system from a Catholic enterprise to a secular one controlled by the state. That work came to fruition during the Third Republic. In the late 1870s and early 1880s, France passed several pieces of legislation called "Ferry Laws," named after Jules Ferry, a prominent politician. These laws gave the state the sole right to grant educational degrees, mandated primary secular education, forbade Catholic orders to teach in state-run schools, and increased state support for teacher training. In 1905, the French passed the Law of Separation of Church and State that completely severed ties with any religion.

In response to these structural opportunities and ideas, Durkheim dedicated a good part of his professional life to education. Durkheim was not only tireless in establishing sociology as a viable field of scholarship, but he was also dedicated to the field of education as a whole. Durkheim saw education as a primary source for socialization of the young and the reorganization of society, much as Comte did. He continued to train teachers throughout his life and published several pieces on education, including *Education and Sociology* and *Moral Education*. "Education . . . remained for Durkheim a privileged applied field where sociology could make its most important contribution to that regeneration of society for which he aimed so passionately" (Coser, 2003, p. 146).

Durkheim's Intellectual World

Montesquieu

Durkheim drew a good deal of his scientific approach from Charles-Louis de Secondat, Baron de La Brède et de Montesquieu (1689–1755). Montesquieu was a French Enlightenment philosopher who is best known for the *Persian Letters* and *The Spirit of Laws*. First published in 1721, the *Persian Letters* is a fictional tale of two foreign travelers who are seeing modern Europe for the first time. Montesquieu used this outsider view to level his own critiques of virtue and self-enlightenment.

The Spirit of Laws (1748) is, perhaps, the first thesis on the sociology of law, which linked expressed laws to the form of government and the nature of the people embraced. Overall, Montesquieu's influence on modernity is felt in the ideas of civil liberty, social equality, and the separation of governing powers.

Durkheim's Latin thesis was entitled *Montesquieu's Contribution to the Rise of Social Science*. His purpose in the thesis was to convince his readers of sociology's validity as a science. "Unmindful of our history," Durkheim (1892/1960a) claimed, "we have fallen into the habit of regarding social science as foreign to our ways and to the French mind" (p. 1). To remind his readers of the French investment in sociology, he reached past Comte and argued that Montesquieu was actually the one to lay down the principles of social science. In making that argument, Durkheim simultaneously gives us insight into how Montesquieu's work influenced him.

Drawing on Montesquieu, Durkheim argued that there are four conditions necessary for sociology to be scientific. First, "before social science could begin to exist, it had to be assigned a *definite subject matter*" (Durkheim, 1892/1960a, p. 3, emphasis added). We saw this issue with Comte and Spencer as well. Before any science can study anything, the thing studied must be seen to objectively, empirically exist. For most of our history, society was seen simply as affiliations of individuals, with no objective existence of its own. That changed with modernity and the work of Comte and Spencer. For Durkheim, sociology's "definite subject matter" is composed of social facts. Durkheim used the term *social fact* to stress the factual or objective existence of social structures and processes. In general, facts exist apart from individual interpretation or action; they are available to everyone and thus empirically substantiated. Durkheim (1895/1938) argued that the category of social facts works the same way: "it consists of ways of acting, thinking, and feeling, external to the individual, and endowed with a power of coercion, by reason of which they control him" (p. 3).

You've experienced this facticity any time you've felt constrained from doing something you shouldn't or felt guilty because you did it. The same is true for doing something you should and feeling pride. That moral force you feel doesn't originate inside you; it's not a psychological entity—it's a social fact that impacts everybody in our society. Durkheim distinguished between material and nonmaterial social facts. *Material social facts* are cultural artifacts, the physical objects that carry cultural meanings, such as flags, clothes, and so on. *Nonmaterial social facts* consist of symbolic meanings and collective sentiments, like love and freedom.

Definition

The concept of *social facts* is the foundation of Durkheim's empirical approach to sociology. Durkheim argues that the facticity of society is established by its objective nature and felt influence. Society exists apart from individuals; it was present before any specific individual was born and will exist after that person's death. Further, society appears to unavoidably influence the individual from the outside—people conform to its demands even if they don't want to. Durkheim used the existence of social facts to argue for society's unique existence and to argue for scientific sociology.

The second characteristic of a science is the ability to *create typologies*. One of the first steps in any scientific understanding is to create a taxonomy, like the Linnaean taxonomy of animal, vegetable, and mineral. Typologies, or taxonomies, are important for a couple of interrelated reasons. First, they are the basic way science understands and explains the world. This is so basic that it's sometimes hard to notice. For example, we speak of oceans, seas, and lakes, yet these are different categories of bodies of water with specific defining characteristics. This principle of classification can be more clearly seen when science reclassifies, as in the example of Pluto. Another reason for classification in science is that it helps scientists think about change. This is key in evolutionary theory—biological evolution is the story of how one type of species evolves to another. We'll see this in the next chapter, introducing Weber's work; he gave us a theory of the evolution of religion from magic to polytheism, pantheism, and eventually to monotheism. Durkheim's most famous typology is that of mechanical and organic solidarity, and he used it to talk about how societies have changed and evolved. We'll talk at length about this typology later.

Hint

It's important for us to see that the use of types isn't restricted to describing societies. Sociologists never understand an individual as a unique person. We theorize about people in terms of types, like race, gender, class, status position, and so on. This isn't as abstract as it might sound—we relate to each other in the same way. For example, you relate to your professor as a student, both of which are social types.

Third, to qualify as a science, sociology must *discover causal laws*. Description through typology isn't the chief goal of classic science—the power of science is its ability to predict, and prediction is based on laws. Society, then, must be seen as subject to sets of laws: "if societies are not subject to laws, no social science is possible" (Durkheim, 1892/1960a, p. 10). The important point to see here is that for us to discover the laws of society, society has to be independent of us. Society can't simply be a mass of individual decisions and actions. Society has to work according to *its own laws*. Sociologists usually argue that this feature of society is an outcome of the fact that different dynamics work at different levels of existence. Think about your body as an example. Knowing how cells work doesn't tell you much about how the circulatory system works, even though the system is nothing more than a bunch of cells. The same sort of thing is true in understanding social things. Knowing about individuals doesn't tell us much about how institutions work. I point this out because Durkheim adds a little something extra to this approach. This little something extra is, I think, one of Durkheim's most profound ideas. We'll talk about it more when we get to Rousseau, in the next subsection.

The fourth condition is *method:* "we must possess a method appropriate to the nature of the things studied and to the requirements of science" (Durkheim, 1892/1960a, p. 13). The scientific method has long been the thread that links diverse sciences together. On the whole, that method is inductive rather than deductive, moving from observations to generalized statements rather than deducing from a priori universals. As Durkheim pointed out, Montesquieu "rightly insisted on the primacy of the inductive method" (p. 64). Durkheim's book entitled *The Rules of Sociological Method* is in many ways an elaboration of what he gleaned from Montesquieu. Social facts exist within a system that is greater than the individuals that comprise it. This system operates according to its own laws and thus constitutes a separate reality. People can discover those laws through empirical observation and induction.

Jean-Jacques Rousseau

One of the most important philosophical influences on Durkheim's theory came from Jean-Jacques Rousseau (1712–1778). Rousseau is best known for *Émile,* a novel in which Rousseau presents his philosophy of education; and *The Social Contract,* Rousseau's philosophy of government. *Émile* and *The Social Contract* were published in the same year (1762). Both books were banned in France due to Rousseau's views on religion and government. By 1789, more than a decade after the author's death, Rousseau's philosophies were at the center of the French Revolution. Durkheim's concern with Rousseau, however, was more focused on the idea of the *general will.*

To understand Rousseau's idea of the general will, Durkheim (1918/1960b) argued that we must first grasp his idea of the state of nature. The *state of nature* was a conceptual device that philosophers used to think about what human beings were like before society. One of the best known of these ideas came from Thomas Hobbes (1588–1679), who argued that human life apart from society was nasty, brutish, and short, and that humankind was perpetually in a state of war.

For Rousseau (1755/1960), the natural state of humankind is encapsulated in one of his most famous quotes: "savage man, destitute of every species of knowledge . . . his desires never extend beyond his physical wants; he knows no goods but food, a female, and rest" (p. 33). First, notice what isn't included in the state of nature: knowledge. Apart from knowledge, humans are like most other animals and guided by sense-certainty alone. Sense-certainty implies that in the natural state, humans didn't have language nor was human behavior moral: "Though natural man is not evil, he is not good; morality does not exist for him. If he is happy, he is not aware of being so" (Durkheim, 1918/1960b, p. 90).

Without knowledge, language, morality, and happiness, what remains for the state of nature? What were humans like before society? Rousseau answered this question with simple physical desires: food, sex, and sleep. In the state of nature, humans lived as other animals: in a balanced give-and-take relationship with their physical environment. Without knowledge and language, "natural man is reduced exclusively to sensations" (Durkheim, 1918/1960b, p. 70). There was no future and no past, no planning or cooperation of any kind. In fact, Rousseau claimed, "It is impossible to imagine why, in this primitive state, a man should need another man" (as cited in Durkheim, 1918/1960b, p. 73).

Why, then, would humans create society? The reason for a *social contract* is easier to see with Hobbes. For Hobbes, people give up self-sovereignty in favor of the guiding power of society because without it, humankind is in a perpetual state of war, due to the fact that one person's judgment is in no way better than another's. For example, if I want something that you have, there is no reason in the natural world that my want is any less valid than yours. I judge the thing as mine; you judge the thing as yours—and without an external authority, the only way to settle it is to fight. If, however, we submit both our wills to an external power (society's laws), then we can live more peacefully, and our lives won't be as brutish and short. But the state of nature Rousseau conceived is almost idyllic and certainly not warlike. So, why did society come into existence? For Hobbes, there was something inside human beings that pushed them to submit to a higher authority. Rousseau, on the other hand, sees a natural harmony of humankind with nature: "Imagine a per-petual springtime on earth. . . . I cannot see how they would ever have surrendered their primitive freedom" (as cited in Durkheim, 1918/1960b, p. 77).

In that statement, Rousseau gives us a hint about the motivation to create society: Perpetual springtime didn't continue. There are a number of ways this could have happened, but at some point the environment changed. It was no longer able to sus-tain the lives of individual human beings. This "resistance men encountered in nature stimulated all their faculties" (Durkheim, 1918/1960b, p. 77). In this struggle, human-kind developed language and the ability to cooperate socially. Social cooperation not only provided for survival, but it also created surplus and new needs. By working together, people were able to create things that didn't exist in nature. There's good evidence that the first artificial product was beer and/or bread. These new goods could only be produced socially, so humankind's new needs made people dependent upon one another. Thus, society emerged and began to develop in complexity and power. At some point, ideas such as ownership and inheritance developed, and these needed additional political power to enforce, which brought alienation, exploitation, and oppression. "Society as it is today is certainly a monstrosity that came into being and continues to exist only through a conjunction of accidental and regrettable cir-cumstance" (Durkheim, 1918/1960b, p. 86).

For Rousseau, this fall from grace not only destroyed the natural state of human-ity, but it also made it impossible to return. We can never go back to the garden; creating society changed our nature—we are now and forever social beings. The problem, then, is how to fashion a new social environment, one that "maintains the same relation to society as natural man to physical nature" (Durkheim, 1918/1960b, p. 93). How would we fashion such a relationship? Who among us, or which polit-ical party, is wise enough to create such a society and lay down its laws? To solve this dilemma, Rousseau argued, we must look to the basic or elementary social form.

Most political philosophy looks at how different types of government come into existence and how they operate. Rousseau, on the other hand, said that there is something more basic we should look at instead—something already existed before the governments that political theory examines. A governing body assumes an already existing social body. In order to help us see this, I'd like you to imagine something for a minute. You're out in a field somewhere, dressed in regal robes, sitting on a gold-encrusted throne, and you're reading the decrees you just wrote that morning. Now scan the surroundings and see that no one else is in the field; it's just you, on your throne, governing nothing. You're either silly or mad because you can't be a monarch without subjects, and this is exactly Rousseau's point. Political theory often neglects the most important element: "the people," the collective that existed before any particular government ruled them. It is that primitive, elementary form of society that Rousseau looked to for answers.

Obviously, this elementary form isn't what we usually think of when talking about society. It's something deeper and more foundational. It came into existence as we were beginning to become social—its existence came as each individual will "vanishe[d] into a common, general will" (Durkheim, 1918/1960b, p. 98). This process created a "corporate body *sui generis*," an entity that exists unto itself. By incorporating each individual will into the general will, humans created a force that was transcendent, something greater under which each individual is equal. This moral force by its very nature creates and imposes inflexible laws that none can use to his or her advantage. Yet the general will implies freedom and equality because no individual's will is subject to another's. With the general will, laws aren't legislated by individuals for special interests; rather, moral laws emerge from the general will. Durkheim said that once this entity *sui generis* came to exist, it needed a foundation in "some being, and since there is no being in nature that satisfies the necessary conditions, such a being must be created. This being is the social body" (p. 103). Drawing again from Rousseau, Durkheim argued that this social body should be "animated by a single, indivisible soul" and that this soul would activate the social body just like the life force in a biological organism, "by a vital force whose synergic action produces the co-operation of the parts" (p. 112).

If this sounds suspiciously metaphysical to you, you're right. But there is a fundamental difference that Rousseau introduced. Most metaphysical philosophies are based on something that exists apart from and prior to human beings. Such a thing is often conceived of as God. This isn't the case with Rousseau (at least not the part of his work from which Durkheim drew). This soul, this transcendent being, didn't create human beings; it emerged from human social interaction. Yet this metaphysical being isn't a false projection, as with Marx and Feuerbach. This general will, which Durkheim called the *collective consciousness,* is an independent vital force.

Definition

The *collective consciousness* is a central theoretical issue in Durkheim's theory of social solidarity. It refers to the collective representations (cognitive elements) and sentiments (emotional elements) that guide and bind together any social group. The collective consciousness varies by four elements: the degree to which culture is shared; the amount of power the culture has to guide individual's thoughts, feelings, and actions; its degree of clarity; and its relative levels of religious versus secular content. Each of these is related to the amount of ritual performance in a collective. According to Durkheim, the collective consciousness takes on a life and reality of its own and independently influences human thought, emotion, and behavior, particularly in response to high levels of ritual.

As you would probably suspect, Rousseau did take his ideas in a religious direction. Rousseau concluded that government and religion should not be separated. Now, be careful here. This isn't related to the "separation of church and state" that we're familiar with. Rousseau argued that the perceived duality between the temporal and spiritual worlds began with Christianity. Looking back at religious history, we can see that early systems of belief didn't see a separation between the physical and spiritual. The gods were the vital force within every physical element and event, such as storms, seasons, plant growth, and so on. Over time and for a variety of reasons—which we'll talk about when we get to Weber—humans began to think of the gods as dwelling in the heavens amongst the stars. Catholicism still teaches that spiritual entities can inhabit material things, as in the Eucharist, and that angels and saints have access to this world. In Catholicism, there is still union between government and religion—the government and religion of the Vatican are synonymous, and the Church has a living disciple in the Pope. It was Protestantism that finally divested the universe of spiritual presence: "The Protestant believer no longer lives in a world ongoingly penetrated by sacred beings and forces" (Berger, 1967, p. 111).

Again, we have to be careful because any religion can be used to legitimate the actions of a government. We can think of this legitimating function of religion as a kind of moral bridge made between the two institutions. That's not what Rousseau had in mind. Government as it evolved over time is the problem; Rousseau thus wants us to reach back to the more fundamental general will that created society before government existed. In that general will, society and religion are one; government, religion, and the people are the same because the general will inhabits each and every person, and every individual will is willingly subjected to the general will. We can clearly see the differences between this kind of religion and what we know of religion today. I've listed the attributes of Rousseau's state religion as follows (as cited in Durkheim, 1918/1960b, p. 134). As you read through it, think about the differences between it and what you know about religion today. In Rousseau's view, civil religion will hold to the following:

The existence of God

The future life

The sanctity of the social contract

The absolute prohibition of any intolerance in matters not covered by the social credo

Zero tolerance of any religion that does not tolerate other religions

Only the state may exclude members from a social group

No church may say that it alone has the way of salvation

Durkheim's Contribution to Functionalism

My source for Durkheim's take on Rousseau comes from a piece published in 1918. The material that Durkheim used came from lectures he delivered between

1887 and 1902 while at the University of Bordeaux. You might remember that I mentioned that period as an extremely productive time for Durkheim. He published *The Division of Labor in Society, The Rules of Sociological Method,* and *Suicide* in those years. The problem that Durkheim was dealing with in two of the books, *The Division of Labor* and *Suicide,* was basically the same: How does increasing structural and social differentiation impact society and people's daily lives? It was clear to Durkheim that in the move from traditional to modern society, the moral base of society had diminished. There was less social "glue" holding people together. If differentiation continued, would it tear society apart? This is what Durkheim was struggling with during this time. Durkheim's strongest answer to this problem didn't come until 1912 with the publication of *The Elementary Forms of the Religious Life,* a work that clearly draws from Rousseau. Notice the title. Like Rousseau's search for the elementary social form that existed before government, Durkheim was after the elementary forms of religion before sectarianism took hold. More to the point, Durkheim explained the sociological base of Rousseau's insight into the basic nature of society—he theoretically explained how the social will, or collective consciousness, was created.

The timing of these works is important not only for understanding how Durkheim's thought developed, but also for seeing what Durkheim added to sociology. In 1981, Jonathan H. Turner, a well-known contemporary theorist, published an article in which he explicated Durkheim's theory of integration in differentiated societies. In it, he says, "Most of these principles are seen to come from Durkheim's *The Division of Labor in Society* and *Suicide.* Other works, while intellectually important, are not viewed as theoretically significant as these" (p. 379). In 1988, Jeffrey C. Alexander published an edited book, *Durkheimian Sociology: Cultural Studies.* Alexander notes that sociologists had focused almost exclusively on *The Division of Labor in Society, The Rules of Sociological Method,* and *Suicide,* which we see in Turner's statement. "But it was only after the completion of these works that Durkheim's distinctively cultural program for sociology emerged" (Alexander, 1988, p. 2).

We will explore more of Durkheim's theory of symbolic culture later in the chapter, but for now I want you to see what Durkheim uniquely added to the functionalist perspective. In our chapter on Spencer, we talked about the formation of functionalism as a perspective in sociology. Explicitly, Spencer built functionalism around the organismic analogy; interrelated systems of structures; requisite needs and functions; and structural differentiation, specialization, and integration. Durkheim elaborates on these and adds an important dimension: the moral base of society and the crucial role that symbols play in our daily lives and social solidarity. Most significantly, Durkheim explains how rituals underlie everything we think, feel, and do.

Durkheim's Sociological Imagination: Cultural Sociology

The concept of culture covers a broad range of issues: ideas, recipes for action, tools, products, norms, values, beliefs, art, and so on. More generally, we can say that culture is made up of symbols and practices that transmit meaning across time and space. Sociologists have been concerned with culture from the beginning. We've seen this with Comte (types of knowledge), Spencer (ceremonial institutions and social categories), and Marx (ideology); and we'll see it with Weber, Simmel, and Du Bois as well. However, each of these theorists sees culture simply as something that comes along with studying society generally. Durkheim, on the other hand, was the first to give culture preeminence. To say this in functionalist terms, Durkheim added the collective consciousness (culture) to Spencer's three requisite functions (regulatory, operative, and distributive). So strong was Durkheim's emphasis on culture that he influenced other disciplines as well, specifically structural anthropology, structural linguistics, and semiotics (Alexander, 1988, pp. 3–6).

In sociology, there are at least three schools of thought concerning culture: culture studies, sociology of culture, and cultural sociology. Cultural studies, more formally known as the Centre for Contemporary Cultural Studies (CCCS), began at the University of Birmingham, England, in 1963 and draws from an eclectic group of sources, most notably literature and a few twentieth century Marxists, such as Antonio Gramsci. As you can imagine, this approach to culture looks for ways in which power is expressed through ideology. The sociology of culture refers simply to studying culture sociologically, in terms of culture's relationship to other social factors. The definition of culture in this case is quite broad: the "(1) *ideas*, *knowledge* . . . and *recipes* for doing things, (2) humanly fabricated *tools* . . . and (3) the *products* of social action that may be drawn upon in the further conduct of social life" (J. R. Hall, Neitz, & Battani, 2003, p. 7). Though not as diverse as CCCS, the sociology of culture draws from a variety of sociological theorists, such as Marx, Weber, and Durkheim.

Cultural sociology is a bit different from sociology's general approach. Where the sociology of culture sees culture as one of many social factors, cultural sociology makes culture primary. Thus, someone who is using a sociology of culture approach would say, "I'm going to study culture," but someone using a cultural sociology perspective would say, "I'm going to study everything in *terms* of culture." Like Durkheim, cultural sociology sees culture as having independent causal force. Seeing culture in this way implies that culture is a social structure, and an analysis of culture must take that structure into account. The most prominent advocate of this approach today is Jeffrey Alexander, who founded the Center for Cultural Study at Yale University (see http://ccs.research.yale.edu/). Alexander argues that the ideas, values, and discourses of a society—its culture—are structured through binary codes of purity and impurity.

Concepts and Theory: Primal Society

I want to begin where others end. Many theory books start with Durkheim's work on the division of labor and suicide. It's there that Durkheim clearly states the problem he wants to address: the structural and social differentiation of society. In today's terms, Durkheim is interested in social diversity. We could phrase his question, then, like this: Can a society have too much diversity? That may be an interesting question, but I think most people in a democratic society would answer "no." Cultural diversity is a good thing, especially in a democracy. However, this question becomes something very different once you grasp how Durkheim understands society.

Like Spencer, Durkheim understands society as a system of interrelated structures that is evolving and becoming more complex over time. Spencer, you'll remember, said that increasing complexity creates problems of coordination and control, and that these issues are directly addressed by increasing the regulatory function (more government for more complex societies) and through interdependency. While Durkheim agrees with Spencer's notion of society as a system and integration through interdependency, he disagrees with Spencer's use of increased regulation. From Durkheim's perspective, Spencer didn't see to the heart of society; he didn't see society's core.

After reading about Rousseau's influence, you know that Durkheim sees a primal social form upon which all the rest of society is based. The fundamental societal need, Durkheim claims, is the collective consciousness. Increasing structural and social differentiation thus impacts the collective consciousness. Here's the question Durkheim wants us to think about: How does structural and social differentiation impact the moral core of society? This is why we are starting with one of Durkheim's last works—we need to understand primal society in order to appreciate the impact of increasing social and structural differentiation. We begin, then, with *The Elementary Forms of Religious Life*. As we'll see, *primal society is religious through and through*.

Durkheim has two major purposes in writing his thesis on religion. His primary aim is to understand the empirical elements present in all religions. He wants to go behind the symbolic and spiritual to grasp what he calls "reality itself" (Durkheim, 1912/1995, p. 22). Durkheim intentionally put aside the issues of God and spirituality. He argues that no matter what religion is involved, whether Christianity or Islam or any other, and no matter what god is proclaimed, there are certain social elements that are common to all religions. To put this issue another way, anytime a god is believed to do something here on Earth, there are certain social elements that are always present. Durkheim is interested in discovering those empirical social elements.

Durkheim's second reason for writing his book is a bit trickier to understand. For Durkheim, religion is the most fundamental social institution. He argues that religion is the source of everything social. That's not to say that everything social is religious, especially today, but Durkheim is convinced that social bonds were first created through religion. We'll see that in ancient clan societies, the symbol that bound the group together and created a sense of kinship (family) was principally a religious one.

Further, Durkheim argues that our basic categories of understanding are of religious origin. Humans divide the world up using categories. We understand things in terms of animal, mineral, vegetable, edible, inedible, private property, public property, male, female, and on and on. Each of these is a category that includes some things and excludes others. Durkheim says that many of the categories we use are of what one might call "fashionable" origin, that is, culture that is subject to change. Durkheim argues that fashion is a recent phenomenon and that its basic social function is to distinguish the upper classes from the lower. There is a tendency for this sort of culture to circulate. The lower classes want to be like the upper classes and thus want to use their symbols. (We want to drive their cars, wear the same kinds of clothes they do, and so on.) This implies, in the end, that fashionable culture is rather meaningless. As he says in an earlier book, "Once a fashion has been adopted by everyone, it loses all its value; it is thus doomed by its own nature to renew itself endlessly" (Durkheim, 1887/1993, p. 87).

Durkheim has little if any concern for such culture (though quite a bit of contemporary cultural theory and analysis is taken up with it). He is interested in primary "categories of understanding." He argues that these categories—time, space, number, cause, substance, and personality—are of social origin, but not the same kind of social origin that fashion has. Fashion comes about as different groups demarcate themselves through decoration, and is purely the work of imagination. The primary categories of understanding, on the other hand, are tied to objective reality. Durkheim argues that the primary categories originate empirically and objectively in society, in what he calls social morphology.

Merriam-Webster (2002) defines *morphology* as "a branch of biology that deals with the form and structure of animals and plants." So, when Durkheim talks about *social morphology,* he is using the organismic analogy to refer to the form and structure of society—in particular, the way in which populations are distributed in time and space. Let's take time, for example. In order to conceive of time, we must first conceive of differentiation. Separate from human symbols, time on Earth appears like a cyclical stream. There is daytime and nighttime, and there are seasons that endlessly repeat themselves. Yet that isn't how we experience time. For us, time is chopped up. For example, at the time I am writing this paragraph, it is February 2, 2012. However, that date and the divisions underlying it are not a function of the way time appears naturally. So, from where do the divisions come? "The division into days, weeks, months, years, etc., corresponds to the recurrence of rites, festivals, and public ceremonies at regular intervals. A calendar expresses the rhythm of collective activity while ensuring that regularity" (Durkheim, 1912/1995, p. 10). The same is true with space. Our spatial divisions, whether east, west, up, down, right, left, and so on are, like time, social constructions.

The important thing to see here is that Durkheim makes the claim that the way in which we divide up time and space and the way we conceive of causation and number are neither a function of the things themselves nor a function of mental divisions. Rather, the way we conceive of these primary categories is a function of the objective form of society—the ways in which we distribute and organize populations and social structures. Our primary categories of understanding come into existence through the way we distribute ourselves in time and space; they reflect our gatherings and rituals. Thus, all of our thinking is founded upon social facts. These social facts, according to Durkheim, originated in religion.

Durkheim wants to make a point beyond social epistemology. He argues that if the four basic categories of understanding have their roots in religion, then all systems of thought, such as science and philosophy, have their roots in religion. Durkheim (1912/1995) extends this theme, stating that "there is no religion that is not both a cosmology and a speculation about the divine" (p. 8). Notice that there are two functions of religion in this quote. One has to do with speculations about the divine—in other words, religion provides faith and ideas about God. Also notice the other function: to provide a cosmology. A cosmology is a systematic understanding of the origin, structure, and space–time relationships of the universe. Durkheim is right: Every religion tells us what the universe is about—how it was created, how it works, what its purpose is, and so on. But so does science, and that's Durkheim's point. Speculations about the universe began in religion; ideas about causation began in religion. Therefore, the social world, even in its most logical of pursuits, was set in motion by religion.

In the end, Durkheim (1912/1995) argues, "Religion gave birth to all that is essential in society." What he means is that "society is not constituted simply by the mass of individuals who comprise it, the ground they occupy, the things they use, or the movements they make, but above all by the idea it has of itself" (p. 425). This idea is born out of the religious. Religion gives us our strongest sense of the collective consciousness; we are most fully aware of our collective soul in religion. Religion gives us our strongest sense of identity, values, norms, symbols, solidarity, and so forth. But more than that, even without religion *per se*, as in modern society, the interactions that are necessary to produce and preserve the collective consciousness are religious in nature.

Enduring Issues

Modernity brought with it the deliberate separation of religion and government. This severance was unique in all of human history—up to this time, government had always been explicitly legitimated by religion. However, in order for democracy to work, there must be a structural separation of church and state. Theocracy—rule by God—is the polar opposite of democracy. In a theocracy, the power to rule comes from the top (God) down; in a democracy, the power to rule comes from the bottom (citizens) up. Yet it is also crystal clear that the founders of modern democracy did not intend for religion to go away. In fact, the strength of ethical and moral conviction that different religions provide is necessary for a healthy democracy. How this relationship works is a continuing issue for social, political, and religious thought.

Defining Religion

But how did religion begin? What is the source of the religious idea or spirituality? Did the idea of religion spring from the fact that humans dream? Did that open the door for us to think about another realm of existence? Or, did religion come out of our fear of death? They say that humans are the only species that knows it's going to die. Did that knowledge push us to create religion? Or, did religion spring from a deep psychological need? Durkheim sees something very different from what any of these questions or approaches implies. He argues that religion originates from an experience, but not the experience of dreams or death. It was the experience of society that gave birth to religion. Humans created religion in the moment they created society.

In order to get at his argument, we will consider the data Durkheim uses, his definition of religion, and—most important—how the sacred is produced. The data that Durkheim employs is important because of his argument and intent. He wants to get at the most general social features underlying all religions. In other words, apart from doctrine, he wants to discern what is common to all religions—not in terms of beliefs, but rather, in terms of social practices and effects. To discover those commonalities, Durkheim contends that one has to look at the most primitive forms of religion.

Using contemporary religion to understand the basic forms of religion has some problems, most notably the natural effects of history and storytelling. You've probably either played or heard of the game where a group of people sits in a circle and whispers a story from one to another. What happens, as you know, is that the story changes in the telling. The same is true with religion, at least in terms of its origins. Basically, what Durkheim is saying is that the farther we get away from the origins of religion, the greater the confusion will be about how religion began in the first place. Also, because ancient religion was simpler, using Durkheim's approach allows us to break the social phenomenon down into its constituent parts and identify the circumstances under which it was born. For Durkheim, the most ancient form of religion is totemism. He uses data on totemic religions from Australian aborigines and Native American tribes for his research.

Conceptually, prior to deciding what data to use, Durkheim had to create a definition of religion. In any research, it is always of utmost importance to clearly delineate what will count and what will not count as your subject. For example, if you were going to study the institution of education, one of the things you would have to contend with is whether homeschooling or Internet instruction would count, or whether only accredited teachers in state-supported organizations constitute the institution of education. The same kinds of issues exist with religion. Durkheim had to decide on his definition of *religion* before choosing his data sources—he had to know ahead of time what counts as religion and what doesn't, especially since he wanted to look at its most primitive form. According to Durkheim (1912/1995):

> *A religion is a unified system of beliefs and practices relative to sacred things, that is to say, things set apart and forbidden—beliefs and practices which unite into one single moral community called a Church, all those who adhere to them.* (p. 44, emphasis original)

Note carefully what is missing from Durkheim's definition: There is no mention of the supernatural or God. Durkheim argues that before humans could think about the supernatural, they first had to have a clear idea about what was natural. The term *supernatural* assumes the division of the universe into two categories: things that can be rationally explained and those that can't. Now, think about this: When was it that people began to think that things could be rationally explained? It took quite of bit of human history for us to stop believing that there are spirits generating almost all natural phenomena. We had gods of thunder, forests, harvest, water, fire, fertility, and so on. Early humans saw spiritual forces behind almost everything, which means that the idea of *nature* didn't occur until much later in our history. The concept of nature—those elements of life that occur apart from spiritual influence—didn't truly begin until the development of science. So, the idea of supernatural is a recent invention of humanity and therefore can't be included in a definition of religion.

Durkheim also argues that we cannot include the notion of God in the definition. His logic here concerns the fact that there are many belief systems that are generally considered religious yet don't require a belief in god. Although he includes other religions such as Jainism, his principal example is Buddhism. The focus of Buddhist faith is the Four Noble Truths, and "salvation" occurs apart from any divine intervention. There are deities acknowledged by Buddhism, such as Indra, Agni, and Varuna, but the entire Buddhist faith can be practiced apart from them. The practicing Buddhist needs no god to pray to or worship, yet we would be hard-pressed to not call Buddhism a religion.

Thus, three things constitute religion in its most basic form: the presence of sacred things, beliefs and practices, and a moral community. The important thing to notice about Durkheim's definition is the centrality of the notion of the *sacred*. Every element of the definition revolves around it. The beliefs and practices are relative to sacred things, and the moral community exists because of the beliefs and practices, which of course brings us back to sacred things. So at the heart of religion is this idea of the sacred.

Hint

How a theorist defines a term is significant. In the first place, the definition determines what needs to be explained. Durkheim has to explain the origin of the sacred, but that would not be a concern for Marx, whose definition of religion is different. Second, theoretical definitions usually include the work a concept does. For Durkheim, religion works to bind people together in a moral community. The implication here is that you usually can't rely on a common dictionary for a definition of a theoretical term.

Creating a Sacred World

What are sacred things? By that I mean, what makes something sacred? For most of us in the United States, we think of the Cross or the Bible as sacred objects. It's easy for us to think they are sacred because they represent something that is truly sacred. Thus, the Cross is sacred because of the death of Christ. However, this assumption falls apart when we look at the vast array of things that have been sacred to humans throughout history: There have been sacred places, animals, flags, and so on. I know of a group that gathers around a sacred vibrating rock. Empirically, humans can make almost anything sacred. The question, then, becomes how do humans do it? How can the U.S. flag and Al Qaeda beliefs both be sacred enough that people will die for them? Let's attempt to explain the power of the sacred using empirical, general terms. Durkheim begins with his consideration of ancient totemic religion.

Totems had some interesting functions. For instance, they created a bond of kinship among people unrelated by blood. Each clan was composed of various hunting and gathering groups. These groups lived most of their lives separately, but they periodically came together for celebrations. The groups were neither related by blood nor connected geographically. What held them together was the name of the totem, or symbol for their clan—all the members of the clan carried the same name, and the members of these groups acted toward one another as if they were family. They had obligations to help each other, to seek vengeance, to not marry, or if they did, to marry one another; and so forth based on family relations. In addition to creating a kinship name for the clans, the totem acted as an emblem that represented the clan. It acted as a symbol both to those within the clan and outside it. The symbol was inscribed on banners and tents and tattooed on bodies. When the clan eventually settled in one place, the symbol was carved into doors and walls. The totem thus formed bonds, and it represented the clan.

In addition, the totem was used during religious ceremonies. In fact, Durkheim (1912/1995) tells us, "Things are classified as sacred and profane by reference to the totem. It is the very archetype of sacred things" (p. 118). Different items became sacred because of the presence of the totem. For example, the clans both in daily and sacred life would use various musical instruments; the only difference between the sacred and mundane instruments was the presence of the totemic symbol. The totem imparted to the object the quality of being sacred.

This is an immensely important point for Durkheim: The totem represents the clan, and it creates bonds of kinship. It also represents and imparts the quality of being sacred. Durkheim then used a bit of algebraic logic: If A = B and B = C, then A and C are equal. So, "if the totem is the symbol of both the god and the society, is this not because the god and the society are the same?" (Durkheim, 1912/1995, p. 208). Here, Durkheim begins to discover the reality behind the sacred and thus, religion. The empirical reality behind the sacred has something to do with society, but what exactly?

One of the primary features of the sacred is that it stands diametrically opposed to the profane. In fact, one cannot exist in the presence of the other. Do you know the story from the Bible of Moses and the burning bush? Moses had to take off his shoes because he was standing on sacred ground. These kinds of stories are repeated over and over again in every religion. The sacred either destroys the profane, or the sacred becomes contaminated by the presence of the profane. So, one of Durkheim's questions is, how did humans come to conceptualize these two distinct realms? The answer to this will help us discover the reality behind sacredness.

Durkheim found that the aborigines of Australia had two cycles to their lives, one in which they carried on their daily life in small groups and the other in which they gathered in large collectives. In the small groups, they would take care of daily needs through hunting and gathering. This was the place of home and hearth. Yet each small group saw itself as part of a larger group: the clan. Periodically, the small groups would gather together for large collective celebrations.

During these celebrations, the clan members were caught up in *collective effervescence,* or high levels of emotional energy. They found that their behaviors changed; they felt "possessed by a moral force greater than" the individual. "The effervescence often becomes so intense that it leads to outlandish behavior. . . . [Behaviors] in normal times judged loathsome and harshly condemned, are contracted in the open and with impunity" (Durkheim, 1912/1995, p. 218). These clan members began to conceive of two worlds: the mundane world of daily existence where they were in control, and the world of the clan where they were controlled by an external force greater than themselves. "The first is the profane world and the second, the world of sacred things. It is in these effervescent social milieux, and indeed from the very effervescence, that the religious idea seems to have been born" (Durkheim, 1912/1995, p. 220).

Definition   Emotional energy or *effervescence* is a key theoretical idea in Durkheim's theory of ritual. It is defined as the level of motivational energy an individual feels after participating in a ritual. It varies by the degree of common focus of attention, common emotional mood, and physical copresence. Emotional effervescence by itself generally dissolves quickly unless it is given a symbol that represents the high level of emotion. These symbols tend to then be seen as sacred (such as national flags and religious emblems) and are used to prompt further ritual enactment through creating a common focus of attention and emotional mood.

In this theory, Durkheim is answering the theoretical questions implicit in Rousseau's work. Rousseau pointed Durkheim to this primal society, the one that forms the bedrock of all things social. Rousseau calls this the "general will," and Durkheim uses "collective consciousness." These concepts, however, only beg these theoretical questions: How did this moral entity come into existence, and how does it work? Durkheim answers those questions here.

For Durkheim, humans are linked emotionally. Undoubtedly these emotions, once established, mediate human connections unconsciously. That is, once humans are connected emotionally, it isn't necessary for them to rationally see or understand the connections, though we will always come up with legitimations. This emotional soup that Durkheim is describing is the stuff out of which human society is built. Initially, emotions run wild and so do behaviors in these kinds of primitive societies. But with repeated interactions, the emotions become focused and specified behaviors, symbols, and morals emerge. Recent work in neuroscience confirms Durkheim's argument about the importance of emotion (for example, see LeDoux, 1996, and Damasio, 1999). Drawing from neuroscience, evolutionary theory, anthropology, and sociology, Jonathan H. Turner (2000) developed a comprehensive explanation of the place emotion has in forming social bonds as well as providing the base for language and rational thought (p. 60).

In general, Durkheim is arguing that human beings are able to create high levels of emotional energy whenever they gather together. We've all felt something like what Durkheim is talking about at concerts, political rallies, or sporting events. We get swept up in the excitement. At those times, we feel "the thrill of victory and the agony of defeat" more poignantly than at others. It is always more fun to watch a game with other people, for example, at a stadium. Part of the reason is the increase in emotional energy. In this case, the whole is greater than the sum of the parts, and something emerges that is felt outside the individual. This dynamic is what is in back of mob behavior and what we call "emergent norms" (and what is called "hooliganism" in Britain). People get caught up in the overwhelming emotion of the moment and do things they normally wouldn't do.

Randall Collins (1988), a contemporary theorist, has captured Durkheim's theory in abstract terms. Generally speaking, there are three principal elements to the kind of interactions that Durkheim is describing (see Figure 4.1): copresence, which describes the degree of physical closeness in space (we either can be closer or farther away from one another); common emotional mood, the degree to which we share the same feeling about the event; and common focus of attention, the degree to which participants are attending to the same object, symbol, or idea at the same time (a difficult task to achieve, as any teacher knows).

Figure 4.1 Durkheim's Ritual Theory

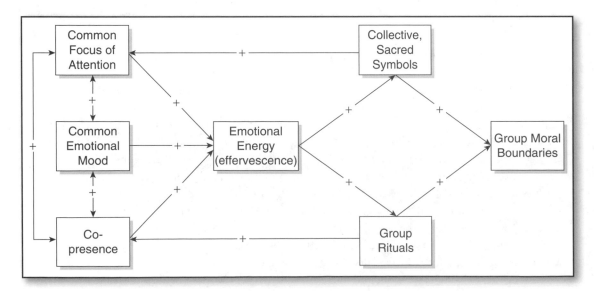

When humans gather together in intense interactions—with high levels of copresence, common emotional mood, and common focus of attention—they produce high levels of emotional energy. People then have a tendency to symbolize the emotional energy, which produces a sacred symbol, and to create rituals (patterned behaviors designed to replicate the three interaction elements). The symbols not only allow people to focus their attention and recall the emotion, but also give the collective emotion stability. These kinds of *rituals* and sacred symbols lead a group to become morally bounded; that is, many of the behaviors, speech patterns, styles of dress, and so on associated with the group become issues of right and wrong.

Groups with high moral boundaries are difficult to get in and out of. Street gangs and the Nazis are good examples of groups with high moral boundaries. One of the first things to notice about these examples is the use of the word *moral.* Most of us probably don't agree with the ethics of street gangs. In fact, we probably think their ethics are morally wrong and reprehensible. But when sociologists use the term *moral,* we are not referring to something that we think of as being good. A group is moral if its behaviors, beliefs, feelings, speech, styles, and so forth are controlled by strong group norms and are viewed in terms of right and wrong. In fact, both the Nazis and street gangs are probably more moral, in that sense, than you are, unless you are a member of a radical fringe group.

Definition

Rituals are the key to Durkheim's theory of social solidarity. In Durkheimian theory, rituals are patterned sequences of behavior that re-create high levels of copresence, common emotional mood, and common focus of attention. In Durkheim's scheme, rituals function to create and reinvigorate a group's moral boundaries and identity. Keep in mind that this definition is specific to Durkheim; other theorists use the concept in different ways.

This theory of Durkheim's is extremely important. First of all, it gives us an empirical, sociological explanation for religion and sacredness. One of the problems that we are faced with when we look across the face of humanity is the diversity of belief systems. How can people believe in diverse realities? The Azande of Africa seek spiritual guidance by giving a chicken a magic potion brewed from tree bark and seeing if the chicken lives or dies. Some Christians drink wine and eat bread believing they are drinking the blood and eating the body of Christ. How can we begin to explain how people come to see such diverse things as real? Durkheim gives us a part of the puzzle. The issues of reality, sacredness, and morality aren't necessarily based on ultimate truth for humans. Our experience of reality, sacredness, and morality are based on Durkheimian rituals and collective emotion. Let me put this another way. Let's say that the Christians are right and the Azande are wrong. How is it, then, that both the Christians and the Azande can have the same experience of faith and reality? Part of the reason is that human beings create sacredness in the same way, regardless of the correctness of any ultimate truth. One of the common basic elements of all religions, particularly during their formative times, is the performance of Durkheimian rituals. These kinds of rituals create high levels of emotional energy that come to be invested in symbols; such symbols are then seen as sacred, regardless of the meaning or truth-value of the beliefs associated with the symbol.

Another reason that this theory is so important is that it provides us with a sociological explanation for the experience that people have of something outside of and greater than themselves. All of us have had these kinds of experiences, some more than others. Some have experienced it at a Grateful Dead or Phish concert, others at the Million Man March, others watching a parade, and still others as they conform to the expectations of society. We feel these expectations not as a cognitive dialogue, but as something that impresses itself upon us physically and emotionally. Sometimes, we may even cognitively disagree, but the pressure is there nonetheless.

Durkheim thus provides us with part of the answer for the moral basis of society. Have you ever wondered about extremist groups? How do they produce such conformity? Or, why is it that people who live in small, rural communities tend to be politically conservative? How is bigotry created and preserved? Or, on a more positive note, how are national pride and identity created? How can Americans, Communists, Nazis, the Ba'ath, the Taliban, and so on all believe they are right?

I generally won't go into such detail about how a theorist goes about creating the theory. However, in this case, I think it's important for a couple of reasons. First, religion is something that most people have strong feelings about, either pro or con, and because Durkheim's theory is powerful, it, too, can elicit strong reactions. So I thought it important for you to see just how and why Durkheim came up with his theory.

Second, Durkheim's theory of religion contains the strongest statements of his general theory and concerns. We see his interest in categories and his assurance that the basis of cognitive categories is the physical form of society. (In the next section, we'll explore the idea of social morphology further.) We also have his strongest statement about how the collective consciousness takes on objective characteristics and a life of its own: The collective consciousness is born out of and maintained by emotion-producing rituals. In addition, we see the importance of symbols for collective life.

Concepts and Theory: Social Diversity and Moral Consensus

Now that we understand the core of society with which Durkheim is concerned, we're in a better place to ask his primary question: How much structural and social diversity can a society have and still maintain consensus around a moral center? Up until modern times, societies stayed together because most of the people in the society believed the same, acted the same, felt the same, and saw the world in the same way. However, in modern societies people are different from each other, and they are becoming more so. What makes people different? How can all these different people come together and form a single society? In order to begin to answer these questions, Durkheim created a typology of societies.

A *theoretical typology* is a scheme that classifies a phenomenon into different categories. We aren't able to explain things directly by using a typology, but it does make things more apparent and more easily explained. Herbert Spencer—the one to whom Durkheim compares his own theory throughout *The Division of Labor in Society*—categorized societies as either industrial or militaristic. In order to construct his typology, he focused on the state. So, for example, when a society is in a militaristic phase, the state is geared toward defense and war and social control is centralized: Information, behavior, and production are tightly controlled by government. But when the same society is in an industrial phase, the state is less centralized and the social structures are oriented toward economic productivity: Freedom in information exchange, behaviors, and entrepreneurship is encouraged. However, Durkheim focuses on something different. His typology reflects his primary theoretical concern: social solidarity.

Mechanical and Organic Solidarity

Social solidarity can be defined as the degree to which social units are integrated. According to Durkheim, the question of solidarity turns on three issues: the subjective sense of individuals that they are part of the whole, the actual constraint of individual desires for the good of the collective, and the coordination of individuals and social units. It is important for us to notice that Durkheim acknowledges three different levels of analysis here: psychological, behavioral, and structural. Each of these issues becomes a question for empirical analysis: How much do individuals feel part of the collective? To what degree are individual desires constrained? And, how are activities coordinated with and adjusted to one another? As each of these varies, a society will experience varying levels of social solidarity.

Definition   *Social solidarity* is Durkheim's term for the level of integration in a society. Generally speaking, integration is the blending and organizing of separate and diverse elements into a more complete, balanced whole. Social solidarity specifically refers to the subjective sense of group membership that individuals have, the constraint of individual behaviors for the group good, and the organization of social units and groups into a single system. Durkheim argues that social solidarity is different in modern societies compared to traditional ones.

Durkheim is not only interested in the degree of social solidarity, but also in the way social solidarity comes about. He uses two analogies to talk about these issues. The first is a mechanistic analogy. Think about machines or motors. How are the different parts related to each other? The relationship is purely physical and involuntary. Machines are thus relatively simple. Most of the parts are very similar and are related to or communicate with each other mechanistically. If we think about the degree of solidarity in such a unit, it is extremely high. The sense of an absolute relationship to the whole is unquestionably there, as every piece is connected to every other piece. Each individual unit's actions are absolutely constrained by and coordinated with the whole.

The other analogy is the organismic one. Higher organisms are quite complex systems, when compared to machines. The parts are usually different from one another, fulfill distinct functions, and are related through a variety of diverse subsystems. Organismic structures provide information to one another using assorted nutrients, chemicals, electrical impulses, and so on. These structures make adjustments because of the information that is received. In addition, most organisms are open systems in that they respond to information from the environment. (Most machines are closed systems.) The solidarity of an organism when compared to a machine is a bit more imprecise and problematic.

Definition	Mechanical and organic solidarity are a part of Durkheim's typology of society. *Mechanical solidarity* refers to high levels of group cohesiveness and normative regulation. *Organic solidarity* is created through mutual dependency, generalized culture, and intermediate (between the level of the state and the level of face-to-face interaction) groups and organizations.

By their very nature, analogies can be pushed too far, so we need to be careful. Nonetheless, we get a clear picture of what Durkheim is talking about. Durkheim says that there are two "great currents" in society: similarity and difference. Society begins with the similarity being dominant. In these societies, which Durkheim terms *segmented*, there are very few personal differences, little competition, and high egalitarianism. These societies experience mechanical solidarity. Individuals are mechanically and automatically bound together. The other current, difference, gradually becomes stronger, and similarity is limited to specific groups and settings. These differentiated social units are held together through mutual need and abstract ideas and sentiments. Durkheim refers to this as *organic solidarity*. While organic solidarity and difference tend to dominate modern society, similarity and mechanical solidarity never completely disappear.

In Table 4.1, I've listed several distinctions between mechanical and organic solidarity. In the first row, the principal defining feature is listed. In *mechanical solidarity*, individuals are directly related to a group and its collective consciousness. If the individual is related to more than one group, there are very few differences and the groups tend to overlap with one another: "Thus it is entirely mechanical causes which ensure that the individual personality is absorbed in the collective personality" (Durkheim, 1893/1984, p. 242). Remember all that we have talked about concerning culture and morality. People are immediately related to the collective consciousness by being part of the group that creates the culture in highly ritualistic settings. In these groups, the members experience the collective self as immediately present. They feel its presence push against any individual thoughts or feelings. They are caught up in the collective effervescence and experience it as ultimately real. The clans that Durkheim studies in *The Elementary Forms of Religious Life* are a good illustration.

Table 4.1 Mechanical and Organic Solidarity	
Mechanical Solidarity	*Organic Solidarity*
Individuals directly related to collective consciousness with no intermediary.	Individuals related to collective consciousness through intermediaries.
Joined by common beliefs and sentiments (moralistic).	Joined by relationships among special and different functions (utilitarian).
Collective ideas and behavioral tendencies are stronger than individual.	Individual ideas and tendencies are strong, and each individual has his or her own sphere of action.
Social horizon limited.	Social horizon unlimited.
Strong attachment to family and tradition.	Weak attachment to family and tradition.
Repressive law: Crime and deviance disturb moral sentiments; punishment meted out by group; purpose is to ritually uphold moral values through righteous indignation.	Restitutive law: Crime and deviance disturb social order; rehabilitative, restorative action by officials; purpose is to restore status quo.

When an individual is mechanically related to the collective, all the rest of the characteristics we see under mechanical solidarity fall into place. In Table 4.1, the common beliefs and sentiments and the collective ideas and behavioral tendencies represent the collective consciousness. The collective consciousness varies by at least four features: (1) *the degree to which culture is shared*—how many people in the group hold the same values, believe the same things, feel the same way about things, behave the same, and see the world in the same way; (2) *the amount of power the culture has to guide an individual's thoughts, feelings, and actions*—a culture can be shared, but then it is not very powerful (A group where the members feel they have options doesn't have a very powerful culture.); (3) *the degree of clarity*—how clear the prescriptions and prohibitions are in the culture (For example, when a man and a woman approach a door at the same time, is it clear what behavior is expected?); and (4) the *content of the collective consciousness varies*. Durkheim is referring here to the ratio of religious to secular and individualistic symbolism. Religiously inspired culture tends to increase the power and clarity of the collective consciousness.

As we see from Table 4.1, in mechanical solidarity, the social horizon of individuals tends to be limited. Durkheim is referring to the level of possibilities an individual has in terms of social worlds and relationships. The close relationship the individual has to the collective consciousness in mechanical solidarity limits the number of possible worlds or realities the individual may consider. In modern societies, under *organic solidarity*, we have almost limitless possibilities from which to choose. Media and travel expose us to uncountable religions and their permutations. Today, you can be a Buddhist, Baptist, or Baha'i, and you can choose any of the varied universes they present. This proliferation of possibilities, including social relationships, is severely limited under mechanical solidarity. One of the results of this limitation is that tradition appears concrete and definite. People in segmented societies don't doubt their knowledge or reality. They hold strongly to the traditions of their ancestors. And, at the same time, people express their social relationships using familial or territorial terms.

But even under mechanical solidarity, not everyone conforms to society. Durkheim acknowledges this and tells us that there are different kinds of laws for the different types of solidarity. The function of these laws is different as well, corresponding to the type of solidarity that is being created. Under mechanical solidarity, *punitive law* is more important. The function of punitive law is not to correct, as we usually think of law today; rather, the purpose is expiation (making atonement). Satisfaction must be given to a higher power, in this case the collective consciousness. Punitive law is exercised when the act "offends the strong, well-defined states of the collective consciousness" (Durkheim, 1893/1984, p. 39). Because this is linked to morality, the punishment given is generally greater than the danger represented to society, such as cutting off an individual's hand for an act of thievery.

Punitive law satisfies moral outrage and clarifies moral boundaries. When we respond to deviance with some form of "righteous indignation," and we punish the offender, we are experiencing and creating our group's moral boundaries. In punishing offenses, we are drawing a clear line that demarks those who are in the group and those who are outside. Punitive law also provides an opportunity for ritual performance. A good example of this principle is the practice of public executions, which have been suspended in what we consider modern states. Watching an execution of a murderer or traitor was a public ritual that allowed the participants to focus their attention on a single group moral norm and to feel the same emotion about the offense. In other words, they were able to perform a Durkheimian ritual that recreated their sacred boundaries. As a result, the group was able to feel its moral boundaries and experience a profound sense of "we-ness," which increased mechanical solidarity.

Organic solidarity, on the other hand, has a greater proportion of *restitutive law*, which is designed to restore the offender and broken social relations. Because organic social solidarity is based on something other than strong morality, the function of law is different. Here, there is no sense of moral outrage and no need to ritualize the sacred boundaries. Organic solidarity occurs under conditions of complex social structures and relations. Modern societies are defined by high structural differentiation, with large numbers of diverse structures necessitating complex interconnections of communication, movement, and obligations. Because of the diversity of these interconnections, they tend to be more rational than moral or familial—which is why we tend to speak of "paying one's debt to society." The idea of "debt" comes from rationalized accounting practices; there is no emotional component, as there would be with moral or family connections. Also, the interests guarded by the laws tend to be more specialized, such as corporate or inheritance laws, rather than generally held to by all, such as "Thou shalt not kill." As an important side note, restitutive and punitive laws are material, tangible items within a society that help us see the nonmaterial organic and mechanical solidarity within a society.

Organic solidarity thus tends to be characterized by weak collective consciousness: Fewer beliefs, sentiments, ideas, and behavioral expectations tend to be shared. There is greater individuality, and people and other social units (like organizations) are connected to the whole through utilitarian necessity. In other words, we need each other in order to survive, just like in an organism. (My heart couldn't do its job for the body without its connection to my lungs and the rest of my biological system.)

The Division of Labor

Earlier, I mentioned that Durkheim says that similarity and mechanical solidarity gradually become channeled and less apparent. That statement gives the impression that the change from mechanical to organic solidarity occurred without any provocation, but that's not the case. The movement from mechanical to organic solidarity, from similarity to difference, from traditional to modern was principally due to increases in the division of labor. The concept of the *division of labor* refers to a stable organization of tasks and roles that coordinate the behavior of individuals or groups that carry out different but related tasks. Obviously, the division of labor may vary along a continuum from simple to complex. We can build a car in our garage all by ourselves from the ground up (as the first automobiles were built), or we can farm out different manufacturing and assembly tasks to hundreds of subcontractors worldwide and simply complete the construction in our plant (as is done today). This example illustrates the poles of the continuum, but there are multiple steps in between.

What kinds of processes tend to increase the division of labor in a society? (Bear with me for a moment; I'm going to put together a string of rather dry-sounding concepts and relationships.) The answer to what increases the overall division of labor is competition. Durkheim sees competition not as the result of individual desires (remember, they are curtailed in mechanical solidarity) or free markets, but rather as the result of what he variously calls *dynamic, moral,* or *physical density.* These terms capture the number and intensity of interactions in a collective taken as a whole. The *level* of dynamic density is a result of increasing population density, which is a function of population growth (birthrate and migration) and ecological barriers (physical restraints on the ability of a population to spread out geographically).

Durkheim's theory is based on an ecological, evolutionary kind of perspective. The environment changes and thus the organism must change in order to survive.

In this case, the environment is social interaction. The environment changes due to identifiable pressures: population growth and density. As populations concentrate, people tend to interact more frequently and with greater intensity. The rate of interaction is also affected by increases in communication and transportation technologies. As the level of interaction increases, so does the level of competition. More people require more goods and services, and dense populations can create surplus workers in any given job category. The most fit survive in their present occupation and assume a higher status; the less fit create new specialties and job categories, thus creating a higher degree of division of labor.

The Problem With Modern Society

There are law-like principles underlying Durkheim's theory. Together, they form a law of human interaction: (1) Humans will use culture to interact; (2) culture is produced to create social meaning and relationships within a specific environment; and (3) the moral strength of any culture is determined by ritual performance. These imply that groups who perform interaction rituals in a stable environment will create a common culture with a good deal of moral strength to it. The inverse is true as well. Changing environments and few or diverse rituals will create particularized cultures with very little moral force. This is essentially what I've already said, but I wanted to lay it out in this way so that you could clearly see the inevitability of the problem that occupied Durkheim's mind.

One of the things that we've seen with all of our theorists so far is that modernity means increasing division of labor. This happens first because of efficiency and second in response to the creation of new and different commodities. The latter factor becomes particularly important as time goes one. You'll see this if you just think of the kinds of job available to you that weren't available to your parents and grandparents. As the division of labor increases, so do the number and types of environments in which people interact. Not only is the structure of the division of labor more complex, but so is the way in which career paths develop. Today, most people don't stay in one job with one company for their entire career, and many careers demand relocation. A friend of mine, for example, is in the process of moving from South Carolina to Germany for a job promotion. Add all of these sorts of factors together and think about them in terms of the laws of human interaction, and I think you'll see why Durkheim is concerned. As people interact with different people, in different circumstances, to achieve different goals, the level of particularized culture increases. This dynamic of course is supercharged as people move from one career to another and from one geographic place to another. Because they contain different ideas and sentiments, the presence of particularized cultures threatens the power of the collective consciousness and social solidarity.

There's an old saying, "Birds of a feather flock together." Well, Durkheim is telling us something slightly different: Birds become of a feather *because* they flock together. In other words, the most prominent characteristics of people come about because of the groups they interact with. As people internalize the culture of their groups, they learn how to think, feel, and behave, and groups become socially distinct from one another. This process is called *social differentiation*. As people become socially differentiated, they, by definition, share fewer and fewer elements of the collective consciousness. This process brings with it the problem of integration. It's a problem that we here in the United States are very familiar with: How can we combine diverse populations into a whole nation with a single identity? But there's more to this problem of modernity. As the division of labor increases, so does the level of structural differentiation—the process through which the needs of society are met via increasingly different sets of status positions, roles, and norms. Spencer made this the central issue for his social theorizing. Here we see the same concern. Durkheim argues, much like Spencer, that structural interdependency creates pressures for integration, but he adds a cultural component: value or culture generalization, which we'll talk about in a moment.

Definition

Social differentiation is the process through which people become different from one another, or culturally diverse. The concept assumes that most distinctions among people are not the result of inborn differences. Rather, people become different from one another in response to increases in the division of labor and structural differentiation, both of which increase the level of particularized culture.

Thus we are confronted with the problem of modernity. Because groups are more closely gathered together, the division of labor increases. In response to population pressures and the division of labor, social structures have differentiated to better meet societal needs. As structures differentiate, they are confronted with the problem of integration. In addition, as the division of labor increases, people tend to socially differentiate according to distinct cultures. As people create particularized cultures around their jobs, they are less in tune with the collective consciousness and face the problem of social integration. I've illustrated these relationships in Figure 4.2.

Figure 4.2   The Division of Labor and Problems of Integration

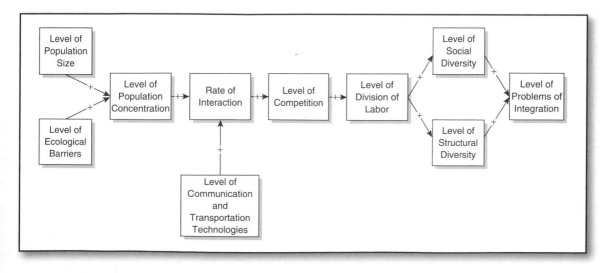

Organic Solidarity and Social Pathology

As social and structural differentiation creates problems of integration, it simultaneously produces social factors that counterbalance these problems: intermediary group formation, culture generalization, restitutive law and centralization of power, and structural interdependency. Together, these factors form organic solidarity. First, structures become more dependent on one another as they differentiate (structural interdependency). For example, your heart can't digest food, so it needs the stomach to survive. It is the same for society as it differentiates. The different structures become dependent upon one another for survival. Further, in order to provide for the needs of other structures, they also must be able to interact with one another. Thus, both structural differentiation and social differentiation also create pressures for a more generalized culture and value system. Let's think about this in terms of communication among computers. I have a MacBook, and my friend Steve has an IBM PC. Each has a completely different platform and operating system. Yet almost every day, my computer communicates with his through e-mail. How can it do this? The two different systems can communicate with one another because there is a more general system (the Internet) that contains values broad enough to encompass both computers.

Societies thus produce more general culture and values in response to the need for different subsystems and groups to communicate. In contemporary structural analysis, this is an extremely important issue. Talcott Parsons (Chapter 9) termed this focus the *generalized media of exchange*—symbolic goods that are used to facilitate interactions across institutional domains. So, for example, the institutional structure of family values love, acceptance, encouragement, fidelity, and so on. The economic structure, on the other hand, values profit, greed, one-upmanship, and the like. How do these two institutions communicate? What do they exchange? How do they cooperate in order to fulfill the needs of society? These kinds of questions, and their answers, are what make for structural analysis and the most sociological of all research. They are the empirical side of Durkheim's theoretical concern.

Of course, in the United States, we have successively created more generalized cultures and values. The idea of *citizen*, for example, has grown from white/male/Protestant/property owner to include people of color and women. Yet generalizing culture is a continuing issue. The more diverse our society becomes, the more generalized the culture must become, according to Durkheim. For example, while we in the United States may say, "In God we trust," it is now a valid question to ask "in which god?" We have numerous gods and goddesses that are worshiped and respected in our society. To maintain this diversity, Durkheim would argue that in the culture there has to be a concept general enough to embrace them all. According to Durkheim, if the culture doesn't generalize, we run the risk of disintegration.

There is, however, a complexity here: *Generalized culture* is often too broad to invest with high levels of emotion, meaning, and importance. For example, it's more difficult to emotionally invest in the idea of "a god" or "higher power" than "The One True God," which is one reason why wars are never fought over the generalized notion of a god—it takes a specific God to motivate people to kill and be killed. As a result of this type of generalized culture, more and more of our relations, both structural and personal, are mitigated by law. This law has to be rational and focused on relationships, not morals. For example, I don't know my neighbors. The reasons for this have a lot to do with what Durkheim is talking about: increases in transportation technologies, increasing divisions of labor, and so on. But if I don't know my neighbors, how can our relationship be managed? Obviously, if I have a problem with their dog barking or their tree limb falling on my house, there are laws and legal proceedings that manage the relationship. Increases in social and structural diversity thus create higher levels of restitutive law (in comparison to restrictive/moral law) and more centralized government to administer law and relations. Of course, we come to value this kind of law and come to believe in the right of a centralized government to enforce the law.

Definition

In functionalist theory, the *generalization of culture*—also called generalized media of exchange—is an important concept in understanding social integration. The generalization of culture is the process through which culture and values become more abstract and are able to transcend specific social groups. As a society experiences greater levels of social diversity, the need for more generalized and transcendent culture and values increases, thus pushing the system to produce more generalized media of exchange. More general or abstract culture facilitates interactions across institutional domains as well as social encounters among diverse individuals.

Both social and structural diversity also push for the formation of *intermediary groups*. Durkheim always comes back to real groups in real interaction: Culture can have independent effects, but it requires interaction to be produced. Thus, the problem becomes, how do individual occupational groups create a more general value system if they don't interact with one another? Durkheim theorizes that societies will create intermediary groups—those between the individual occupational groups and the collective consciousness. Such groups are able to simultaneously carry the concerns of the smaller groups as well as the collective consciousness.

For example, I'm a sociology faculty member at the University of North Carolina. Because of the demands of work, I rarely interact with faculty from other disciplines (like psychology), and I only interact with medical doctors as a patient. Yet I am a member of the American Sociological Association (ASA), and the ASA interacts with the American Psychological Association and the American Medical Association, as well as many, many other groups. And all of them interact with the U.S. government as well as other institutions. This kind of interaction among intermediary groups creates a higher level of value generalization, which, in turn, is passed down to the individual members. As Durkheim (1893/1984) says,

A nation cannot be maintained unless, between the state and individuals, a whole range of secondary groups are interposed. These must be close enough to the individual to attract him strongly to their activities and, in so doing, to absorb him into the mainstream of social life. (p. liv)

As we've seen, structural interdependency, culture generalization, intermediary group formation, restitutive law, and centralization of power together create organic solidarity (Figure 4.3). Increasing division of labor systemically pushes for these changes: "Indeed, when its functions are sufficiently linked together they tend of their own accord to achieve an equilibrium, becoming self-regulatory" (Durkheim, 1893/1984, p. xxxiv). As a side note, part of what we mean by functional analysis is this notion of *system pressures creating equilibrium.* Notice the dynamic mechanism: It is the system's need that brings about the change. It's kind of like pulling your car into the gas station because it needs gas. Society is a smart system, regulating its own requirements and bringing about changes to keep itself in equilibrium. (Remember, for Durkheim, society can act on its own.) However, if populations grow and/or become differentiated too quickly, the system can't keep up, and these functions won't be sufficiently linked together. Society can then become pathological or sick.

Figure 4.3 Organic Solidarity

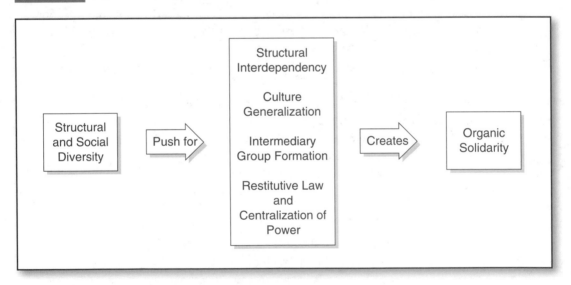

Durkheim elaborates two possible *social pathologies.* (He actually mentions three—anomie, forced division of labor, and lack of coordination—but he clearly elaborates only the first two.) The first is *anomie*—social instability and personal unrest resulting from insufficient normative regulation of individual activities. Durkheim argues that social life is impossible apart from normative regulation. Without norms to regulate interactions, cooperation is impossible. If it were necessary to "grope de novo for an appropriate response to every stimulus from the environing situation, threats to its integrity from many sources would promptly effect its disorganization" (Durkheim, 1903/1961, p. 37). As we've seen, the production of norms requires interaction. The problem occurs when the combination of overly rapid population growth with excessive division of labor and social diversity hinders groups from interacting. Links among and between the groups cannot be formed when growth and differentiation happen too quickly. The result is anomie.

Definition

Anomie is a concept that is used by both Durkheim and Georg Simmel. Anomie literally means to be without norms or laws. Because human beings are not instinctually driven, they require behavioral regulation. Without norms guiding behavior, life becomes meaningless. Anomie is a pathology of modernity and tends to occur under high levels of structural differentiation and division of labor when the culture of society does not generalize quickly enough. High levels of anomie can lead to anomic suicide.

The other pathology that Durkheim considers at some length involves *class inequality* and *the forced division of labor*. Durkheim argues that the division of labor must occur "spontaneously"; that is, it should emerge naturally from the social system in response to population growth and density. The division of labor should not occur due to a powerful elite driven by profit motivations. According to Durkheim (1893/1984), it is dysfunctional to force people to work in jobs for which they are ill suited:

> We are certainly not predestined from birth to any particular form of employment, but we nevertheless possess tastes and aptitudes that limit our choice. If no account is taken of them, if they are constantly frustrated in our daily occupation, we suffer, and seek the means of bringing that suffering to an end. (pp. 310–311)

This condition is both similar to and different from what Marx talks about. The ideas are similar in that the division of labor is forced; there is a sense of separation or suffering due to that division of labor, and this sense of suffering can lead to the fragmentation of social solidarity. But the source of separation is different. For Marx, alienation is a state of existence that the individual may or may not subjectively experience. Alienation for Marx is defined by separation from species-being, and people only become aware of it through critical or class consciousness. Durkheim, on the other hand, assumes that humans are egotistical actors without an essentially good nature from which to be alienated. Durkheim's alienation comes about only as a result of a pathological form of the division of labor—forced by capitalist greed rather than organic evolution. It exists as a subjective state; people are always aware of alienation when it occurs.

Durkheim's solution for this pathological state is "justice" enforced by the state—specifically, price controls by useful labor (equal pay for equal work) and elimination of inheritance. Two of the most powerful tools in producing a structure of inequality are ascription and inheritance. Ascription assigns different status positions to us at birth, such as gender and race. Apart from legislation guaranteeing equal pay for equal work, ascription strongly influences inequality. For example, studies done in the United States consistently show that women earn about 70 percent of what men make for the same job, even though they have the same qualifications. In addition, inheritance is an obvious way to maintain structural inequality; estimates are that during the 55-year period from 1998 through 2052, at least $41 trillion will be inherited in the United States (Havens & Schervish, 2003). Both ascription and inheritance make the perpetuation of class differences appear natural: Our position is ascribed to us, and we either are wealthy or poor (or somewhere in between) by birth. America may be the land of opportunity, but it is not the land of *equal* opportunity. It matters if you are male or female, if you are black or white, if you go to school in Harlem or Hollywood, and so on. On the other hand, if inheritance is done away with and laws are implemented that bring equal pay for equal work, structured inequality would have a difficult time surviving.

Concepts and Theory: Individualism

In early traditional societies, the human self was an utterly social self. People were caught up in and saw themselves only in terms of the group. The self was an extension of the group just as certainly as your arm is an extension of your body. Yet as societies differentiated both structurally and socially, the self became more and more isolated and took on the characteristics of an individual. Thus, in modern societies, the individual takes on increasing importance. We can think of many benefits from this shift. We have increased freedom of choice and individual expression under conditions of organic solidarity. But there are some dysfunctional consequences as well. Although Durkheim didn't phrase it in this way, in addition to the two pathologies of modern society previously mentioned (anomie and forced division of labor), we can add suicide. Durkheim's study of suicide is one of the most provocative works in early sociology. He took what most people think of as a purely personal decision and demonstrated that a society's suicide rate is a function of social factors: the level of group attachment and the regulation of a person's actions. From these two factors, Durkheim explains four different types of suicide.

Suicide

There are two critical issues for the individual in modern society: the level of group attachment and that of behavioral regulation. People need a certain level of *group attachment*. We are social creatures, and much of our sense of meaning, reality, and purpose comes from having interpersonal ties (both in terms of number and density) and a sense of "we-ness" or collective identity. As we'll see when we get to George H. Mead, our sense of self arises from social interactions. Certainly, meaning and reality are social attributes, as is the sense of purpose and motivation that comes along with social identities (such as "student"). Of course, because we are human we can create utterly individualistic realities and meanings that do not directly relate to the social group around us. However, we usually consider those realities (and the people having those realities) to be either strange or crazy. Most of us are aware, and even unconsciously convinced, that our personal meanings and realities have to be linked in some way to the social group around us. This is why, when group attachment is too low, Durkheim argues that *egoistic suicide* is likely: Low group attachment leads to extreme individualism and the loss of a sense of reality and purpose.

However, extremely high group attachment isn't a good thing for the individual either. High attachment leads to complete fusion with the group and loss of individual identity, which can be a problem, especially in modernity. Under conditions of high group attachment, people are more likely to commit *altruistic suicide*—giving one's life for the sake of the group. The Kamikaze pilots during WWII are a good example. Some contemporary examples include religious cults such as the People's Temple and Heaven's Gate, and the group solidarity the U.S. government fosters in military boot camps. Under conditions of high group attachment, individual life becomes less meaningful and the group's reality the most significant.

It's important to note here that modern societies are characterized by the presence of both mechanical and organic solidarity. It is certainly true that, in general, the society is held together by organic means—general values, restitutive law, and dependent opposites—yet it is also true that pockets of very intense, particularized culture and mechanical solidarity exist as well. These kinds of group interactions may in fact be necessary for us. This need may explain such intense interaction groups as rock music and sports fans. Both of these groups engage in the kind of periodic ritual gatherings that Durkheim describes in *The Elementary Forms of Religious Life.* During these gatherings (shows or events), rock and sports fans experience high levels of emotional energy and create clear group symbols. Yet extremes of either organic or mechanical solidarity can be dangerous, as Durkheim notes.

The other critical issue for individuals in modern society is *behavioral regulation.* Because in advanced, industrialized nations we believe in individualism, the idea of someone or something regulating our behavior may be objectionable. But apart from regulation, Durkheim argues, our appetites would be boundless and ultimately meaningless. The individual by him- or herself "suffers from the everlasting wranglings and endless friction that occur when relations between an individual and his fellows are not subject to any regulative influence" (Durkheim, 1887/1993, p. 24). Further, our behaviors must have meaning to us with regard to time. Time, as we think of it, is a function of symbols. (The past and future only exist symbolically.) Symbols, of course, are a function of group membership.

Thus, there are a variety of reasons why the regulation of behaviors is necessary. Behaviors need to be organized according to the needs and goals of the collective, but the degree of regulation is important. Under conditions of rapid population growth and diversity, anomie may result if the culture is unable to keep pace with the social changes. Under these conditions, it is likely there will be an increase in the level of *anomic suicide.* The lack of regulation of behaviors leads to a complete lack of regulation of the individual's desires and thus an increase in feelings of meaninglessness. On the other hand, overregulation of behaviors leads to the loss of individual effectiveness (and thus increases hopelessness), resulting in *fatalistic suicide.*

I've listed the suicide types in Table 4.2. It's important to keep in mind that the motivation for suicide is different in each case, corresponding to group attachment and behavior regulation. Also note that the kinds of social pathologies we are talking about here are different from the ones in the previous section. Here, we are seeing how modernity can be pathological for the *individual,* in the extremes of attachment and regulation. In the previous section, we looked at how modernity can be pathological for society as a whole and its solidarity.

Table 4.2 Suicide Types

	Group Attachment	Behavior Regulation
High	Altruistic Suicide	Fatalistic Suicide
Low	Egoistic Suicide	Anomic Suicide

The Cult of the Individual

Both egoistic and anomic suicide can be seen as a function of high levels of individuality, but individuality is not the problem. Individualism is in fact necessary in modernity. Before we go on, I need to make a distinction between this idea of individuality and what might be called egoistic hedonism or materialism (what most of us think of when we hear the term individualism: "I gotta be me"). From a Durkheimian view, individuals who are purely and exclusively out to fulfill their own desires can never form the basis of a group. Group life demands that there be some shared link that motivates people to work for the collective rather than for individual welfare. As we've seen in our discussion of Durkheim, this kind of group life and awareness is dependent upon a certain degree of collective consciousness. The idea of the individual that we have in mind here is therefore not pure ego. Rather, what is at stake might better be understood as the idea of "individual rights" and how it has progressed historically.

To do the concept justice, we should really go back to early Greek and then Roman times. But in actuality, we need only go back to the founding of the United States. What does the following sentence mean? "We hold these truths to be self-evident, that all men are created equal, that they are endowed by their Creator with certain unalienable Rights, that among these are Life, Liberty, and the pursuit of Happiness." The history of the United States is the tale of working out the meaning of that line. Obviously, the central struggle concerns the term "all men." Initially, "all men" referred only to white, property-owning, heterosexual, Protestant males. Somewhere along the line, the U.S. government decided (often in response to fierce civil struggle, such as the fight for women's suffrage) that you didn't have to own property and that you didn't have to be male, white, or Protestant (although you still have to be heterosexual in most states) in order to have full access to civil rights. The reason there has been struggle over this sentence is the basis that is given for these rights. According to the quoted line from the U.S. Declaration of Independence, the basis for civil rights is simply being human. These rights can neither be earned nor taken away. They are yours not because of anything you've done or any group membership you have, but simply because of your birth into the human race. Thus, what matters isn't your group membership; rather, it's you as an *individual* human being that matters. It is this moral idea of individualism that Durkheim has in mind and that has the potential for creating social solidarity.

Durkheim calls this new moral basis for society *the cult of the individual.* The individual, as he or she is historically separated from the group, becomes the locus of social concern and solidarity. The *individual* becomes the recipient of social rights and responsibilities, rather than castes or lineages. Today, we see the individual as perhaps the single most important social actor. Even our legal system here in the United States is occupied with preserving the civil rights of the perpetrator of a crime because it is the individual that is valued. The individual becomes the focus of our idea of "justice," which Durkheim sees as the "medicine" for some of the problems that come with pathological forms of the division of labor. For example, in a society such as the United States, the problems associated with labor issues, poverty, deviance, and depression (all of which can be linked to Durkheim's pathologies) are generally handled individually through the court system or counseling. Remember that Durkheim sees culture as the unifying force of society, so the importance of these kinds of cases for Durkheimian sociology has more to do with the culture that the practices create and reproduce than the actual legal or psychological effects. In this way, the individual becomes a ritual focus of attention, the symbol around which people can seek a kind of redress for the forced division of labor, inequality, and anomie that we noted previously. Today, the individual has taken on a moral life. But it is not the particular person per se, with all of her or his idiosyncrasies, who has value. Rather, it is the ethical and sacred *idea* of the individual that is important.

This notion of the individual has become perhaps the collective symbol of modernity and postmodernity. What moral factor can link together extremely diverse groups? What can street gangs and medical doctors hold in common? Perhaps it's the belief that individuality is supremely important. This cultural idea of the individual becomes the focus of both public and private rituals and part of the moral fiber that holds modern society together. As Durkheim (1957) says,

> From being limited and of small regard, the scope of the individual life expands and becomes the exalted object of moral respect. . . . If he be the moral reality, then it is he who must serve as the pole-star for public as well as private conduct. (p. 56)

> This cult, moreover, has all that is required to take the place of the religious cultures of former times. The belief in individualism has taken on sacred qualities and thus serves as well as religion to bring about the communion of minds and wills that is a first condition of any social life. (p. 69)

Thinking About Modernity and Postmodernity

Durkheim's perspective is and always has been central to sociology. The issues of culture, group cohesion, and religion have never left the sociological landscape, although the importance we grant them has varied. As a result of the social unrest in the 1960s and early 1970s, the ideas of culture and functionalism took a backseat to Marx and conflict theory. However, since the mid-1980s, culture has become more and more important for sociological thinking, due in large part to postmodernism. Culture is at the heart of the postmodern critique, and sociologists, whether they agree with the critique or not, have had to consider the issue of culture. As Hall and Neitz (1993) state, "At the end of the twentieth century, culture once again has become a central focus of sociological theorizing" (p. 1).

Grand Narratives, Doubt, and Civil Religion

Durkheim's theory is motivated by his concerns about culture in modernity. Durkheim sees modernity as being driven by increasing levels of population density and division of labor. These two dynamics create distinctly modern societal effects: increasing importance of the state and restitutive law, decreasing importance of religion and moral solidarity, increasing structural complexity and interdependency, increasing generalization of culture and media of exchange, and increasing levels of individualization. We have also seen that Durkheim is concerned about possible pathological states coming from these effects, particularly anomie and the forced division of labor. I believe that Durkheim had an additional lingering doubt about social cohesion in modernity that we haven't talked about yet. It has to do with symbolic unification.

The kinds of symbols that Durkheim is interested in are religious. He isn't arguing that religion is needed; rather, he is asserting that only the kinds of symbolic meanings that religion produces are needed. "Thus there is something eternal in religion that is destined to outlive the succession of particular symbols in which religious thought has clothed itself" (Durkheim, 1912/1995, p. 429). Pay particular attention here, because this is the point at which Durkheim expresses some doubt about purely organic solidarity and where postmodernism comes into play. Durkheim says that no replacement for these religious-type symbols has yet been created, which he explains in this way: "If today we have some difficulty imagining what the feasts and ceremonies of the future will be, it is because we are going through a period of transition and moral mediocrity" (1912/1995, p. 429). Nevertheless, he is convinced that the kinds of symbols and symbolic activities that society needs will be created: "A day will come when our societies once again will know hours of creative effervescence during which new ideals will again spring forth and new formulas emerge to guide humanity for a time" (p. 429).

There is therefore a gap in Durkheim's theory and confidence. Even though he argues for organic solidarity, he nonetheless senses that society needs something more. It is at this very point that postmodern theory finds its niche. Many postmodernists argue that the need for, or legitimacy of, cultural unification has passed, in particular the need for what is termed *grand narratives*. A narrative is a story, and a grand narrative is an abstract, symbolic story that transcends and links together diverse cultures.

One of the earmarks of modernity is the presence of or belief in grand narratives—overarching themes that make things look monolithic. The social world of modernity is defined primarily through capitalism, science and technology, and the nation-state. A specific culture arose that served to reinforce these social factors: a secular and liberal grand narrative of equality and freedom, given hope by the advances of science and embodied in Western colonialism. The social world of postmodernity, on the other hand, is characterized by increasing doubt concerning the validity of these grand narratives. There is a way in which we could talk about this facet of postmodernity as the *institutionalization of doubt*. Part of the doubt originates in the failure of Western colonialism and disappointment in the inability of technically advanced nations to create true equality. The ubiquitous doubt of postmodernity is also facilitated by universal education in science and the commodification of religion, as explained subsequently.

Western colonialism was based on the belief in one best system. The foundation stone of equality was faith in the cultural melting pot. We no longer believe in forcing a dominant culture on others in our society, and we no longer think that a melting pot that puts forth one national identity is the way to go. But notice what these changes imply: There are few if any unifying themes on which we can agree. But perhaps we shouldn't agree on any culturally unifying themes, because to do so would be to deny or silence the voices of others. This is postmodernism's point exactly: Postmodernism advocates the legitimation of *polyvocality* (many voices).

This doubt about any kind of ultimate story is also, interestingly enough, facilitated by science and religion. Science teaches us to doubt. Prior to science, knowledge was held to be absolute. Science, however, is based on knowledge testing and accumulation: All information is held in doubt, tested, and replaced. Since the scientific model informs most of our schooling, we come away with a culture of doubt. Further, religion has also contributed to our doubt about grand narratives, as it has come under the influence of mass media and marketing, which we saw when we considered Luckmann's theory in Chapter 2. Religions use billboards, TV spots, newspaper ads, and so on to convey their message. They create sayings and logos that capture their mission, differentiate them from other churches, and make them easy to relate to. Thus, as with everything that is subjected to the trivializing effects of commodification and advertising, the religious individual is faced with a marketplace of fragmented ultimate meanings, competing churches and sects, and substitute religious communities (such as TV evangelists). According to some postmodernists, this kind of meaning market results in an individual consciousness that is unstable, that is doubtful about general legitimating myths, and that is only concerned with immediate sensations and emotions.

On the other hand, there are those who argue from a Durkheimian perspective that we are not postmodern and that the solidarity functions of religion have moved to the sphere of the state. This point of view is in keeping with Durkheim's hope and is championed by Robert Bellah (1970, 1975, 1980; Bellah, Madsen, Sullivan, Swidler, & Tipton, 1985, 1991). Bellah acknowledges the contemporary debate over the cohesiveness of national culture, but it's his position that moral debates over the dysfunctions of institutions are an intrinsic element of the culture of any society. Societies have always argued over the economy, education, religion, and so forth. He makes the case that these debates actually constitute culture. He sees culture as a dynamic conversation or argument, something like a Durkheimian ritual. Thus, for Bellah, the debate over the condition of culture is a vehicle through which a kind of collective consciousness is produced.

However, Bellah also argues that religion has evolved in modernity to the point where the dualism between the sacred and the profane that was crucial to past religions and morals has collapsed. The collapse of the sacred is due to the loss of an institutional sphere where collective morals are self-consciously formed—institutional differentiation and size have grown to the point where individuals have lost the ability to collectively inform the general moral culture and have become passive recipients of institutional control.

Bellah decries the results of capitalistic exploitation and the colonization of everyday life and the public sphere by institutions that have failed to live up to their moral obligation. He makes the same point as many postmodernists: Capitalist greed and exploding advertising have eroded the collective moral base. According to Bellah and his colleagues (1991), what we need is to "move away from a concern for maximizing private interests" and to move toward "justice in the broadest sense—that is, giving what is due to both persons and the natural environment" (p. 143). This can be accomplished through the cultivation of face-to-face communities where people are "involved in the creation of regional cultures in some degree of harmony with the natural environment, where individuals, families, and local communities could grow in moral and cultural complexity" (p. 265). This argument clearly echoes Durkheim's concern with the creation of intermediate moral groups and the need to interact in face-to-face situations to produce moral culture.

Enduring
Issues

Most people who were thinking about democracy during the Enlightenment saw that democracy required an institutional space that sat between the state on the one hand and daily life and local associations on the other. This space was talked about as *civil society*, the "dynamic side of citizenship, which, combining as it does achieved rights and obligations, finds them practised, scrutinized, revamped, and redefined" (Marshall, 1998, p. 74). For early sociological thinkers such as Alexis de Tocqueville, Harriet Martineau, and Jane Addams, civil society initially included the press, capitalism, voluntary or moderate religion, and education. Together, these provided the structural and cultural bases for democracy. Bellah's notion of civil religion fits here as well. Jeffrey Alexander (2006) argues that the original idea of civil society doesn't work for democracy in the twentieth century. Alexander terms this new space the *civil sphere* rather than civil society and defines it as "a world of values and institutions that generates the capacity for social criticism and democratic integration at the same time" (p. 4).

These interactions will revolve around civil rather than religious discourses. Central to these civil moral discourses are *representational characters*, a kind of symbol. They are a way through which people define what is good and legitimate in a society. In the United States, Abraham Lincoln, Helen Keller, Eleanor Roosevelt, John Kennedy, Jacqueline Kennedy Onassis, Rosa Parks, and Martin Luther King Jr. are examples of such characters. They have assumed mythic significance and have the ability to generate a degree of social solidarity. Today, we might also add such media and sports figures as Oprah Winfrey and Lance Armstrong. What's important about these characters is that they provide us with a focus for intense interactions around central themes in our society. As such, they are the touch point for Durkheimian rituals that produce social solidarity.

Central themes (such as, "Americans as the chosen people, the champions of freedom") along with representative characters can form a civil religion. In civil religion, sacred qualities that in previous times were the custody of religion are attached to certain civil institutional arrangements, historical events, and people. The theme of Bellah's (1975) work is a call to create such a civil religion because "only a new imaginative, religious, moral, and social context for science and technology will make it possible to weather the storms that seem to be closing in on us in the late 20th century" (p. xiv).

It is certainly the case that the United States has a great deal of cultural diversity and institutional differentiation. But are we left without a unifying story or grand narrative that can give us a collective identity? Or are we instead a nation with a civil religion that can bring social solidarity? Or are there layers of each, like Durkheim's continuum of mechanical and organic solidarity? If there are pockets of each, how do they relate to each other? How do they form a collective consciousness? If we are indeed as diverse and differentiated as we seem, what is holding us together? What values do we hold in common? How can large-scale collective projects, such as war, be carried out? Or is the diversity only skin deep? Are we modern or postmodern? I don't know the answers, but these are good Durkheimian (and important) questions.

> For if society lacks the unity that derives from the fact that the relationships between its parts are exactly regulated, that unity resulting from the harmonious articulation of its various functions assured by effective discipline, and if, in addition, society lacks the unity based upon the commitment of men's wills to a common objective, then it is no more than a pile of sand that the least jolt or the slightest puff will suffice to scatter. (Durkheim, 1903/1961, p. 102)

Summary

- Durkheim is extremely interested in what holds society together in modern times. In order to understand this problem, he constructs a perspective that focuses on three issues: social facts, collective consciousness, and the production of culture in interaction. Durkheim argues that society is a social fact, an entity that exists in and of itself, which can have independent effects. The facticity of society is produced through the collective consciousness, which contains collective ideas and sentiments. The collective consciousness is seen as the moral basis of society. Though it may have independent effects, the collective consciousness is produced through social interaction.

- Durkheim argues that the basis of society and the collective consciousness is religion. Religion first emerged in society as small bands of hunter-gatherer groups assembled periodically. During these gatherings, high levels of emotional energy were created through intense interactions. This emotional energy, or effervescence, acted as a contagion and influenced the participants to behave in ways they normally wouldn't. So strong was the effervescent effect that participants felt as if they were in the presence of something larger than themselves as individuals, and the collective consciousness was born. The emotional energy was symbolized and the interactions ritualized so that the experience could be duplicated. The symbols and behaviors became sacred to the group and provided strong moral boundaries and group identity.

- Because of high levels of division of labor, modern society tends to work against the effects of the collective consciousness. People in work-related groups and differentiated structures create particularized cultures. As a result, society has to find a different kind of solidarity than one based on religious or traditional collective consciousness. Organic solidarity integrates a structurally and socially diverse society through interdependency, generalized ideas and sentiments, restitutive law and centralized power, and through intermediary groups. These factors take time to develop. If a society tries to move too quickly from mechanical to organic solidarity, it will be subject to pathological states, such as anomie and the forced division of labor.

- At the center of modern society is the cult of the individual. The ideal of individuality, not the idiosyncrasies of individual people, becomes one of the most generalized values a society can have. However, the individual can also be subject to pathological states, depending on the person's level of group attachment and behavioral regulation. If a society produces the extremes of either of these, then the suicide rate will tend to go up. Suicide due to extremes in group attachment is characterized as either egoistic or altruistic. Suicide due to extremes in behavioral regulation is characterized as either anomic or fatalistic.

- Durkheim's argument for social solidarity is a moral one. He holds the hope that one day a religious-like grand narrative will once again provide the firm center for society. Some thinkers, like Robert Bellah, agree with Durkheim and argue that civil religion can provide just such a base. However, postmodernists are unsure that such a grand narrative is possible or even desirable. Postmodernists point to the pervasive doubt that has come about through the decentering of religion, the failure of Western colonialism, the effects of mass media and markets, and the doubt intrinsic within science as the reasons why a Durkheimian grand narrative is unlikely. Further, grand narratives are based on privileged knowledge: In order to produce a grand narrative, a particular kind of knowledge has to be seen as more important than all others. Grand narratives are thus morally objectionable, under postmodern values.

BUILDING YOUR THEORY TOOLBOX

Learning More—Primary and Secondary Sources

Primary sources: Almost every book of Durkheim's is worth reading. The following are indispensable:

- Durkheim, É. (1938). *The rules of sociological method* (S. A. Solovay & J. H. Mueller, Trans.; G. E. G. Catlin, Ed.). Glencoe, IL: The Free Press. (Original work published 1895)

- Durkheim, É. (1951). *Suicide: A study in sociology* (J. A. Spaulding & G. Simpson, Trans.). Glencoe, IL: The Free Press. (Original work published 1897)

- Durkheim, É. (1984). *The division of labor in society* (W. D. Halls, Trans.). New York: The Free Press. (Original work published 1893)

- Durkheim, É. (1995). *The elementary forms of the religious life* (K. E. Fields, Trans.). New York: The Free Press. (Original work published 1912)

Durkheim secondary sources:

- Alexander, J. C. (Ed.). (1988). *Durkheimian sociology: Cultural studies.* Cambridge, UK: Cambridge University Press. (Excellent collection of contemporary readings concerning Durkheim's contribution to cultural sociology)

- Giddens, A. (1978). *Émile Durkheim.* New York: Viking Press. (Brief introduction to Durkheim's work by one of contemporary sociology's leading theorists)

- Jones, R. A. (1986). *Émile Durkheim: An introduction to four major works.* Newbury Park, CA: Sage. (Part of the Masters of Social Theory series; short, book-length introduction to Durkheim's life and work)

- Lukes, S. (1972). *Émile Durkheim, his life and work: A historical and critical study.* New York: Harper & Row. (The definitive book on Durkheim's life and work)

- Meštrovic, S. G. (1988). *Émile Durkheim and the reformation of sociology.* Totowa, NJ: Rowman & Littlefield. (Unique treatment of Durkheim's work; emphasizes Durkheim's vision of sociology as a science of morality that could replace religious morals)

Seeing the Social World (knowing the theory)

- Write a 500-word synopsis of Durkheim's sociological imagination, making certain to include not only how he sees the world (perspective) but also how that perspective came about (history, social structures, and biography).

- After reading and understanding this chapter, you should be able to define the following terms theoretically and explain their theoretical importance to Durkheim's theory: social facts,

(Continued)

(Continued)

society sui generis, collective consciousness, religion, sacred and profane, ritual, effervescence, social solidarity (mechanical and organic), punitive and restitutive law, the division of labor, social differentiation, cultural generalization, intermediary groups, social pathologies, anomie, suicide (altruistic, fatalistic, egoistic, and anomic), the cult of the individual.

- After reading and understanding this chapter, you should be able to answer the following questions (remember to answer them theoretically):

  1. Explain the organismic analogy and use it to analyze the relations among and between social structures.

  2. Define social facts and explain how society exists sui generis.

  3. Explain how society is based on religion.

  4. Discuss how Durkheimian rituals create sacred symbols and group moral boundaries.

  5. Define collective consciousness, social solidarity, and mechanical and organic solidarity.

  6. Explain the problem of modernity and describe how organic solidarity creates social solidarity in modernity.

  7. Describe how organic solidarity can produce certain social pathologies.

  8. Define the cult of the individual and explain its place in producing organic solidarity.

Engaging the Social World (using the theory)

- I'd like for you to go to a sporting event—football, basketball, or hockey would be best. Analyze that experience using Durkheim's theory of rituals. What kind of symbols did you notice? What kinds of rituals? Did the rituals work as Durkheim said they would? What do you think this says about religious rituals in contemporary society? Can you think of other events that have the same characteristics?

- There are a lot of differences between gangs and medical doctors, but there might also be some similarities. Using Durkheim's theory and perspective, how are gangs and medical doctors alike?

- Explain this event and the reactions from a Durkheimian perspective: On September 11, 2001, hijacked jetliners hit the World Trade Center in New York and the Pentagon outside Washington, D.C. News headlines around the world proclaimed "America Attacked," and people in places such as San Diego, CA; Detroit, MI; and Cornville, AZ, wept openly. What Durkheimian processes must have been in place for such an event to happen? And, how would Durkheim explain the fact that Americans had such a strong emotional reaction to the loss of people unknown to them? Also, explain the subsequent use of flags and slogans, and the "war on terrorism," using Durkheim's theory.

- Often in theory class, I will take the students on a walk. We walk through campus, through a retail business section, past a church, and through a residential area. Either think about such a walk or go on an actual walk yourself. Based on Durkheim's perspective (not necessarily his theory), what would he see? How would it be different from what Marx would see?

Weaving the Threads (building theory)

- Compare and contrast Spencer, Marx, and Durkheim on the central features of modernity. What do they say makes a society modern? What problems do they see associated with modern society and modern lives?

- Compare and contrast Spencer's and Durkheim's theories of social change and the problems of integration. How can they be integrated?

- Analyze Marx's position on religion using Durkheim's theory. I don't want you to argue for or against Marx; I want you to understand and convey Marx's position using Durkheim.

Rationality and Organization

Max Weber

As we've seen, capitalism in modernity was intended to facilitate equality. Rather than social position being defined by birth, capitalism promised that each person could progress as far as his or her talents and efforts would take the individual. However, Marx points out, capitalism is built upon exploitation and creates a class structure that limits, rather than facilitates, social mobility. Weber's primary concern is rationality. On the surface, reason and rationality appear to be the hope humankind has been looking for. Through reason and rationality, we could discover the secrets of the universe and use them to better humanity. Through reason and rationality, we could create social organization free of favoritism, treating all people equally. We could apply rationality to all facets of our lives, lifting them out of the dark ages of superstition and uncertainty and placing them on the sure pillar of reason. Like capitalism, the inexorable push toward rationality has brought benefits; and, like capitalism, reason and rationality have brought unanticipated and, at times, negating consequences.

Before we get into considering Weber's theory of rationalization, I want you to be aware that Weber is one of sociology's most intricate thinkers. Part of this complexity is undoubtedly due to the breadth of his knowledge. Weber was a voracious reader with an encyclopedic knowledge and a dedicated workaholic. In addition, Weber was in contact with a vast array of prominent thinkers from diverse disciplines. As Lewis Coser (2003) comments,

> In leafing through Weber's pages and notes, one is impressed with the range of men with whom he engaged in intellectual exchanges and realizes the widespread net of relationships Weber established within the academy and across its various disciplinary boundaries. (p. 257)

This social network of intellectuals in diverse disciplines helped create a flexible mind with the ability and tendency to take assorted points of view.

Another reason why Weber's writings are complex is due to the way he views the world. Weber sees that human beings are animals oriented toward meaning, and meaning is subjective and not objective. Weber also understands that all humans are oriented toward the world and each other through values. Further, Weber sees the primary level of analysis as the social action of individuals; for Weber, individual action is social action only insofar as it is meaningfully oriented toward other individuals. Weber sees these meaningful orientations as produced within a unique historical context. This orientation clearly sets him apart from Comte, Spencer, Durkheim, and Marx, who were much more structural in their approaches. It also means that Weber's (1949/1904) explanations are far more complex and tentative: "There is no absolutely 'objective' scientific analysis of culture [because] . . . all knowledge of cultural reality . . . is always knowledge from particular points of view" (pp. 72, 81). At the same time, as we will see, Weber believes that we can create objective knowledge. Yet, while it is possible to create objective knowledge about how things came to be historically so and not otherwise, Weber is extremely doubtful about prediction. But I'm getting ahead of myself.

THEORIST'S DIGEST

Central Sociological Questions

It's difficult to narrow Weber down to just a few central concerns. However, it's safe to say that Weber wrote the most about these three subjects: religion, the state, and the economy. But it would give a false impression to simply say that he is interested in those institutions, per se. His interests are more fundamental: Weber viewed the evolution of religion, politics, and economics as a way of understanding the process of rationalization and its uneasy relationship with cultural values. Weber is also centrally concerned with how stratified systems of class, status, and power produce social change.

Simply Stated

For Weber, life becomes more rational as mystery, emotion, and tradition recede—and as technical efficiency, mathematical and logical calculation, and material and social control increase. Over time, efficiency, calculation, and control have increased in tandem with technology, ethical monotheism, early Protestantism, bureaucratized government, and rational capitalism. Social change that is fueled by stratification comes as groups question the legitimacy of the existing system and meet the technical conditions of group organization (leadership, goals and ideology, and communication).

Key Ideas

Ideal types, action, rationalization, magic and religion, professionalization, symbolism, traditional and rational capitalism, class, status, power, legitimation, authority (charismatic, traditional, rational-legal), routinization, bureaucracy.

The Sociological Imagination of Max Weber

Weber's Life

Karl Emil Maximilian "Max" Weber was born on April 21, 1865, in Erfurt, Germany (Prussia). His father, Max, was well to do, practiced law, and held several political posts. He was an important member of the National Liberal Party, which supported Bismarck's foreign policies (explained in subsequent section), and served on its executive committee. He also served in the Prussian Chapter of Deputies and the German Reichstag. In temperament, the senior Weber was like many German men of the time: strong, assertive, self-indulgent, and patriarchic. Weber's mother, Helene Fallenstein, was in many ways her husband's opposite. She was deeply concerned with the plight of the working class in Berlin and she was a devout Calvinist, and thus focused on inner righteousness. For a good portion of her family life, she was greatly troubled about the worldliness of her husband. Her concern eventually included her son, as young Weber identified with his father. The tensions between his father and mother would later have a profound impact on Weber's life.

Weber's home life also had an impact on his intellectual development. It's obvious that Weber was naturally brilliant, but his home provided an atmosphere that enriched his mind at an early age and perhaps created a sense of self-sufficiency about education. In their biographical sketch, Gerth and Mills (1948) conclude, "The intellectual companions of the household and the extensive travels of the family made the precocious young Weber dissatisfied with the routine instruction of schools" (pp. 3–4). In addition to this home environment, Weber suffered from meningitis at an early age, which predisposed him to reading alone. As a result, Weber had even as a young boy extensive knowledge of the Greek classics and was fluent in such philosophers as Kant, Goethe, and Spinoza before entering college.

In 1882, Weber entered the University of Heidelberg, where he studied law, history, and economics. After a year's military service in 1883, Weber returned to school at the University of Berlin. He lived with his parents while completing his education and saw firsthand how the marriage had broken down after the death of their daughter, Max's sister. Weber also saw his father's treatment of his mother in a new light: "he came to resent his father's bullying behavior toward her" (Coser, 2003, p. 237). In 1889, Weber completed his dissertation in law and went on for his "habilitation," a kind of postgraduate degree, which he completed in 1891. The title of his habilitation thesis was *Roman Agrarian History and Its Importance for State and Civil Law.*

In 1893, Weber married Marianne Schnitger—the marriage proved significant for both their lives. Marianne was born to Eduard Schnitger and Anna Weber. Anna's last name isn't coincidental: Her grandfather was Max Weber's brother. Two years after Marianne was born, her mother died giving birth to a second child. After his wife's death, Eduard shipped Marianne off to live with his mother and sister. While Eduard didn't live in the home, his two brothers did. Both brothers went insane and were in and out of mental asylums. Eduard suffered mental illness as well, but Marianne didn't experience that firsthand as she did with her uncles. Marianne eventually was sent to finishing school, after which she became acquainted with the Weber side of her family. She and Max's mother, Helene, became close friends and confidantes, and in 1892, Marianne and Max became engaged.

The pair moved to Freiburg in 1894 where Weber taught economics. Weber took a position at Heidelberg University where he was professor of political science. Heidelberg was a university city in every sense, and the Weber home soon became a gathering place for intellectual and academic discussions; many of the participants went on to strongly influence twentieth century thought. Marianne was active in these meetings, which at times became significant discussions of gender and women's rights. Georg Simmel, whom we'll meet in the next chapter, frequently attended. It was in Heidelberg that Marianne began her own intellectual and activist career. She studied with Heinrich Rickert, a leading neo-Kantian, and became active with feminist groups. But after just 3 years of intellectual and professional growth, it came to a sudden end.

In July of 1897, Max's parents came to spend a few weeks with the couple in Heidelberg. Everyone had hoped that Max's father would stay home; Helene especially had hoped to be allowed to schedule time with her son and his wife, but Max Sr. insisted on going. The father took the occasion of being at his son and daughter-in-law's to chastise Helene for her independence. At this point, the younger Weber had had enough. He told his father that he would "break all ties with him unless he agreed to a pact that guaranteed his mother certain rights to make visits without his father's permission" (R. Collins, 1986b, p. 15). The elder Weber refused and was told to leave. The parents left, and less than 2 months later, Max's father was dead; the shattered relationship between father and son never mended. Max soon suffered a complete emotional and psychological breakdown that prevented him from speaking in public or teaching his classes. He tried taking leaves of absence and stayed in several sanatoriums, but Weber had to resign his position with the university.

You can imagine how this impacted Marianne; she had watched her two uncles go insane and knew of her father's insanity. Yet Marianne devotedly cared for Max over the next almost 6 years of recovery. Amazingly, Marianne managed to publish two scholarly, political papers during this time: "Fichte's Socialism and Its Relation to Marxist Doctrine" and "Politics and Women's Movement." Equally surprising is how Marianne and Max saw the breakdown impacting their marriage. Marianne devoted herself to Max's care. She saw her childhood experiences as giving her a special compassion and ability to care for those who are suffering. The illness also let her experience a husband's need in a way that "Weber's sovereign self-sufficiency had occasionally made her wonder." She discovered that "the fact that he needs me is a never-ending source of happiness" (Weber, 1926/1975, pp. 236, 237). For his part, Max told Marianne that the illness "reopened the purely human side of living." He had been driven by a compulsion to work, an "inner treadmill" that he experienced as a need to feel overwhelmed by the workload. The illness changed that for Weber. He wanted, rather, "to live a full personal life with my *Kindele* [baby] and to see her as happy as it is given to me to make her" (p. 236). Under Marianne's care, Max slowly recovered. In 1903, he was able to work for a few weeks at a time before relapsing into the "dreariness of the dead period" (p. 268). Weber received an invitation to speak in St. Louis, Missouri, at an international scholarly congress. After the presentation in 1904, the Webers took an extended vacation and traveled the United States. The trip triggered a wellspring of new energy for Max. Marianne said that she "had the feeling that she was bringing home a man restored to health" (p. 302).

Weber began working again, but his focus had changed. While convalescing, he "devoted himself to thinking about thought and the logical and epistemological problem of his science" (Weber, 1926/1975, p. 306). Weber ended up publishing several essays on these issues, the three best known of which were published together in *The Methodology of the Social Sciences* (1949). The first of these, "'Objectivity' in Social Science and Social Polity," was written because he had just taken over editorship of *The Journal of Social Science and Social Policy;* the essay set out the type of methodology that would be accepted in the journal. The last of the three, "The Meaning of 'Ethical Neutrality' in Sociology and Economics," was published during World War I. As in every war, many academics (as well as religious leaders) were using their scholastic authority to legitimate the nation's participation in war. This essay was born out of "Max Weber's pressing need to know the grounds for his own actions and his strong belief that man's dignity consists in his capacity for rational self-determination" (Shils, 1949, p. v).

The years 1904 through 1914 marked the most productive period of Weber's life. In that 10-year period, he wrote some of the foundational documents of sociology: *The Protestant Ethic and the Spirit of Capitalism, The Sociology of Religion,* and *The City.* He also began work on what was to become his magnum opus: *Economy and Society: An Outline of Interpretive Sociology.* In 1909, Weber helped form the first German Society for Sociology. The mental breakdown had obviously changed Weber, but how? How did this part of Weber's personal biography influence his sociological imagination? Personally, I don't feel qualified to even hazard a guess. For an interesting analysis, I would refer you to Randall Collins's (1986b) *Max Weber: A Skeleton Key,* pp. 14–30. Collins's conclusion is that Weber's mother and father represented opposing political and moral forces of modernity. These conflicting forces not only led to the breakdown, but also influenced Weber's work, which is in many ways fragmented and certainly complex. According to Collins, "Weber was truly in the crucible between contending forces, and his multisided intellectual work is the product" (p. 28).

Randall Collins (1986b) characterizes Weber's works as "somewhat schizophrenic, concluding that "they don't add up to one grand synthesis" (p. 11). Weber certainly had a lot to say about many different things, and Weber's writings don't collectively add up to one single theory. However, there may be a different reason for this than Weber's personal problems. Marianne Weber (1926/1975) tells us, "Weber did not care about the systematic presentation of this thinking" (p. 309). She explains that Weber was interested in solving multiple concrete social problems. Once he began to tackle an issue, the information flowed quickly from his mind to pen. "And yet he wanted to be done with it quickly and express himself as briefly as possible, because new problems . . . constantly crowded in upon him" (p. 309). This drive from problem to problem wasn't simply the result of an overactive mind.

Max Weber tells us that more than anything, he wanted to be involved in politics because that is the sphere in which practical solutions could be applied to social problems. He saw his work as a scholar "as a product of free hours, as a sideline of life," one where "the scientifically most useful *new* ideas . . . have always come to me when I was laying on the sofa with a cigar in my mouth" (as cited in Marianne Weber, 1926/1975, p. 187). The best or most useful part of being a scholar for Weber was "on the practical-pedagogical rather than on the really 'scholarly' side" (p. 187). Weber saw multiple social issues that needed practical solutions, and he thus saw teaching as having greater value than his research. I don't think we should be surprised in the least that Weber's work covers a vast social landscape but doesn't ever come together in a single theoretical perspective.

The period after 1904 also marked a return to work for Marianne. Like her husband, Marianne was interested in scholarship and political involvement. On one of the Webers' trips to the United States, Marianne met Jane Addams, who became the first American woman to win the Nobel Peace Prize in 1931. Marianne admired Addams because she saw in her "the work of a woman who has the courage of her convictions" (Weber, 1926/1975, p. 287). Marianne also published several papers during this period, some of which were critical of Charlotte Perkins Gilman, whom we will cover in Chapter 7. Marianne argued that Perkins's work advocated women's workforce participation and "it cannot be the goal of our movement to replace the mother at home by the working wife" (Roth, 1990, p. 65). During this time, Marianne also published a significant study on *Marriage, Motherhood, and the Law.* The Weber home once again because a hub of intellectual life, and both Marianne and Max were at its center. Describing a typical evening, Marianne Weber (1926/1975) wrote,

> Enveloped in cigar smoke, one intellect caught fire from another—an attractive spectacle when the knowledge they had acquired in long discipline burst forth as vibrant streams with a personal coloration, and in their union gave birth to new insights. (p. 369)

At the onset of World War I, Max volunteered for service with the Reserve Military Hospitals Commission; he founded and oversaw the administration of nine military hospitals in Heidelberg. Weber was initially caught up in the nationalism surrounding the war, but gradually became more critical of Germany's role. He specifically stood against submarine warfare, which he felt would surely bring the United States into the conflict, and he became critical of the German political system generally and argued in favor of a more parliamentary system. After the war, Weber was involved in the peace negotiations (he was against the treaty, thinking it was biased and unfair), joined the German Democratic Party, and ran for office, but party politics worked against him. Marianne, on the other hand, became the first woman elected to the German State Parliament and served as president of the Federation of German Women's Organizations. During this period, Max finished his multivolume work on the sociology of religion: *The Religion of China, The Religion of India, and Ancient Judaism.*

In the midst of this productivity and political activism, Max Weber died of pneumonia, June 14, 1920. At that time, Marianne backed off from her own work and dedicated her time and effort to preserving the work of her husband: "she insisted on bringing Weber back to life for herself and posterity" (Roth, 1990, p. 65). In addition to writing the authoritative work of her husband's life, *Max Weber: A Biography,* she prepared a number of Weber's writings for publication. Marianne saw this time in her life as essential for both preserving and promoting Max's work, whom she felt was never fully appreciated during this life. So significant was Marianne's work during this time, Guenther Roth concludes, "that without Marianne Weber, her husband's work might not have gained its later importance for the course of social science" (p. 63). (Note: A good deal of the information about Marianne's life I've used was drawn from Lengermann and Niebrugge-Brantley's [1998] excellent book introducing the women founders of sociology. I highly recommend it.)

Weber's Social World

The period between 1815 and 1848 were years of political tension and economic stagnation for Germany. Under the Holy Roman Empire, there had been well over 300 German political entities, the most powerful of which were Austria and Prussia.

After the defeat of Napoleon in 1813, the Congress of Vienna (1815) created the German Confederation, a loose economic affiliation of less than 40 German-speaking territories. As I noted in the chapter on Marx, these territories practiced protectionist economic policies, which in turn hampered the building of railroads and the process of industrialization. Provincial politics and ethnocentrism hampered any move toward nationalism, and while Christianity in the German Confederation was divided between Catholics and Protestants, overall they were united in their attitude against Jews. This was the world into which Karl Marx was born.

The Congress of Vienna did one other thing that would affect the future of Germany and eventually the world: It granted an edge to Prussia in an age-long rivalry between Prussia and Austria. One significant cause of this tension goes back to the Thirty Years War. Beginning in about 1600, Protestantism began to move into the northern sections of Europe. The wave of conversion quickly moved to military action by 1618, which pitted Protestant Prussia against Catholic Austria. The Treaty of Westphalia of 1648 ended the bloodshed but not the rivalry between Prussia and Austria. So strong was this tension that historians often talk about this relationship as the German Dualism or the German Question. As I mentioned, the 1815 Congress of Vienna helped tip the balance in favor of Prussia by increasing its territory and population. More significantly, "Prussia also was poised to outstrip Austria in economic development, a major factor in the century of industrialisation" (Fulbrook, 2004, p. 105). The German Question was forever settled by the Austro–Prussian War (1866) and the Franco–Prussian War (1870), which ended with a united Germany under Wilhelm I of Prussia.

Wilhelm I appointed Otto von Bismarck (1815–1898) Imperial Chancellor of the German Empire. Under Bismarck's guidance, Germany began a period of four decades of unprecedented economic growth. Bismarck did two things that spurred this growth. First, he created the cartel system. A *cartel* is a formal agreement among manufacturers to fix prices and production. This state-sponsored system encouraged industry to pool resources in research, development, and production techniques, as well as monopolize markets. The second thing Bismarck did was to create the first social state; today we call this the welfare state. Under Bismarck, Germany committed itself to the economic, physical, and social welfare of its workers. The state created insurance programs for health, accidents, old age, and unemployment.

Where the German territories lagged behind England and France, the political and economic structures of unified Germany surged ahead through three waves of industrialization. The first, as I hope you would now expect, was the expansion of a national railway system between 1877 and 1886. Beginning in 1887, Germany began expansion into dye manufacturing—by 1900, Germany dominated the world market with three major and five smaller firms producing 90 percent of the world's dye. Germany's industrializing tide included a wave of chemical innovation and production between 1897 and 1902, and electrical engineering between 1903 and 1918.

Max Weber thus lived and worked in an environment noted for unprecedented economic growth and state expansion. Weber was deeply concerned with the continued growth of German nationalism and was thus was politically active for most of his life. This political activism was different from Marx's, first because of the new political structure and second because of Weber's family standing (as we'll soon see, Weber family was well situated politically, culturally, and economically). The political and economic environment also informed his sociology. Germany's economic background provided a basis for Weber's distinction between traditional and rational capitalism. *Traditional capitalism* was motivated by immediate material concerns. Depending on their economic position, people would sell goods in order to make a living or to maintain a specific lifestyle—people thus earned money to spend money. *Rational capitalism* has an entirely different motive: the acquisition of money (capital) in order to make more money—people thus earned money to make money. Weber saw firsthand the potential for rapid economic growth that industrialized rational capitalism brought.

Weber also experienced the rapid development and extension of state bureaucratic organization. Germany was a newly united state struggling not only with bringing the nation's economy up to par, but with international legitimacy as well. Out of this historical, structural context, Weber developed a theory of the state and global relations that far exceeded that of any theorist up to his time and still forms the basis of sociological analysis of national and international politics. While the state bureaucracy had been growing in Germany since 1770, the centralization of German government into a single state along with Bismarck's social and political programs supercharged the process. Thus, state bureaucracy grew exponentially in Weber's time. In response, Weber created a theory of modern bureaucracy so robust that it continues to influence the sociological study of organizations today. In addition, the predominance of German bureaucracy in his lifetime undoubtedly oriented Weber toward the emphasis on rationalization as the major cultural, historical force moving society toward modernity.

Another indicator of the power of the social context under Bismarck is the Nobel Prize. Alfred Nobel (1833–1898) was a Swedish chemist who made his money through the invention and production of such explosives as nitroglycerin and dynamite. The story goes that Nobel read of his own death in a French newspaper. He obviously wasn't dead—the French paper had mistakenly reported his death instead of his brother's—but the story got him thinking about his legacy. He changed his will and left provision for the Nobel Foundation to award individuals whose work gave the greatest benefit to humankind. The first prizes were awarded in 1910, and by 1920 (the year of Weber's death), Germany had won more Nobel Prizes than any other nation on earth.

This early German dominance of the Nobel Prize owes a debt to a pre-Bismarck structural feature, one that is significant in understanding Weber's sociological imagination: the German university system. We've seen that one of the effects of the Enlightenment generally was the secularization of education, which influenced or created every academic discipline we now have. A significant event in this educational shift was the founding of the Humboldt University of Berlin in 1810. In many ways, the University of Berlin was the first modern university (R. Collins, 1998, pp. 638–646). Prior to the modern university, creative and intellectual work was usually funded in two ways: patronage and publication. In the patronage system, a wealthy individual (patron) would support the artist or scholar. Those artists and scholars not fortunate enough to have a patron made a living through publications and public lectures. All that began to change under the Humboldt system.

There were a number of social and structural factors in back of the changes in universities. There was of course the demand for secular teachers—as we saw in our last chapter, this structural demand was an important factor in Durkheim's work. In Germany, yet another structural demand was created. Beginning in 1770 and officially extended in 1804, Prussia became the first Western society to require 3 years of university education and a civil examination for government employment (R. Collins, 1998, p. 642). One of the effects of these structural demands was the creation of the research PhD as we know it today: "The German university revolution created the modern research university, where professors were expected not only to teach the best knowledge of the past but also to create new knowledge" (p. 643). This granted a great deal of independence to the intellectual. Rather than being dependent upon either a patron or book sales, the scholar was free to search for truth for truth's sake. This shift also institutionalized the conflictual yet detached nature of modern scholarship. Secular intellectual progress occurs through the abstract scientific eye and through doubt and debate.

The German university structure impacted Weber in a number of ways. First, Weber grew intellectually and professionally in an already well-institutionalized education system. This meant that he entered into academic debates that were ongoing discussions, most notably debates about idealism, history, and social science methods. The university and its ideas were foundational to Weber in a way that wasn't true for Comte, Marx, Spencer, or even Durkheim. Not only did the university influence the ideas Weber entertained, but it also affected the way in which he built up his theoretical work: incrementally in the context of respectful debates and established professional relations. This effect is especially seen in the fact that his home "became a gathering ground for the flower of Heidelberg's academic intellectuals, and Weber, though still quite young, came to be seen as the central figure in an extended network of colleagues and like-minded scholars" (Coser, 2003, p. 238).

Thus far in thinking about Weber's sociological imagination, I've emphasized the tremendous economic, political, bureaucratic, and intellectual growth of Germany. Yet Weber saw that in spite of all this development, Germany had entered a time where there were new problems on the horizon, and "he viewed the emerging trends of German society with mounting despair" (Bendix, 1960, p. 53). In a speech given to German economists when he was just 29 years of age, Weber insightfully observed that the problems facing his generation were very different than for those who worked toward unification (pp. 53–54). Like Durkheim, Weber saw that great society-wide tasks, such as war or unification, required an "appeal to the collective sentiments of the whole nation" (p. 53). And like Durkheim, Weber knew that under peaceful or "functional" conditions, particularized interests would arise. However, that is where Durkheim and Weber part company.

With Marx as an intellectual background, Weber saw particularized interests working against class consciousness (Marx) or collective consciousness (Durkheim). In a speech given in 1893, Weber refers directly to specific sorts of groups that he later (1922/1968) generalizes to class, status, and party. As we'll see, Weber uses these different social categories to argue that the kind of solidarity needed to create a social revolution or overt conflict is difficult to achieve in modern society. I also think that we can see shades of Weber's theory about charismatic, traditional, and rational legal authority in this speech. One of the central points in that theory is that the major task after any significant social change is routinization: Society must make the changes routine so that they can be carried out efficiently. In speaking to the generation that brought about German unification, he said, "we confront tasks of a different kind" (Bendix, 1960, p. 53). The tasks of the previous generation demanded and allowed for a strong sense of unity and collective consciousness, which he called "tremendous illusions." Gone were the high ideals and utopian visions. Weber's time was the time of routinization: the time to build the bureaucratic machine that would systematize and preserve the changes brought about by the old guard.

There is one more historical factor I need to include: World War I. Weber was politically and actively involved during the war. More important for understanding his sociological imagination are some of his speeches and publications from that time, most notably, "Politics as a Vocation" (Weber, 1921/1948a) and "Suffrage and Democracy in Germany" (1917/1994). These strike me as important because they give us a clearer grasp of what Weber intended with some of his better-known ideas, and it perhaps helps us see why Weber's theories are more pessimistic than any we've seen thus far. In general, Weber argued that there are structural barriers with participatory democracy.

Weber argued that modern government is different from others that came before it. The modern state is defined by the "*monopoly of the legitimate use of physical force* within a territory" (Weber, 1921/1948a, p. 78). The key to Weber's theory of the state is legitimacy. *Legitimacy* is our belief in the right of others to rule; legitimacy explains when and why people obey. There are three kinds of legitimacy: charismatic, traditional, and rational-legal. We'll go into more depth with these later, but the important thing for us now is that charismatic authority is "personal devotion and personal confidence in . . . qualities of individual leaders" (p. 78). It implies the idea of "*calling* in its highest expression" (p. 79). Today, we might call this a cult of personality. Weber also argued that the allocation of power through political parties is another trait of the modern state. This is extremely clear in the United States where we have two political parties that vie for power.

There is a tendency in party systems for charismatic leaders to rise to the fore. This implies that people in a modern democratic state vote for people based on personalities rather than actual political issues and democratic principles. Writing during World War II, Talcott Parsons (1942), arguably the most influential sociologist of the twentieth century, said that Weber's idea of charismatic leadership works itself out in party politics when "one party claims that it alone is 'fit' to govern the country, and that the policies of its rival are 'Unamerican'"—in such a condition "there must be a decision as to which is *the* legitimate leader" (p. 157).

In an insight that was far ahead of his time, Weber wondered about the influence of newspaper conglomerates, advertising, and biased journalism on the political process and swaying public opinion toward party politics and personality. To put this in context, the First Amendment to the Constitution of the United States guarantees freedom of the press. The reason for this was tied up with civil society and the need for citizenry to have accurate information in their exercise of freedom of speech, assembly, and the right to "petition the Government for a redress of grievances." As a result of seeing what the German press did in World War I, Weber wondered if such ideals held true. He saw that through advertising, business influenced "the press politically in a grand style" (Weber, 1921/1948a, p. 97). Weber thus saw these three factors—the cult of personality, party politics, and capitalist-controlled newspapers—as a danger to participatory democracy: "Since the time of the constitutional state, and definitely since democracy has been established, the 'demagogue' has been the typical political leader" (p. 96).

Weber talked about yet another threat to participatory democracy. This threat comes from a completely different direction: rational-legal authority. As I've noted, one of Weber's important theories concerns bureaucracies, the epitome of rational-legal systems. We'll soon see that bureaucracies are organizations that are built around expert knowledge systems and rational-legal control. Weber argued that modern states operated as an organization in the same way as a factory—as a bureaucracy.

This type of organizational form is historically specific. While ancient systems such as those found in China had bureaucratic elements, their core was still based in familial relations. The actual power rested in monarchs and emperors whose position was given by right of lineage. Underneath party politics and personality, the core and practices of the modern state are bureaucratic. Every American president, for example, has established or legitimated policies based on expert knowledge. In such a system, "there is simply no place in which non-expert citizens could make a contribution" (Barbalet, 2010, p. 209). Bureaucratic state management thus implies that "the ballot slip is the *only* instrument of power which is at all *capable* of giving the people who are subject to bureaucratic rule a minimal right of co-determination" (Weber, 1917/1994, pp. 105–106).

Weber's Intellectual World

As we've seen, Weber was well situated in the German intellectual community and had numerous professional and personal associations with academics. Further, if you read any of Weber's work, which I encourage you to do, you'll find that he makes frequent reference to other people's work. In fact, in the second paragraph of his introduction to *Economy and Society* (1922/1968), Weber references six academics: Karl Jaspers, Heinrich Rickert, Georg Simmel, Friedrich Gottl, Ferdinand Tönnies, and Rudolf Stammler (pp. 3–4). Thus, any account of Weber's intellectual context will either need to be quite long or an informative summary. Our story will tend toward the latter and will focus on Weber's concerns with German idealism and historicism. Lewis Coser (2003) captures this approach well:

> All in all, [Weber] attempted to direct the German idealistic position into a closer relation with the positivist tradition . . . and still retained what seemed to him the distinctive achievements of the German tradition: the emphasis on the search for subjective meanings that impel the action of historical actors. (p. 247)

In doing this, Weber is going to give us our first real critique of the positivist school that Comte founded. Positivist sciences assumes that the universe is empirical, it operates according to law-like principles, and that humans can discover those laws and use them to improve the human condition. Social science, then, seeks universal theories: theories that explain social phenomena without reference to a specific time, place, or subjective individual experience. This is the approach we saw in Spencer, Marx, and Durkheim. The prime movers in those theories are large-scale social structures. Weber's perspective, on the other hand, is a cultural one that privileges individual social action within a historically specific cultural milieu. Weber clearly opens up the possibility that culture and meaning can trump social structures: "Not ideas, but material and ideal interests directly govern men's conduct. Yet very frequently the 'world-images' that have been created by 'ideas' have, like switchmen, determined the tracks along which action has been pushed by the dynamic of interest" (Weber, 1915/1948c, p. 280). What this means for Weberian sociology is that society isn't always subject to universal laws, and people's actions aren't always predictable.

Enduring Issues

The predictability of human action is one of the thorniest and most persistent issues in the social sciences. As we saw in Chapter 1, science assumes that the universe operates according to universal laws, and Comte saw the discovery of those laws in society as the business of sociology. Weber casts doubt on both universal laws and predictability, but he doesn't do away with them entirely. He is, however, proposing a different kind of sociology from what we've thus far seen, and Weber is the first of many to do so. If you continue your studies, you'll find this is a continually debated issue: How predictable is human action? If it is predictable, then where's free will? If it isn't predictable, then how do we account for the tremendous social order we see?

As Coser (2003) notes, Weber built his perspective in dialogue with German idealism. More than any other, Kant was responsible for situating the German Enlightenment in idealism. You remember our discussion of Immanuel Kant from Chapter 3: Kant argued that the mind and its creation couldn't be reduced to general laws of positivism. The mind transcends the objective physical world. The categories of thought that we use to see the world around us are present *before observation* (a priori). The thing-in-itself (Kant's term for the physical world) isn't the goal or intent of our observations of the world—it represents the *limit of observation*. Universal laws of nature are thus valid not because they are intrinsic to the material world, but, rather, because they exist within our minds and thus within subjective experience. This idea of Kant's had important implications for the social and behavioral (human) sciences and formed the basis of a great deal of debate in Germany regarding the methods of those disciplines.

The Problems of Values and Meaning

There are two strands of neo-Kantian thought that were particularly important to Weber: the work of Windelband and Rickert as well as that of Wilhelm Dilthey.

Wilhelm Windelband (1848–1915) and Heinrich Rickert (1863–1936) formed what's called the Baden School and proposed an *axiological epistemology* for the human sciences. Axiological means something that is based on or influenced by values. Windelband and Rickert thus made a distinction between the natural and human sciences methodologies. Because the natural sciences deal with objective facts, their methodologies are *nomothetic* (able to produce universal laws).

However, because knowledge about human beings always implies value (axiological), the human sciences are restricted to *ideographic* methodologies. The word ideographic is based on the Greek word *ideo* and means "idea." We can understand the "graphic" part of the term through synonyms—something that is graphic is picturesque, vivid, or pictorial (we could also compare regular novels with graphic novels). Taken together, something is ideographic if it conveys an idea graphically, like with a picture. Thus, Windelband and Rickert were arguing that while the natural world can be studied and expressed mathematically, the human world must be studied and expressed meaningfully. Human behavior, then, is historically specific, and the task of the human sciences is to understand the meaningful relationship between the historical, cultural context and individual subjective experiences.

There are a number of ways that Windelband and Rickert's work (they aren't considered sociologists) influenced Weber's. First, Weber came to argue that humans are oriented toward culture, meaning, and values. Human action is based in meaning: "We shall speak of 'action' insofar as the acting individual attaches a subjective meaning to his behavior" (Weber 1922/1968, p. 4). Second, like Windelband and Rickert, Weber saw that this orientation toward meaning holds implications for social research because culture implies values. Value here is simply a preference for one thing over another. For example, I value my dogs more than my lawn or my truck, but I value my wife more than my dogs. In other words, my wife is more meaningful to me than my dogs, lawn, or truck. This feature of our lives implies that "empirical reality becomes 'culture' to us because and insofar as we relate it to value ideas" (Weber, 1904/1949, p. 76).

This implies that social science cannot be objective to the extent that the natural sciences can. Weber recognized that to ask a question about society or humans is a cultural act: *It requires us to place value on something.* In other words, we choose problems to study in sociology "on a basis of the *significance* which the concrete constellations of reality have for us" (Weber, 1904/1949, p. 80). For example, it would have been almost impossible for us to study spousal abuse 300 years ago. (It would have been difficult even 50 years ago.) It isn't that the behaviors weren't present; it is simply that the culture would not have allowed us to define them as abusive, at least not very easily. Further, one sociologist might choose to study spousal abuse and another how the level of bureaucratization influences the freedom school administrators have to make decisions based on pedagogical concerns. Both choices are based on the significance—the value—each of these issues has for the individual researcher.

Thus, sociology is deeply intertwined with subjectivity, and subjectivity is a problem for scientific research because it is supposed to be objective. Weber argued that while the selection of a research topic is subjective, the sociologist must constantly work to keep his or her own value judgments out of the research process and findings. Theories should be chosen, data gathered and analyzed, and findings presented without reference to the value judgment of the sociologist. This approach is difficult and more complicated than it might seem. Remember, as sociologists we are studying people, and people act on the basis of *their values*. So, the data we gather is itself infused with value by the subjects we're studying. As scientific sociologists, we can't pass judgment on their values: "to *judge* the *validity* of such values is a matter of *faith*" (Weber, 1904/1949, p. 55). Nor can a scientific sociologist "tell anyone what he *should* do" (p. 54). However, sociological research and theory can help a person see "what he *can* do—and under certain circumstances—what he wishes to do" (p. 54).

Hint

As mentioned in Chapter 1, there are several assumptions sociologists make in order to do their work (this is true of every discipline). One of the assumptions is the purpose of sociological knowledge. We saw that Comte "invented" sociology in order to have a scientific basis for reorganizing society. In his own way, Marx made a similar assumption: He said the purpose is to change the world, not simply to understand it. Durkheim felt that sociology would be a science of morals, a way to discern the best ethical guides for society. Weber's assumption about the purpose of sociological knowledge is clearly different from Comte, Marx, or Durkheim. When possible, it's important to know what assumptions a theorist is making and to understand how they influence what he or she does.

Windelband and Rickert also informed Weber's approach to the kind of data to be gathered. Think back to what we said about ideographic methodology. This approach implies that the gathered data should be historical and culturally detailed. Weber (1904/1949) argued that our aim as sociologists is the

> understanding of the characteristic uniqueness of the reality in which we move. We wish to understand on the one hand the relationships and the cultural significance of individual events in their contemporary manifestations and on the other hand the causes of their being historically *so* and not *otherwise.* (p. 72)

Notice two things in this quote: First, Weber talks about the "uniqueness of the reality." Again, there is the implication that invariant laws don't drive the human world. Our moment in time is different from all others before it, as was Weber's (which is why we've spent time considering the historically specific factors that influenced each theorist's sociological imagination). What we think and feel, the values that we hold, and the meanings of our world are unique.

Second, notice that Weber did not say that causal explanations aren't possible. Sociology may not, in Weber's view, be able to predict how society will be in 20 years. However, sociology can discern the factors and processes that caused a society, a culture, and its people to be the way they are. The third thing I'd like for you to see is that Weber's approach demands a full and rich understanding of a society and its moment. In Weber's sociology, the data is what we would today call "thick descriptions," layer upon layer of detailed historical accounts of social structures, culture, and people. Specifically, in his sociology, Weber was interested in explaining the relationships among cultural values and beliefs (generally expressed in religion), social structure (overwhelmingly informed by the economy), and the psychological orientations of the actors. And remember, none of these elements is particularly determinative for Weber.

Specific Methods: Ideal Types

As I've said, Weber's sociological imagination puts a lot of emphasis on subjective meaning. The question, then, is how to create objective knowledge, which is what science is interested in, out of subjective meaning. Plus, I don't know if you thought of this yet, but it's extremely difficult to make sense out of the amount of historical data in which Weber was interested. Heinrich Rickert provided an answer to both these issues in the concept of ideal types. *Ideal types* are analytical constructs that don't exist anywhere in the real world. They simply provide a logical touchstone to which we can compare empirical data. Ideal types act like a yardstick against which we can measure differences in the social world. These types provide objective measurement because they exist outside the historical contingency of the data at which we are looking. "Historical research faces the task of determining in each individual case, the extent to which this idea-construct approximates to or diverges from reality" (Weber, 1904/1949, p. 90).

There are two main kinds of ideal types: historical and classificatory. *Historical ideal types* are built up from past events into a rational form. In other words, the researcher examines past examples of whatever phenomenon he or she is interested in and then deduces some logical characteristics. Weber uses this form in *The Protestant Ethic and the Spirit of Capitalism* (1904–1905/2002). In that work, Weber constructs an ideal type of capitalism (traditional and rational) in order to show that the capitalism in the West is historically unique. Classificatory ideal types, on the other hand, are built up from logical speculation. Here the researcher asks him- or herself, what are the *logically possible* kinds of _____ (fill in the research interest)? A good example of this in Weber's work is his ideal type bureaucracy.

Definition

Ideal types are a part of Weber's theoretical methodology used to create systematic, objective knowledge in the face of the subjectivity of culture. Ideal types are purely conceptual constructions that are used as standards of comparison in the empirical world. There are two kinds of ideal types: historical and classificatory.

Specific Methods: Verstehen

The other important influence on Weber's sociological imagination that I want us to consider is Wilhelm Dilthey (1833–1911). Dilthey was interested in unpacking personal experience, the expressions of that experience, an individual's self-awareness, and the awareness of others. His approach was to focus on the objective products of the historically situated human mind. In contrast to the psychology of the time, Dilthey argued that the mind can't be understood through introspection or self-examination alone. In a way that is a little reminiscent of Marx, Dilthey contended that it's the *products of the mind,* such as art, language, literature, and actions, that provide an objective gateway to understanding. The mind is manifested, made clear, through such cultural and social expressions, and it's there, among the objective manifestations, that the mind can be interpreted. Further, those objective manifestations arise out of specific historical contexts. Thus, to understand ourselves, we must understand our history.

When we talk about this aspect of Weber, we tend to use the German word *verstehen,* which simply means understanding. The reason we do, I think, is because "understand" is used so commonly and not in the way it is intended here. As a social science methodology, it is probably most familiar to you in the context of ethnographic studies. An ethnographer is one who steeps him- or herself in a culture deeply enough to understand it, yet maintains the outsider perspective sufficiently to tell the story of an unknown culture to others.

Weber (1922/1968) said there are two types of understanding (*verstehen*). The first "is the direct observational understanding of the subjective meaning of a given act as such" (p. 8). The "subjective meaning" is the meaning it has for the person involved in the action. "As such" means it's apparent and taken simply for what it is, as when we hear or read 2 + 2 = 4. Actions "as such" can also be understood through direct observation, as when someone chops wood or lights a cigar. Emotions can be understood this way as well. We can see anger on someone's face or joy in his or her laughter. But the mind's manifestation generally has another, deeper level that is of far greater interest.

This second understanding is "in terms of *motive* the meaning an actor attaches . . . which consists in placing the act in an intelligible and more inclusive context of meaning" (Weber, 1922/1968, p. 8). This understanding seeks to know why an actor does what he or she does. A person may write 2 + 2 = 4, if she is a teacher instructing students; or it may be part of a ritual a person performs if he or she suffers from obsessive-compulsive disorder. The woodcutter may be getting a cord of wood ready to sell, or he may be letting off steam because of a recent fight with his partner. Weber noted that motivations might be rational or irrational.

Weber explained that there are three contexts that are important for understanding. The *historical approach* seeks to understand the intended meaning of individual actors. I want to pause here and point out that this is why I've included the sociological imagination of our theorists in this book. Most of us see ourselves as simply doing what we do as a result of individual decisions. Let's imagine I see you in a local coffee shop reading a book (the action as is). I ask you why you're reading that specific book (seeking your motivation). You'd probably answer something like, "It's for a class I'm taking." That sort of explanation is all most of us see, and it's precisely why Dilthey said that the mind couldn't be known simply through self-examination. In talking about the sociological imagination, C. Wright Mills (1959) said, "the individual can understand his own experience and gauge his own fate *only by locating himself within his period*" (p. 5, emphasis added). We cannot understand our own motivations unless and until we understand them historically. If 200 years ago I had asked someone why he or she were reading a book in a public space, the person would not—and likely could not—give an answer based on a college education being used as a credential to enter the workforce (which, by the way, is what "It's for a class I'm taking" more probably means).

The next context for understanding is *sociological mass phenomena.* Here, Weber (1922/1968) intended "the average of, or an approximation to, the actually intended meaning" (p. 9). The best-case scenario for this type of understanding exists in "cases of mass phenomena which can be statistically described and unambiguously interpreted" (p. 10). This sounds straightforward to our sociological ears, because a good portion of our research is cast in this mold. We use statistics and standard interpretive schemes to make sense of them. But for Weber, things weren't this clear-cut: "processes of action which seem to an observer to be the same or similar may fit into exceedingly various complexes of motive in the case of the actual actor" (p. 10). Though Weber seemed more confident in the third context, "the meaning appropriate to a *scientifically formulated pure type* (ideal type)" (p. 9, emphasis added), he was in the end equally tentative about it as well.

Where does Weber's emphasis on understanding bring us? Weber saw meaning as central to humankind and its interpretation (*verstehen,* understanding) as fundamental to sociology. Yet by its very nature, interpretive understanding is tricky business because humans are extremely complex beings. We need to understand the historical, cultural context; we need to look for sociological mass phenomena; and we need to use ideal types as a measuring rod to sort things objectively, but in the end, "there remains only the possibility of comparing the largest possible number of historical or contemporary processes" (Weber, 1922/1968, p. 10). Where we end up is Weber's comparative sociology. The Weberian sociologist is never certain and always delving deeper and searching for comparative cases.

Weberian sociologists, then, will see the world in terms of ideal types (abstract categorical schemes), broad historical and cultural trends, or from the point of view of the situated subject (interpretive sociology). Weber's causal explanations have to do with understanding how and why a particular set of historical and cultural circumstances came together, and his general explanation is always subject to case-specific variations. There are a couple of things that this implies. First, Weber rarely gives us highly specified relationships among his concepts. He is more concerned with providing the historical preconditions for any phenomenon. Thus, it isn't as easy to create a dynamic model of Weber's theory as it is for Durkheim's, nor is it as true to the theorist's thinking to do so. What we *can* do is create a picture of the general historical processes that tend to produce an environment conducive to whatever issue it is we are trying to explain. The second implication of this Weberian approach is that the general theory will hold if and only if there are no mitigating circumstances. In other words, Weber will give us a general theory, but it will only hold if nothing gets in the way. So, with religion, for example, Weber may give us a general account of how ethnical monotheism evolved from magic, but we can see histories where it didn't work the way the theory implies. That is one of the frustrating things about reading Weber's *The Sociology of Religion* (1922/1993); however, it is also what makes the account more accurate than many simpler sociologies.

Weber's Sociological Imagination

You might notice that the above heading doesn't say anything specific about Weber's sociological imagination, as headings for previous theorists did. The reason for that I imagine is fairly obvious: Weber is a bit too complex to capture in a single heading. We've seen how his personal biography both prepared him for academic life and helped create the complexities of his mind. He was intimately aware of politics (his father) and religion (his mother) and the tension that exists between the two in modernity. It's also interesting to discover that scholastic work was a sideline for this man; he wanted to be involved in politics and saw teaching as a substitute. It's also a powerful insight into Weber to know that he wanted so badly to be involved in politics yet was so adamant about keeping values out of sociology. He had very clear standards and kept them rigorously.

Weber's complex sociological imagination was also strongly influenced by the German intellectual climate of his day. German idealism had a pronounced impact on his work, as he gave prominent place to the inner experience and awareness of the individual. Yet this was tempered by German historicism, which placed the individual's mind in its historically specific cultural context. From these sources, Weber developed a unique methodology that informed all of his sociological research. Ideal types and *verstehen* are the tools he used to investigate religion, bureaucracy, human action, capitalism, conflict and social change, stratification, states and political power, and the ongoing process of rationalization.

It's much easier to encapsulate all of Comte, Spencer, Durkheim, and even Marx under single headings. One reason is that in many ways, Weber's methodology and empirical research have had independent effects—thus, there isn't a single Weberian school. There are multitudes of studies that fall under interpretive sociology or historical-comparative sociology that find their base in Weber's work yet don't look at the same empirical phenomena as he, and because of that, his perspective is less easily seen in his methodological offspring. I think his influence, and thus his perspective, is more straightforwardly observed in the empirical studies that have followed in his footsteps, most notably conflict and stratification studies, organizational studies, action theory, and phenomenological studies of meaning.

Weber's work in stratification and conflict was formulated in an intellectual conversation of sorts with Marx. Marx, you'll remember, argued that stratification was based solely on class; thus, conflict and social change would arise around class issues. As we'll see, Weber gives us first a more complex understanding of class by adding the middle class to Marx's owners and workers dualism. Weber's understanding of stratification is also more complex because he adds status groups, like race and gender, and political parties to the issue of class. This "tripartite model of class, status group, and power group ('party'), as independent although interacting forms of stratification, has become the new paradigm" (R. Collins, 2009, p. 14).

Weber also expanded our understanding of Marx's notion of ideology by showing that all knowledge systems are legitimated. Having dismissed Marx's ties to economic materialism and sensitized by Kant's distinction between the mind and the thing-in-itself, Weber was able to help us see that culture is socially constructed in all its forms and always carries with it legitimating stories validating the culture. This insight helped form the foundation for the sociology of knowledge. One of the most powerful uses of this perspective is Peter Berger and Thomas Luckmann's *The Social Construction of Reality* (1966); in that book, they see legitimation as one of two primary mechanisms through which we objectify and make real our culture. Berger and Luckmann's work is part of a sociological perspective called social phenomenology. That perspective was founded by Alfred Schutz (1932/1967), who based his entire theory on Weber's idea of meaningful social action. In addition to Berger and Luckmann, Schutz's phenomenology is the inspiration behind Harold Garfinkel's (1967) school of ethnomethodology, which upended traditional sociological approaches for 20-some odd years.

Weber is also the primary foundation for the sociology of organizations. All studies of bureaucracy come back to Weber. One of the earliest and best-known applications is found in Peter M. Blau and Marshall W. Meyer's *Bureaucracy in Modern Society* (third edition, 1987). An interesting and popular application of Weber's bureaucracy is George Ritzer's (2004) *The McDonaldization of Society*.

Another significant area that Weber has influenced is action theory. Action theory has been an interest of philosophy since the time of Aristotle; in sociology, action theory begins with Weber. Both disciplines are concerned with theories that explain the complex process of intentional or willful action. Weber argues that action takes place when a person's behavior is meaningfully oriented toward other social actors, usually as types of instrumental or value rationality.

The most pervasive application of Weber's action theory comes from Talcott Parsons (Chapter 9). Parsons's first major publication was *The Structure of Social Action* (1937), and in it he proposed his voluntaristic theory of social action. We'll talk a bit more about that theory and Parsons's influence in Chapter 9, but let me give you a hint of it by quoting from Oxford University Press's dictionary of sociology: "For some twenty to thirty years after the Second World War, Talcott Parsons was the major theoretical figure in English-speaking sociology, if not in world sociology" (Marshall, 1998, p. 480).

As you might have already picked up, Weber probably inspired our thinking in more areas than any other person. If we are going to study religion, bureaucracy, culture, politics, conflict, war, revolution, the subjective experience of the individual, historic trends, knowledge, or the economy, then we have to incorporate Weber. In my opinion, he is in many ways our most seminal thinker.

Concepts and Theory: The Process of Rationalization

I presented a version of modernity in Chapter 1 that emphasized it as a unique time period dominated by a specific way of thinking (reason and rationality) and a particular form of knowledge (objective science). I also gave the impression that these things sprang to life out of the Enlightenment and that previous epochs were dominated by faith and fatalism. I said one further thing that I hope you remember: The story I gave you was just one of many possibilities, and I wanted to use it as a touchstone, a way to organize our thinking. In retrospect, you might now know that what I gave you is a historical ideal type of modernity, one constructed from past events and made into a rational idealized form. Remember, it's idealized in the sense that its historical accuracy isn't a major concern—it's not idealized in the sense that it represents a goal that theorists and societies should strive to attain. As Weber intended, I'm using it simply and only as a way of understanding our theorists.

So far, Weber has varied more from the ideal type than the others. We've seen that variation in his emphasis on subjective meaning and tentativeness about invariant laws. In fact, Weber has been characterized as anti-positivist. I'm not sure I'd go that far, if for no other reason than it puts Weber in too clear a category. Weber also varies from my ideal type in that he doesn't see reason and rationality rising out of the dark ages like a brilliant morning sun. I mean this in two ways: First, Weber argues that the process of rationalization has been around for a long time; in other words, it isn't specific to modernity. Weber would certainly agree that rationality came to fruition in modern times and it is modernity's defining feature, but he would also point to the long historical buildup that preceded modernity. The second way that Weber deviates from my brilliant sun analogy is that he finds little to be hopeful or happy about in the modern era. In the rest of this chapter, we'll clearly see rationalization as a historical process; I also think you'll get a sense of Weber's gloomy outlook.

Weber uses the term *rationalization* in at least three different ways: He uses it to talk about means-ends calculation, in which rationality is individual and specific. Rational action is action based on the most efficient means to achieve a given end. Second, Weber uses the term to talk about bureaucracies. The bureaucratic form is a method of organizing human behavior across time and space. Initially, we used kinship to organize our behaviors, using the ideas of extended family, lineages, clans, moieties, and so forth. But as the contours of society changed, so did our method of organizing. Bureaucracy is a more rational form of organization than the traditional and emotive kinship system.

Definition

Rationalization is a central concept in Weber's theory of modernity and social change. It is the process through which affective ties, spirituality, and tradition are replaced by rational calculation, efficiency, and control. Rationalization is most associated with increasing levels of bureaucratization.

Last, Weber uses the term rationalization in a more general sense. One way to think about it is to see rationalization as the opposite of enchantment. Specifically, an enchanted world is one filled with mystery and magic. Disenchantment, then, refers to the process of emptying the world of magical or spiritual forces. Part of this, of course, is in the religious sense of secularization. Peter Berger (1967) provides us with a good definition of *secularization:* "By secularization we mean the process by which sectors of society and culture are removed from the domination of religious institutions and symbols" (p. 107). Thus, both secularization and disenchantment refer to the narrowing of the religious or spiritual elements of the world. If we think about the world of magic or primitive religion, filled with multiple layers of energies, spirits, demons, and gods, then in a very real way, the world has been subjected to secularization from the beginning of religion. In Western Europe, the birthplace of modern science, the number of spiritual entities has generally declined from many, many gods to one. In addition, the presence of a god has been removed from being immediately available within every force (think of the gods of thunder, harvest, and so on) to being completely divorced from the physical world, existing apart from time (eternal) and space (infinite). In our more recent past, secularization, demystification, and rationalization have of course been carried further by science and capitalism.

This general process of rationalization and demystification extends beyond the realm of religion. Because of the prominence of bureaucracy, means-ends calculation, science, secularization, and so forth, our world is emptier. Weber sees this move toward rationalization as historically unavoidable; it has become above all else the defining feature of modernity: "The fate of our time is characterized by rationalization and intellectualization and, above all, by the 'disenchantment of the world'" (Weber, 1919/1948b, p. 155). As Weber implies, rationalization leads inexorably to an empty society, at least in terms that have traditionally mattered to people. The organizational, intellectual, and cultural movements toward rationality have emptied the world of emotion, mystery, tradition, and affective human ties. We increasingly relate to our world through economic calculation, impersonal relations, and expert knowledge. Weber tells us that as a result of rationalization, the "most sublime values have retreated from public life" and that the spirit, "which in former times swept through the great communities like a firebrand, welding them together," is gone (p. 155). Weber not only sees this as a condition of the religious or political institutions in society, but also sees the creative arts, like music and painting, as having lost this vital spirit as well. Even our food is subject to rationalization, whether it is the "McDonaldized" experience (Ritzer, 2004) or the steak dinner that is subjected to the "fact" that it contains in excess of 2,000 calories and 100 grams of fat. Thus, for Weber, the process of modernization brings with it a stark and barren world culture.

Types of Social Action

One of Weber's influences on sociology comes in the form of social action. As we've seen, Weber defines action as something to which an individual attaches subjective meaning. A person's action becomes social "insofar as its subjective meaning takes account of the [action] of others and is thereby oriented in its course" (Weber, 1922/1968, p. 4). Weber created an ideal type of social action in which he talks about the different kinds of meaning social action can have.

Hint It's important to note the implicit distinction Weber is making with the concept of action—action is distinct from behavior. All animals behave; they move, acquire nutrients from the environment, react to stimuli, and so on. This is true about humans as well, but humans can also *act*. Behavior for humans becomes action when we assign meaning to it. In our next chapter, George Mead will give us another quality of action: intentional choice. This quality of human action creates what's known as the problem of social order: If humans can freely choose their actions, why do they continually create patterns of repetitious actions?

Remember that Weber's intent is to create an objective measure that exists outside of the content—the ideal type helps us understand actions in ways that self-reflection cannot. Weber asked himself, what are the possible different kinds of human action? He came up with four. In Figure 5.1, these ideal types are pictured as a block to which the variety of complex behaviors can be compared. When we use this ideal typology, we are able to offer an objective explanation of human behavior.

Figure 5.1 Social Action Typology

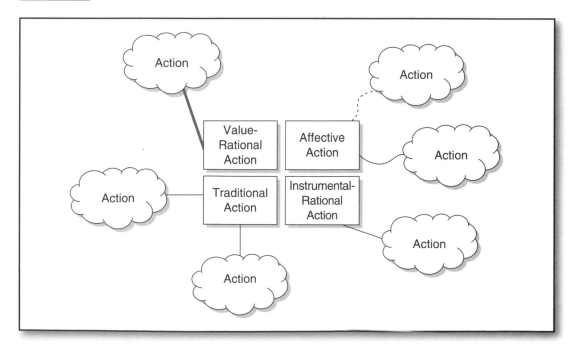

Instrumental-rational action is behavior in which the means and ends of action are rationally related to each other. So, your action in coming to the university is instrumental-rational in that you see it as a logical means to achieve an "end," that is, a good job or career. *Value-rational action* behavior is based upon one's values or morals. If there is no way you could get caught paying someone to write your term paper for you, then it would be instrumentally rational for you to do so. It would be the easiest way to achieve a desired end. However, if you don't do that because you think it would be dishonest, then your behavior is being guided by values or morals and is value-rational. *Traditional action* is action that is determined or motivated by habit or time-honored beliefs and meanings; *affective action* is determined by people's emotions in a given situation. All of these different types of behavior can become social action insofar as we take into account the behaviors and subjective orientations of other actors, whether nearby or absent, and whether in the past, present, or future.

We can tease out an important element in Weber's thinking from this ideal typology. Value-rational behavior is distinguished from affective action because Weber sees some semblance of self-aware decision-making processes in value-rationality. Values are emotional, but when we consciously decide to act because of the logic of our values, it is rational behavior. Emotion, on the other hand, is irrational and proceeds simply from the feelings that an actor may have in a given situation. According to Weber (1922/1968), both affective and traditional behaviors lay "very close to the borderline of what can justifiably be called *meaningfully oriented action*" (p. 25, emphasis added) because they are not based on explicit decisions.

The important thing I want you to see here is Weber's emphasis on rationality. The two types that are most meaningful both involve rational choices; the difference is what those choices are based on (efficiency or value) and not on the decision process itself. Less meaningful actions are those that are nonrational. Here, rationality is restricted because of the basis for action: tradition or emotion. These two clusters—rational/nonrational and meaningful/meaningless—vary by history and culture. Traditional and affective actions were more characteristic of times and places outside the influence of Western European Enlightenment. Ancient Chinese society, for example, was organized around tradition and strong, elaborated family ties. However, as contemporary China is becoming Westernized, traditional and affective social actions are receding, and rational action, both instrumental and value rational, is increasing. The issue for Weber (1904/1949) is "the causes of their being historically *so* and not *otherwise*" (p. 72).

My problem at this point is how to organize the historical picture that Weber gives us. As we know, Weber himself didn't bring together his different studies in a cohesive whole, so we're left to our own devices. I'm going to start with Weber's work in religion. My reason for doing so isn't that religion is fundamental to his theory, as is the case with Durkheim. It's simply that Weber's theory of capitalism is based on his study of Protestantism; since Weber's analysis of religion is historical, and Protestantism came late to the religious scene, it simply seems convenient. Weber's theories of class, status, and power fit best following capitalism because class is historically specific to a particular kind of capitalism. With these three clustered together in this way, we'll end our discovery of Weber's theory with bureaucracy.

Concepts and Theory: The Evolution of Religion

Historically, religion in general has moved from magic to polytheism, to pantheism, to monotheism, and to ethical monotheism. There isn't really any debate about that historical line; there is, however, debate about why religion changed. There are two basic approaches to explaining this movement: progressive revelation and social evolution. *Progressive revelation* argues that there has always been one God who is concerned with the way we behave. But because of our inability to receive the full revelation of God, He (and these gods are always male) had to progressively reveal truth to us. Implicitly or explicitly, this perspective posits a God who is bound to human history. In traditional Christianity, for example, humankind had to reach bottom through the revelation of the law before it was possible for God to reveal grace. *Social evolution,* on the other hand, is empirically based and doesn't assume a metaphysical being guiding the process. The only link that is explored is the one between religion and society (at least in sociological terms). Weber gives us such a model of religious change.

However, *these two models are not necessarily exclusionary.* As I mentioned, most religions see God working through history. Further, it is possible that part of humankind's preparedness to receive God's revelation is linked to society, since humans are social beings. The reason I say this is to let you know that you aren't faced with an either/or decision. If you are religious, you don't have to reject your religion in order to see value in Weber's theory. Weber is only concerned with the empirical elements of religious change. We can either believe that there are spiritual forces behind those changes or not. In either case, our beliefs are based on assumptions we make about how the universe works. And, as is always the case, assumptions are never tested or proven.

From Magic to Religion

Weber is interested in explaining how ethical monotheism came about. Monotheism, of course, is the belief in one God. *Ethical monotheism* is the belief in one God that cares about human behavior. In terms of history, such a God is a recent occurrence. For much of our history, we believed in many gods and goddesses, and we didn't feel that they cared much about how we behaved. What's more, most of the gods and goddesses were quite risqué in their own actions. Weber's approach is to look through history and see what kinds of social factors were associated with changes in the way people thought about God.

Weber first recognizes that the movement from magic to religion is the same as the change from naturism to symbolism. *Magic* is the direct manipulation of forces. These forces are seen as being almost synonymous with nature. So to assure a good harvest, a magician might perform a fertility ritual, because seeds and fertility are related. *Religion,* on the other hand, is more symbolic. Let's take Christian communion, for example. In Protestant churches, the bread and grape juice symbolize the body and blood of Christ. They represent not only the atonement, but also the solidarity believers have in sharing the same body. The Catholic example is a bit more interesting. While there is quite a bit of symbolism in the Eucharist, there is also what is known as transubstantiation. In transubstantiation, the wine and bread literally become the blood and body of Christ. Taking the sacrament brings about union with Christ. Through it, venial sin and punishment are remitted.

Definition Magic and religion are ideas Weber uses to understand the evolution of spiritual forms of action and social organization. *Magic* is based on a naturalistic view of the universe and is concerned with direct manipulation of natural forces; *religion*, on the other hand, is symbolic and is concerned with rituals that dramatically enact abstract truths. Evolution toward more symbolic forms is associated with professionalization, bureaucratization of government, and technology.

The reason I compare the Protestant and Catholic versions of communion is to bring out this point of Weber's. In Protestant communion, there are *only symbolic* elements present. In the Catholic Eucharist, there are actual elements present that are effectual in their own right. Weber would argue, then, that there are magical elements present in the Eucharist. This is not meant as a slight against Catholicism (or a slight against Protestantism, depending on your perspective). It is simply meant to point out the distinction between more symbolic and naturalistic elements in religion. With the Eucharist, there is a direct manipulation of forces to accomplish some purpose, which is Weber's definition of magic. We can also see the complexity of the social world in this example. Weber argues that, *in general,* there is a movement from magic to religion, from physical manipulation to symbols. However, real life is more complex. Catholicism shows us that due to a religion's specific history, it may contain both symbolic and magical elements.

Generally speaking, humanity moved from magic to religion due to economic stability and professionalization. To understand these effects, we can picture a small hunter-gatherer society. In such a group, everyone lives communally, the division of labor is low, and life is uncertain. The tribe is subject to the whims of nature: Game may or may not be there, and the weather may or may not be good enough for the berries to grow. In an attempt to make their world more stable, tribe members perform rituals before the hunt and at season changes. They perform rituals at childbirth or when they need to find water. These rituals constitute magical manipulations, and the people performing the rituals would be the same as those who are involved in the actual work. For example, we can imagine the men gathering together before a hunt, perhaps painting themselves with animal blood or putting on masks and acting out the hunt.

Over the years, the tribe learns about horticulture and they begin to plant food. They become sedentary and tied to the land, and life becomes a bit more predictable. After some time, they learn how to use metal, plow the land, and irrigate. From these small technological advances come surplus, population growth, power, and a different kind of division of labor. Rather than living communally, certain people are able to work at specialized jobs. Thus, the uncertainty of life is further diminished.

What we want to glean from this little story is how the type of economy can influence the development of religion. Early in our story, many people were involved in the practice of magic. Because people were tied to nature, their view of how things worked was naturalistic and not symbolic. They were concerned with day-to-day existence: What mattered was getting food, water, and shelter. It was necessary for them to have a method of control that each person could use. But as life became more predictable, people did not have to be as concerned with immediate sustenance. Tasks became more specialized. Some people tilled the soil; others planted the seeds. Some were in charge of the irrigation; others transported the harvested grain to the grinders. And those who had a knack for it became what Weber characterizes as the "oldest of all vocations," professional necromancers (magicians).

Three things happened as a result of the *professionalization* of magic: Individuals could devote all of their time to experiencing it, they increased their knowledge surrounding the experience, and they acquired vested interests (that is, their livelihood became dependent upon an esoteric knowledge). The experience factor is important in this case. Weber argues that ecstatic experiences are the prototypical religious experience. Religion always involves transcendence, the act of going above or outside of normal life, and ecstasy is a primal form of transcendence. Like Durkheim, Weber (1922/1993) argues that "ecstasy occurs in a social form, the orgy, which is the primordial form of communal religious association" (p. 3). For laymen, the experience is only available on occasion, but because the professional wizard is freed from daily concerns, she or he can develop practices that induce ecstatic states almost continually.

Definition

Professionalization is a concept in Weber's theory of religious evolution. In its simplest form, professionalization occurs when a person or group is paid to do specific work. This move frees these people from basic concerns of sustenance. Professionals are thus able to elaborate (become more abstract and complex) and control their specific fields, thus protecting their vested interests.

These experiences, along with increasingly available time, prompted the professional to develop complex belief systems. As a result, what had once been seen as something everyone could practice (magic) became part of secret lore and available to only a select few. Because the professional magician could afford to engage in continual transcendent experiences, he saw more behind the world than did others. The world thus became less empirical and more filled with transcendent beings, spirits that controlled and lived outside of daily life. The magician developed beliefs and ritual systems that reflected this growing complexity. In addition, professionals could consider their own thoughts and beliefs; the very act of reflexive thought will tend to make ideas more abstract and complex. Reflexive thought by its nature is abstract because it isn't thinking about anything concrete. And because it isn't thinking about anything concrete, it can go anywhere it wants. Further, the connections among ideas that are "discovered" through this kind of thinking tend to be systematized around abstract ideas.

Let's think about sociology as an example of professionalization. In a very real way, everybody is a sociologist. We all have an understanding about how society works and what is involved in getting around in society. That's one of the funny things about teaching an introduction to sociology course—most of the students have an "Oh, I knew that" response to the material we talk about. But what do professional sociologists do? Well, because we get paid to do nothing but sit around and think about society, we have made things very abstract and complex. We use big words (like *dramaturgy* and *impression management*) to talk about pretty mundane things (picking out clothes to wear). And we've developed these ideas into some pretty complex theories that take years of education to understand.

Further, think about the book you are reading right now. It is the result of my thinking about other people's thinking (Weber's, for example). But it's really more than that. This chapter could not have been written right after Weber. What we have here is not only my reading of Weber, but also my reading of other people's reading of Weber. And because I get paid to do this kind of thing, I can weave all this complexity into a systemic whole. More than that—and this is an important point for Weber—we have made it so that you can't understand sociology without the help of a professional. You had to pay to get this book, and you had to pay to get another professional to stand up in front of the class and explain this explanation of Weber. Thus, professionalization pushes for symbolic complexity because of practice, reflexive thinking, and vested interests. (I have to make sociology complex to protect my job.) This kind of process is exactly what created symbolic religion.

I bring the ideas of professionalization and economic changes together in Figure 5.2. As technology allows people to be less tied by direct relations to the environment, they are less dependent upon magic for manipulating nature. As the economy produces surplus and creates more complex divisions of labor, necromancers can be relieved of other duties and paid simply to practice magic—they become professionals. These two factors work together to increase the level of abstract symbols, which in turn moves magic toward religion.

From Polytheism to Ethical Monotheism

In the beginning, religious practice was oriented around local gods and goddesses. These gods lived in specific locations and were connected to specific collectives. The next step in religious evolution was to place these local gods into organized pantheons. Organized pantheons are not simply clusters of various gods and goddess. Rather, in a pantheon, each deity is given a specific sphere of influence, and the activities and deities are related to one another.

Professionalization and symbolism play important roles in the religious evolution to pantheons. Professionals affected religion primarily through increasing the level of abstraction (analogous thinking, religious stereotyping, and the use of symbols in ritual rather than actual things—Weber points out that the oldest use of

Figure 5.2   Evolution to Symbolism

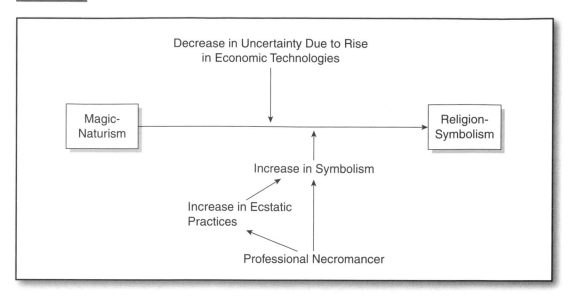

paper money was to pay off the dead, not the living) and creating coherent systems of knowledge out of localized beliefs. In short, as religion became more professionalized, it tended to become more rationally and abstractly organized.

At this point in our evolution, politics has come to play a very important role as well. One of the things the idea of God does is unite different groups into a single community. A kin- or tribal-based god provides the symbolic ties that small groups need, but at some point in our history, we began to bring these different collectives together through conquest or voluntary association. Different groups, each with its own god, began to form larger collectives. In order to link these groups, a more abstract and powerful god was needed—one that could be seen above all other gods, thus linking all the people. The gods that were highest on the polytheistic hierarchy began to be seen as less and less attached to the earth and more as part of the heavens. As populations with even greater diversity came together, more abstract symbols were needed to link them. These universal gods became seen as more powerful than all other gods. They became the god of gods and lord of lords.

This development toward monotheism was aided by the need of monarchs to break the hold that the priesthood had on the populace through the "multiplicity of sacerdotal gods" (*sacerdotal* meaning "pertaining to priests"). As long as there was a multiplicity of gods still linked to the daily needs of the people, there would be need for specific rites to approach those gods. In order to consolidate power, it was necessary for the king to eliminate those various paths, because each one represented a power that he didn't control. Monarchs thus began to monetarily and politically privilege the universal god and his priests. And the idea of the one God was thus born.

Yet one more aspect needs to be added here: morality. The gods had typically never been concerned with the behaviors of their cults. Weber argues that ethical monotheism came about in response to the increasing rationalization of the state and the social control of human behavior. Because of the changes in the way government was organized, from traditional to rational-legal authority, a culture developed that proclaimed that individual behaviors should be controlled. Of course, this was in response to the needs of a large population under a centralized state: The larger and more diverse a population, the greater the need for rational, centralized control. And, according to Weber (1922/1993), this need was reflected in religion: As the state became more interested in the actions of the populace, so did the ruling god. "The personal, transcendental and ethical god is a Near-Eastern concept. It corresponds so closely to that of an all-powerful mundane king with his rational bureaucratic regime that a causal connection can scarcely be overlooked" (p. 56).

In Figure 5.3, we still see the influence of professionalization. This remains an important factor in all areas of modern life. But rather than symbolization, rationalization now plays an important role in the movement toward an ethical god. As state governments become more and more bureaucratized and rational law becomes the way in which relationships are managed, people begin to see their lives as subject to rational control. Of course, today we feel this extended to almost every area of our lives. We not only sense that our bodies, emotions, and minds can be rationally controlled, we believe they should be controlled.

Religion contributed to this belief through the role of the prophet. Weber categorizes prophecy as two kinds: exemplary and ethical. The exemplary type is found in India and other Eastern civilizations. This kind of prophet shows the way by being an example. The emphasis in this case is on a kind of self-actualization. There is no real sense of right and wrong, but only of a better way. Neither is there a god to whom the individual or collective owes allegiance. The Buddha is a good example of this. The ethical prophet, on the other hand, is vitally concerned with good and evil and with bringing a wayward people back to the right relationship with the dominant god. Just as a centralized state needs and creates a single national identity and story, so the ethical prophet presents the cosmos as a "meaningful, ordered totality" (Weber, 1922/1993, p. 59). Thus, God and the state come together to rationally control human behavior.

Figure 5.3 Polytheism to Ethical Monotheism

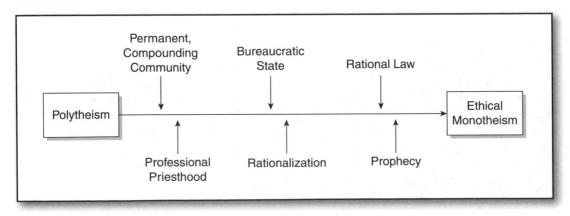

Concepts and Theory: The Rise of Capitalism

For Weber, there are three main factors that influenced the rise of capitalism as an economic form: religion, nation-states, and transportation and communication technologies. But, as with most of Weber's work, it is difficult to disengage the effects of these factors. There is not only quite a bit of overlap when he talks about these issues, but he is also interested in explicating the preconditions for capitalism rather than determining a causal sequence. So, what we have are a number of social factors that overlap to create the bedrock out of which capitalism could spring, but did not necessarily cause capitalism. Picking up from the last section, we'll begin with religion.

The Religious Culture of Capitalism

The Protestant Ethic and the Spirit of Capitalism (1904–1905/2002) is probably Weber's best-known work. It is a clear example of his methodology. In it, he describes an ideal type of the spirit of capitalism, he performs a historical-comparative analysis to determine how and when that kind of capitalism came to exist, and he uses the concept of *verstehen* to understand the subjective orientation and motivation of the actors. Weber had three interrelated reasons for writing the book. First, he wanted to counter Marx's argument concerning the rise of capitalism—Weber characterizes Marx's historical materialism as "naive." The second reason is very closely linked to the first: Weber wanted to argue *against brute structural force* and argue for the effect that cultural values could have on social action.

The third reason that Weber wrote *The Protestant Ethic* was to explain why rational capitalism had risen in the West and nowhere else. Capitalism had been practiced previously, but it was traditional, not rational capitalism. In *traditional capitalism,* traditional values and status positions still held; the elite would invest but would spend as little time and effort as possible doing so in order to live as they were "accustomed to live." In other words, the elite invested in capitalistic ventures in order to maintain their lifestyle. It was, in fact, the existence of traditional values and status positions that prevented the rise of rational capitalism in some places. *Rational capitalism,* on the other hand, is practiced to increase wealth for its own sake and is based on utilitarian social relations. Though Marx didn't make distinctions, this is the type of capitalism he analyzed. It is the pursuit of money in order to use it as capital to gain more money.

Weber's argument is that there are certain features of Western culture that set it apart from any other system, thus allowing capitalism to emerge. As we talk about this culture, it is important to keep in mind that Weber is describing the culture of capitalism in its beginning stages. In many ways, the United States is now experiencing a form of late capitalism, and some of the spirit that Weber is describing has been lost to one degree or another. Also keep in mind that what he describes is an ideal type.

Weber's first task was to define the *spirit of capitalism*. The first thing I want you to notice is the word *spirit*. Weber is concerned with showing that a particular cultural milieu or mind-set is required for rational capitalism to develop. This culture or mind-set is morally infused: The spirit of capitalism exists as "an ethically-oriented maxim for the organization of life" (Weber, 1904–1905/2002, p. 16). This culture, then, has a sense of duty about it, and its individual components are seen as virtues.

Definition

The *spirit of capitalism* is a key factor in Weber's explanation of rational capitalism. The spirit of capitalism refers to the cultural values and beliefs that make rational capitalism possible: the belief that life should be rationally organized, that economic work is the most valued of all action, and that quantification is the true estimate of value and worth.

As the previous quote indicates, modern capitalism contains principles for the way in which people organize or live out their lives. For example, you woke up this morning and you will go to sleep tonight. In between waking and sleeping, you will live your life, but how will you use that time? From an even broader position, you were born and you will die. What will you do with your life? Does it matter? According to Weber, under capitalism, it does matter. How you live your life is not simply a matter of individual concern; the culture of modern capitalism provides us with certain principles, values, maxims, and morals that act as guideposts telling us how to live. In my reading of Weber, I see three such prescriptions in the spirit of capitalism: the value of money for its own sake, vocational calling and duty to work, and bottom-line legitimation.

Weber begins his consideration of the spirit of capitalism with a lengthy quote from Benjamin Franklin:

> Remember, that time is money . . . that credit is money . . . that money is of the prolific, generating nature. . . . After industry and frugality, nothing contributes more to the raising of a young man in the world than punctuality and justice in all his dealings. . . . The sound of your hammer at five in the morning, or eight at night, heard by a creditor, makes him easy six months longer. . . . [K]eep an exact account for some time both of your expenses and your income. (Franklin, as cited in Weber, 1904–1905/2002, pp. 14–15)

From these sayings, Weber gleans the first maxim of rational capitalism: Life is to be lived with a specific goal in mind. That is, it is good and moral to be honest, trustworthy, frugal, organized, and rational because it is *useful for a specific end:* making money, which has its own end, the acquisition of money, and more and more money. The culture of modern capitalism says that money is to be made but not to be enjoyed. Immediate gratification and spontaneous enjoyment are to be put off so that money can be "rationally used." That is, it is invested to earn more money. The making of money then becomes an end in itself and the purpose of life.

The second prescription is that each of us should have a vocational calling. Of course, another word for vocation is job, but Weber isn't simply saying that we should each have a job or career. The emphasis is on our attitude toward our vocation, or the way in which we carry out our work. There are two important demands to this attitude. First is that we are obligated to pursue work: We have a duty to work. In the spirit of capitalism, work is valued in and of itself. People have always worked. But generally speaking, we work to achieve an end. The spirit of capitalism, however, exalts work as a moral attribute. We talk about this in terms of a work ethic, and we characterize people as having a strong or weak work ethic. We can further see the moral underpinning when we consider that the opposite of a strong work ethic is being lazy. We still see laziness as a character flaw today. In the culture of capitalism, work becomes an end in itself, rather than a means to an end. More than that, it becomes the central feature of one's life, overshadowing other areas such as family, community, and leisure.

We not only have a duty to work, but we also have a duty *within* work. Weber (1904–1905/2002) says that "*competence and proficiency* is the actual alpha and omega" of the spirit of capitalism (p. 18). Notice the religious reference: In the New Testament, Jesus is referred to as the alpha and omega. Weber is again emphasizing that this way of thinking about work is a moral issue. Our duty within work is to organize our lives according to *scientific* vantage points. That is, under the culture of capitalism, we are morally obligated to live our lives rationally.

The rationally organized life is one that is not lived spontaneously. Rather, all actions are seen as stepping-stones that bring us closer to explicit and valued goals. Let me give you a contrary example to illustrate this point. A number of years ago, a (non-Hawaiian) friend of mine managed a condominium complex on the island of Maui. In the interest of political and cultural sensitivity, he hired an all-Hawaiian crew to do some construction. He tells of the frustrations of having crews come to work whenever they got up, rather than at the prescribed 8 AM. What's more, periodically during the day, the crew would leave at the shout of "Surf's up!" The work got done and it was quality work, but the Hawaiians he supervised did not organize their lives rationally. They lived and valued a more spontaneous and playful life. In contrast, most of us have Day Runners or smart phones that guide our lives and tell us when every task is to be performed in order to reach our lifetime goals.

The third prescription or value of the spirit of capitalism, according to Weber, is that life and actions are legitimized "on the basis of strictly quantitative calculations" (Weber, 1904–1905/2002, p. 35) Weber makes an interesting point with regard to legitimation or rationalization: Humans can rationalize their behaviors from a variety of ultimate vantage points. And we always do legitimize our behaviors. We can all tell stories about why our behaviors or feelings or prejudices are right, and those stories can be told from various religious, political, or personal perspectives. Weber's point here is that the culture of capitalism values quantitative legitimations. That is, capitalist behaviors are legitimized in terms of bottom-line or efficiency calculations. So, for example, *Roger and Me*, a film by Michael Moore (1989) depicts the closing of the General Motors (GM) plants in Flint, Michigan, resulting in the loss of over 30,000 jobs and the destruction of Flint's economy. The film asks about

GM's social responsibility; from GM's position, however, the closing was legitimated through bottom-line, financial portfolio management.

These cultural directives find their roots in Protestant doctrine and practice. But Weber doesn't mean to imply that these tenets of capitalist culture are religious—far from it. What we can see here is how culture, once born, can have unintended effects that are independent from its creating group. Protestantism did not directly produce capitalism; however, it did create a culture that, when cut loose from its social group, influenced the rise of capitalism.

The most important religious doctrine behind the spirit of capitalism is Martin Luther's notion of a calling. Prior to Luther and the advent of Protestantism, a calling was seen as something peculiar to the priesthood. Men could be called out of daily life to be priests and women to be nuns, but the laity was not called. Luther, however, taught that every individual can have a personal relationship with God; people didn't need to go through a priest. Luther also taught that individuals were saved based on personal faith. Prior to this, the church taught that salvation was a property of the church—people went to heaven because they were part of the Bride of Christ, the Church. With this shift to the individual also came the concept of a calling. If each person stood before God individually, and if priests aren't called to intercede, then the ministry belongs to the laity. Each person will stand before God on Judgment Day to give an account of what he or she did in this life. That means that God cares about what each individual does and has a plan for each life. God's plan involves a *calling*. Luther argued that every person in the Church is called to do God's will. So it is not simply the case that ministers are called to work for God—carpenters are called to work *as carpenters* for God as well. The work of the carpenter is just as sacred as the work of the minister or pastor. One calling is not greater than another; each is a religious service. This doctrine, of course, leads to a *moral organization* of life.

This idea of a calling was elaborated upon and expanded by John Calvin. Calvin took the idea of God's omniscience seriously: If God knows something, then God has always known it. Calvin also took seriously the "sin nature" of humanity. According to the sin nature doctrine, every human being is born in sin, reckoned sinful under Adam. That being the case, there is nothing anyone can do to save him- or herself from hell. We are doomed because of our very nature. Salvation, then, is from start to finish a work of God. Taking these ideas together, we come up with the Calvinistic idea of *predestination*. People are born in sin, there is nothing they can do to save themselves (neither faith nor good works), salvation is utterly a work of grace, and God has always known who would be saved. We are thus predestined to go to heaven or hell.

This doctrine had some interesting effects. Let's pretend you're a believer living in the sixteenth century under Calvin's teaching. Heaven and hell are very real to you, and it is thus important for you to know where you are headed. But there isn't anything you can do to assuage your fears. Because salvation is utterly of God, joining the Church isn't going to help; neither is being baptized or evangelizing. Confessing faith isn't going to help either, because you are either predestined for heaven or you are not. And you can't go by whether or not you "feel" saved—feelings were seen as promoting "sentimental illusions and idolatrous superstition" (Weber, 1904–1905/2002, p. 60).

"*Restless work in a vocational calling* was recommended as the best possible means to *acquire* the self-confidence that one belonged among the elect" (Weber, 1904–1905/2002, p. 66). Good works weren't seen as a path to salvation; rather, they were viewed as the natural fruit: If God has saved you, then your life will be lived in the relentless pursuit of His glory. In fact, only the saved would be able to dedicate their entire lives in such a way. The emphasis was thus not on singular works, but on an entire life organized for God's glory.

Thus, diligent labor became the way of life for the Calvinist. Every individual was called to a job and was to work hard at that job for the Lord. Even the rich worked hard, for time belonged to the Lord, and glorifying Him was all that mattered. Everything was guarded, watched, and recorded. Individuals kept journals of daily life in order to be certain that they were doing good works. Their entire lives became defined rationally and were thus systematically ordered. Further, if one was truly chosen for eternal salvation, then God would bless that individual, and the fruits of his or her labor would multiply. In other words, in response to your labor, you could expect God to bless you economically. Yet asceticism became the rule, for the world was sinful and the lusts of the flesh could only lead to damnation. The pleasures of this world were to be avoided. So the blessings of God were reinvested in the work to which God had called you. Taken together, then, the Protestant ethic commits each individual to a worldly calling, places upon her or him the responsibility of stewardship, and simultaneously promises worldly blessings and demands abstinence. This religious doctrine proved to be fertile ground for rational capitalism: a money-generating system that values work, rational management of life, and the delay of immediate gratification for future monetary gain.

Structural Influences on Capitalism

Thus, Weber argues that rational capitalism in the West found a seedbed in a culture strongly influenced by Protestantism. Yet there are other preconditions for the emergence of capitalism. I've illustrated all of these preconditions in Figure 5.4. They are divided into institutional, structural, and cultural influences, but these demarcations are not clear-cut. I've pictured them as rather shapeless, overlapping spheres, as that's how Weber talks about them. All of these processes mutually reinforce one another. With capitalism in particular, there is movement back and forth between culture and structure in terms of causal influence.

On the far left of the figure, you see the changes in institutions that were necessary for rational capitalism to arise. We've talked about Protestantism, and we'll cover the state shortly. These institutional shifts came along with and independently influenced various structural and cultural factors, which are depicted in the other two spheres. I spent more time explaining the cultural influences of Protestantism because they are more difficult to pick out; plus, by this time you have a good deal of experience thinking about social structures such as the state and these sorts of issues. But bear in mind that these effects are equally important, and you must understand how each of them works in order to comprehend the base for rational capitalism.

Figure 5.4 Preconditions of Capitalism

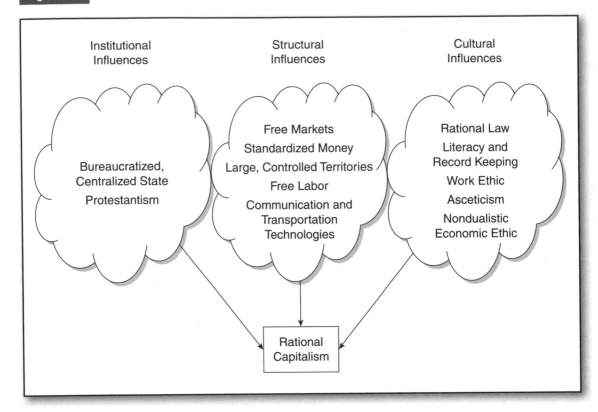

As we've seen, nation-states are relatively recent inventions. Up until the nineteenth century, the world was not organized in terms of nation-states. People were generally organized ethnically, with fairly fluid territorial boundaries. They didn't have nations as we think of them today. A *nation* is a collective that occupies a specific territory, has a common history and identity, and sees itself as sharing a common fate. The widespread use of the idea of a nation for organizing people was necessary for capitalism. Nations are responsible for controlling large territories, standardizing money, organizing social control, and facilitating free and open markets. All of these factors allow for easier exchanges of goods and services involving large populations. These kinds of exchanges are, of course, necessary for rational capitalism to exist. Labor also has to be free from social and structural constraints. Capitalism depends on a labor force that is free to sell its labor on the open market. Workers can neither be tied down to apprenticeships or guild obligations nor attached to land obligations as in feudalism.

Along with increases in communication and transportation technologies, nation-states and Protestantism helped form an objectified, rationalized culture, wherein written records were kept, people practiced a strong work ethic coupled with asceticism, and the traditional dualistic approach to economic relations broke down. This latter issue is particularly important for Weber. In traditional societies, groups usually had two different kinds of ethics when participating in exchanges: On one hand, there were restrictions having to do with ritual and fairness when dealing with group members, but also people outside the group could be exploited without measure. Both of these frameworks needed to be lifted in order for capitalism to flourish. All business and loans needed to be rationalized so that there could be continuity.

Concepts and Theory: Class, Authority, and Social Change

As we can see with capitalism, social change for Weber is a complex issue, one involving a number of variables coming together in indeterminate ways. Social change always involves culture and structure working together, and it involves complex social relations. We turn our attention now to stratification and change. Again, we will see the interplay of culture and structure as well as complex social categories.

Weber's understanding of stratification is more complex than Marx's. Marx hypothesized that in capitalist countries there will be only one social category of any consequence—class. And in that category, Marx saw only two roles or types: owners and workers. For Weber, status and power are also issues around which stratification can be based. Most important, class, status, and power do not necessarily covary. That is, a person may be high on one of those dimensions and low on another (like a Christian minister, typically high in terms of status but low in terms of class). These crosscutting life circumstances or affiliations can prevent people from forming into conflict groups and bringing about social change. For example, in the United States, a black man and a white man may both be in the poor class, but their race (status) may prevent them from seeing their life circumstances as being determined by similar factors.

Weber also argues that all systems are socially constructed and require people to believe in them. Marx did say that capitalism has a cultural component that holds it together. (If not for ideology, the proletariat would immediately overthrow the system.) True communism, however, requires no such cultural reinforcement because it corresponds to our species-being nature. For Weber, legitimation is the glue that holds not only society together, but also its systems of stratification. Thus, for Weber, issues of domination and authority go hand in hand. (In fact, Weber used just one German word to denote them both—*Herrschaft.*) Therefore, people must have some level of belief in the authority (culture) of those who are in charge, and they must cooperate with the system to some degree in order for it to work. Although the concept may sound a bit like an oxymoron, we could call this aspect of stratification cooperative oppression.

Class

Let's begin our consideration of these issues by taking a closer look at Weber's complex idea of stratification. Weber's definition of class is different from Marx's.

Marx defines class around the ownership of the means of production. Weber (1922/1968), on the other hand, says that

[A] class situation exists where there is a "typical probability of
   1. procuring goods
   2. gaining a position in life and
   3. finding inner satisfaction,
   a probability which derives from the relative control over goods and skills and from their income-producing uses within a given economic order. (p. 302)

In other words, Weber defines class as being based on your ability to buy or sell goods or services that will bring you inner satisfaction and increase your life chances (how long and healthfully you will live).

Weber also sees class as being divided along several dimensions as compared to Marx's two. While Marx acknowledges that there are more than two class elements, he argues that the other classes (such as the *lumpenproletariat* or petite bourgeoisie) become less and less important due to the structural squeeze of capitalism. Weber also speaks of two main class distinctions, yet they are constructed around completely different issues, each with "positively privileged," "negatively privileged," and "middle class" positions. The property class is determined by property differences, either owning (positively privileged) or not owning (negatively privileged). Rentiers (people who live off property and investments) are clear examples of the positively privileged (owners), whereas debtors (those who have more debt than assets) are good examples of the negatively privileged (not owners). The middle property classes are those that do not acquire wealth or surplus from property, yet they are not deficient either.

Definition   For Marx, *class* is the only structure that matters, and it is specifically defined by the ownership of the means of production. In capitalist societies, class bifurcates into owners (bourgeoisie) and workers (proletariat). For Weber, class is one of three systems of stratification, status and power being the other two. Weber also defines class more complexly than does Marx. According to Weber, one's class is defined by the probability of acquiring the goods and positions that are seen to bring inner satisfaction. Class position is determined by the control of property or market position, both of which may be positively, negatively, or medially possessed.

The commercial class is determined by the ability to trade or manage a market position. Those positively related to commercial position are typically entrepreneurs who can monopolize and safeguard their market situation. Negatively privileged commercial classes are typically laborers. They are dependent on the whim of the labor market. The in-between or middle classes that are influenced by labor market positions are those such as self-employed farmers, craftspeople, or low-level professionals who have a viable market position yet are not able to monopolize or control it in any way.

In Figure 5.5, I've given us a picture of Weber's ideas about class. We can see that there are two axes to class: property and market position. People can have a positive, negative, or middle position with respect to each of these issues. I've conceptualized this as a typology, because Weber spoke of people holding a position on both issues. I have also provided some contemporary examples in this typology. Today, those in upper management, such as CEOs, are not only paid large salaries, but also are given stock options that translate into ownership, which places them high in both the property and market dimensions. (They hold a monopoly with the skills they have.) For example, *Forbes* magazine (Forbes.com, 2003) lists Jeffrey C. Barbakow of Tenet Healthcare as having been the highest-paid American CEO in 2003. He received about $5.4 million in salary, plus he had stock gains of $111 million. Weber's typology of class also allows us to conceptualize certain kinds of knowledge as a resource or market position and thus a class indicator. Those who monopolize such skills and knowledge in our society include medical doctors and other such professionals. Stock owners as well as Weber's rentiers use the control of property to attain wealth, and we find them in the upper-right corner of the picture. Those who have little or no control over property and market are the poor.

Weber sees that holding a specific class position does not necessarily translate into being a member of a group. Unless and until people develop a similar identity that focuses on their class differences, they remain a statistical aggregate. In other words, the census bureau may know you are in the middle class, but you may not

Figure 5.5 Weber's Concept of Class

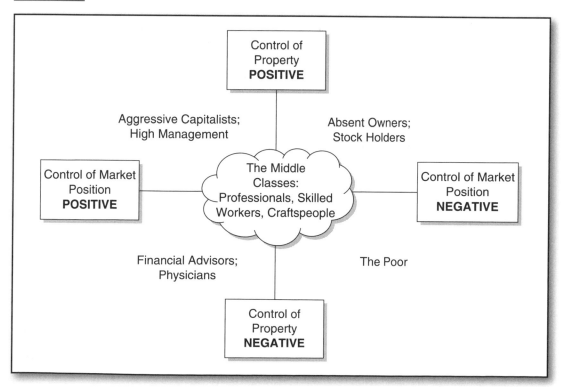

have a group identity around your class position. So, in addition to the property and commercial classes, Weber also talks about social class. His emphasis here seems to be on the social aspect—the formation of unified group identity. Weber identifies at least three variables for the formation of an organized social class:

1. *It is organized against immediate economic groups.* There must be some immediate market position or control issue, and the other group must be perceived as relatively close. Thus, in today's economy, workers are more likely to organize against management rather than ownership.

2. *Large numbers of people are in the same class position.* The size of a particular group can give it the appearance, feel, and unavoidability of an object.

3. *The technical conditions of organization are met.* Weber recognizes that organization doesn't simply happen; a technology is needed to organize people. People must be able to communicate and meet with one another, there must be recognized and charismatic leadership, and there must be a clearly articulated ideology in order to organize.

Keep in mind that these are all variables. A group may have more or less of any of them. As each of these increases, there will be a greater likelihood of the formation of a social class.

Status and Party

Weber sees stratification as more complex than class. Money isn't the only thing in which people are interested. People also care about social esteem or honor. Weber termed this issue status. *Status,* for Weber (1922/1968), entails "an effective claim to social esteem in terms of positive or negative privileges" (p. 305). Thus, status groups are hierarchically ranked by structural or cultural criteria and imply differences in honor and privilege.

Definition

Status is one of three systems of stratification in Weber's theory. Status is social honor or esteem that is hierarchically arranged. It is generally associated in modern societies with different levels of education or career, distinctive lifestyles, or family traditions and history. Examples of status groups include religion and race.

Status may be based or founded on one or more of three things: a distinct lifestyle, formal education, or differences in hereditary or occupational prestige. Being a music fan may entail a distinctive lifestyle—people who listen to jazz are culturally different from those who listen to rock. And there are positive and negative privileges involved as well: You can get a degree in jazz at my university but not in rock. Homosexuals can also be seen in terms of lifestyle status groups. One's education level can provide the basis of status as well. Being a senior at the university is better, status-wise, than being a freshman, and you have greater privileges as a senior as well. Professors are also examples of those who acquire prestige by education. They are good examples because they usually rate high on occupational prestige tables but low in class and power. Race and gender are also examples of status founded on perceived heredity. Gender is a social construction that is built upon biological distinctions (sex). Believing that heredity is destiny is one of the main ways that gender discrimination has been legitimated. So, it's the *perception* that gender is a matter of birth that helps gender inequality to continue. The really interesting thing is that even the sex categories we use are just a *perception* as well. Anne Fausto-Sterling, Brown University professor of biology, makes the case that there are at least five biological sex categories. So "seeing" only two is a social construction, and, thus, a perception.

Status groups maintain their boundaries through particular practices and symbols. Boundary maintenance is particularly important for status groups because the borders are more symbolic than actual. The *status* differences between jazz and rock music, for example, are in the ear of the listener. For example, at my university you could get a degree in music performance with an emphasis in jazz but not rock, even though some rock performances are just as demanding (if not more) than jazz, and both forms are based on the same earlier form: the blues. (You can't get a degree in blues performance either.) Generally speaking, Weber argues, we maintain the boundaries around status through marriage and eating restrictions, monopolizing specific modes of acquisition, and different traditions—all of which are described in greater detail next.

The norm of marriage restriction is fairly intuitive; we can think of religious and ethnic groups who practice "endogamy" (marriage within a specific group as required by custom or law). But eating restrictions may not be as evident. It may help us to realize that most religious groups have special dietary restrictions and practice ritual feasting. Feasts are always restricted to group members and are usually seen as actively uniting the group as one. (The traditional Jewish Passover Seder is a good example.) We can also think of special holidays that have important feast components, like Thanksgiving and Christmas here in the United States. And, if you think back just a few years, I'm sure you can recall times in junior high when someone your group didn't like tried to sit at your lunch table. For humans, eating is rarely the simple ingestion of elements necessary for biological survival. It is a form of social interaction that binds people together and creates boundaries.

We also maintain the symbolic boundaries around status groups through monopolizing or abhorring certain kinds and modes of acquisition, and by having certain cultural practices or traditions. The kinds of things we buy obviously set our status group apart. That's what we mean when we say that a BMW automobile is a status symbol. Status groups also try and guard the modes of acquisition. Guilds, trade unions, and professional groups can function in this capacity. And, of course, different status groups have different practices and traditions. Step concerts, pride marches, Fourth of July, Kwanzaa, Cinco de Mayo, and so on are all examples of status-specific traditions and cultural practices.

When sociologists talk about Weber's three issues of stratification, they typically refer to them as class, status, and power. However, power is not the word that Weber uses; instead, he talks about *party*. What he has in mind, of course, is within the sphere of power, "the chance a man or a number of men to realize their own will in a social action even against the resistance of other who are participating in the action" (Weber 1922/1968, p. 926). Yet, it is important to note that power always involves social organization, or, as Weber calls it, party. Weber uses the word party to capture the social practice of power. The social groups that Weber would consider parties are those whose practices are oriented toward controlling an organization and its administrative staff. As Weber puts it, a party organizes "in order to attain ideal or material advantages for its active members" (p. 284). The Democratic and Republican Parties in the United States are obvious examples of what Weber intends. Other examples include student unions or special interest groups like the tobacco lobby, if they are oriented toward controlling and exercising power.

Crosscutting Stratification

One of Weber's enduring contributions to conflict theory is this tripartite distinction of stratification. Marx rightly argues that overt conflict is dependent upon bipolarization. The closer an issue of conflict gets to having only two defined sides, the more likely that the conflict will become overt and violent. But Weber gives us an understanding of stratification that shows the difficulty in achieving bipolarization.

Every individual sits at a unique confluence of class, status, and party. I have a class position, but I also have a variety of status positions and political issues that concern me. While these may influence one another, they may also be somewhat different. To the degree that these issues are different, it will be difficult to achieve a unified perspective. For example, let's say you're white, gay, male, the director of human resources at one of the nation's largest firms, and Catholic. Some of these social categories come together, like being white, male, and in upper management. But other categories are not related, such as being gay and being Catholic. If this is true of you as an individual, then it is even more so in your association with other people. You'll find very few white, gay, male, Catholic upper managers to hang around with. And if you add in other important status identities, such as Southern Democrat, it becomes even more complex. Thus, Weber is arguing that the kind of conflict that produces social change is a very complex issue. Different factors have to come together in unique ways in order for us to begin to formulate groups and identities capable of bringing about social change.

Enduring Issues

Social stratification and inequality are two of sociology's primary sub-fields. Yet the more we study it, the more complicated it appears to be. Of course, Weber was the first to point out the complexity of inequalities in modern society. More recently, Patricia Hill Collins (2000) uses the idea of *intersectionality*. The basic idea is that systems of race, gender, class, nationality, sexuality, and age crisscross one another at specific social locations that form a matrix of domination. The analytical issue, then, is to empirically discover how these various systems come together for any specific person or group. Collins is specifically concerned with how these systems shape black women's experiences and are, in turn, influenced by black women.

Authority and Social Change

Because of Weber's emphasis on subjective meaning, he was able to see that all systems of stratification or power are based on legitimation. *Legitimation* refers to the process by which power is not only institutionalized, but also is given moral grounding. Legitimations contain discourses or stories that we tell ourselves that make a social structure appear valid and acceptable. The strongest legitimations will make social structure appear inevitable and beyond human control—a good contemporary example is the use of religion to legitimate denying marriage rights to homosexuals. A legitimated order creates a unified worldview and is based on a complex mixture of two kinds of legitimations: subjective (internalized ethical and religious norms) and objective (having the possibility of enforced sanctions from the social group [conventions] or an organizational staff [law]). Weber indicates that subjective legitimacy is assumed in the presence of the objective.

The function of legitimation is to render power into authority. Part of the reason behind this need is the cost involved in the use of power. If people don't believe in authority to some degree, they will have to be forced to comply through coercive power. The use of coercive power requires high levels of external social control mechanisms, such as monitoring (you have to be able to watch and see if people are conforming) and force (because they won't do it willingly). To maintain a system of domination not based on legitimacy costs a great deal in terms of technology and manpower. In addition, people often ultimately respond to the use of coercion by either rebelling or dropping out entirely—the end result is thus contrary to the desired goal.

Definition

Legitimation is a theoretical concept Weber uses to describe the effects that specific stories, histories, and myths have in granting ethical or moral authorization to social power and relations. Legitimations are both *subjective* (existing within the self of the individual because of socialization) and *objective* (available to everyone "like us" and enforceable through informal and formal norms). Legitimations turn coercive force into acceptable authority.

Authority, on the other hand, implies the ability to require performance that is based upon the performer's *belief in the rightness of the system.* Because authority is based on socialization—the internalization of cultural norms and values—authority requires low levels of external social control. We can thus say that any structure of domination can exist in the long run if and only if there is a corresponding culture of authority. Weber identifies three ideal types of authority. (Keep in mind that these are ideal types and may be found in various configurations in any society.) These ideal types of authority are as follows:

1. *Charismatic authority*—belief in the supernatural or intrinsic gifts of the individual. People respond to this kind of authority because they believe that

the individual has a special calling. (Examples of people who convey this type of authority include Susan B. Anthony, Adolf Hitler, Martin Luther King Jr., John F. Kennedy, Golda Meir, and Jesus. Notice that it is people's belief in the charisma that matters; thus, we can have Hitler and Jesus on the same list.)

2. *Traditional authority*—belief in time and custom. People respond to this kind of authority because they honor the past and they believe that time-proven methods are the best. (Good examples of this type of authority are your parents and grandparents, the Pope, and monarchies.)

3. *Rational-legal authority*—belief in procedure. People respond to this kind of authority because they believe that the requirements or laws have been enacted in the proper manner. People see leaders as having the right to act when they obtain positions in the procedurally correct way. (A good example of this type of authority is that of your professor—it does not matter who the professor is, as long as he or she fulfills the requirements of the job.)

These diverse types of authority interact differently in the process of social change. According to Weber, the only kind of authority that can instigate social change is charismatic. Traditional and rational-legal authorities bring social stability—they are each designed to maintain the system. Charismatic individuals come to bring social change, yet charismatic authority is also inherently precarious. Because charisma is based on belief in the special abilities of the individual, every instance of charismatic authority will fail when that person dies—the gifts die with the person. Thus, every charismatic authority will someday have to face the *problem of routinization* (making something routine and thus predicable). Every social movement based upon charismatic authority—and Weber argues that all social movements are—routinizes the changes by either using traditional authority or rational-legal authority.

Definition

Routinization is the process through which something is made habitual or routine. The concept is most closely associated with Weber's theory of bureaucratization—the purpose of bureaucracy is to routinize actions. The problem of routinization, however, is somewhat different. It occurs after the charismatic leader dies. Because charismatic authority is rooted in an individual, when that individual is gone, the authority must be made routine, either through traditional or rational-legal authority.

The case of the Christian religion might give us insight. Jesus was a charismatic leader. When he died, the movement was faced with the problem of continuing his leadership. Over time, the routinization of Christ's authority took two forms, one in Catholicism and the other in Protestantism. All authority in the Catholic Church is derived from the Pope, which is a form of traditional authority based on a line of succession going back to the Apostle Peter. Protestant denominations don't have such a line of succession, and most denominations require educational credentials (rational-legal authority) to enter the ministry. In actuality, things are a bit more complex because the Catholic Church requires priests to be educationally accredited, and both Christian forms tend to see their pastors as called (charismatic authority) by God. Here is where the power of ideal types becomes clear. Because ideal types aren't hampered by the contingencies or complexities of real life, they give us an abstract and objective way of understanding and comparing actual instances. While both Catholics and Protestants say their pastors are called by God, we can confidently claim that neither are pure incidences of charismatic authority because the ideal type is based solely on the individual's gifts and calling; plus, charismatic authority brings change, and in neither case do we see that as a defining feature. We can also say that the core of Catholic authority rests in tradition, whereas the core of Protestant authority does not. Weber understood the complexities of modern life like few before him and he used ideal types to make sense of the multifaceted existence that is ours.

Combining Weber's ideas of class, status, and party with his argument concerning authority and social change, we can put together a theory of conflict and social change. I've depicted this theory in Figure 5.6; I encourage you to follow the model as I describe each of the variables. From Weber's point of view, the possibility of social change depends on how much people question the legitimacy of the stratification system. Questioning begins in response to the two factors listed in the top box in Figure 5.6. People are more likely to question the system when there are clear breaks between stratified groups (*perceived group boundary*) or when the three scarce resources, class, status, and power, all appear to belong to a single social group (*degree of correlation*).

As legitimacy is questioned, *the technical conditions of group conflict* tend to be met. These include charismatic leadership, clearly articulated goals and ideology, and the ability to meet and communicate with one another. How much *social change* results depends on the level of each of the technical conditions. In other words, as the group becomes organized around a set of goals, and has access to places to meet and communication technologies, and as group members are motivated by charismatic leadership and inspiring ideology, the possibility of significant social change increases. Notice that greater social changes will tend to create greater needs for *routinization*. Routinization can be achieved through traditional or rational-legal authority; but given the level of bureaucratization in modern society, there will be greater pressure for rationally, legally established routines.

Notice that Figure 5.6 is circular. The reason for this is that Weber argues that routinization will in the long run lead to a system of stratification around class, status, and power that again sets up the conditions for conflict and change. Marx's dialectical theory has an end point: communism. It's utopian in that sense and is based on a specific view of human nature. Weber, however, makes no such assumption. He thus sees conflict and change as constant, universal features of every human collective. For Weber, social systems will forever move cyclically through routinization and charismatic change.

Figure 5.6 Weber's Theory of Social Change

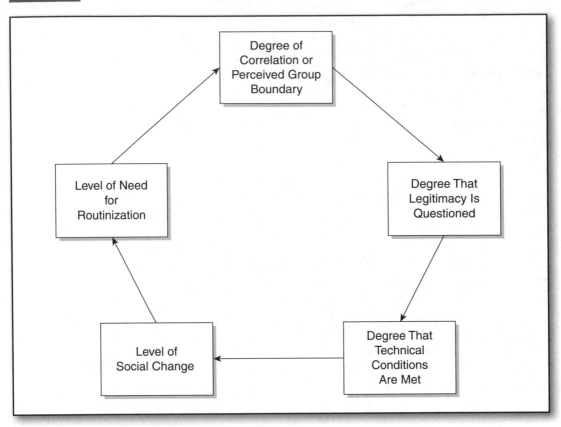

Concepts and Theory: Rational-Legal Organization

To be human is to be social, and to be social is to be organized. Thus, organization is primary to what it means to be human. Human beings have been organized over time by different social forms. These different organizational forms have clear consequences for us, not only socially but personally as well. In the not-too-distant past, we organized much of our lives around affective (emotionally based) systems, such as kinship. Kinship linked people through blood and marriage. In such organizations, people saw each other in terms of familial obligations and rights, which they felt as emotional ties. But today, we are organized through mostly bureaucratic means. From the time you were put in school at the age of 5 or 6, bureaucracies have been organizing your life. You know how to stand in line, how to use time and space, and how to relate to people (the grocery clerk, the teacher, the minister, the insurance agent) as a result of spending most of your life in a bureaucracy.

Bureaucracies tend to rationalize and routinize all tasks and interactions. Tasks and interactions are rationalized in terms of means-ends efficiency. They are routinized in the sense that they may be carried out without thought, planning, or dependency on individual talents. Thus, each person and task is met in the same efficient and equal manner; personal issues and emotions do not have a place in a bureaucracy. While bureaucracies have been around for quite awhile, for much of their history, they were not the primary way in which humans were organized. Certain social factors came about that simultaneously pushed out affective systems of organization (traditional) and set the stage for systems based more on reason than emotion. As these different features lined up, they created an environment ripe for rational organization.

Definition   *Bureaucracy* is a system of organizing people and their behavior that is characterized by the presence of written rules and communication, job placement by accreditation, expert knowledge, clearly outlined responsibilities and authorities, explicit career ladders, and an office hierarchy. The purposes of the bureaucratic form are to rationalize and routinize behavior. Unintended consequences include the bureaucratic personality and the iron cage of bureaucracy.

According to Weber (1922/1968, pp. 217–226; 956–958), there are at least six factors that pushed for bureaucratic organization in modernity; they don't necessarily occur in any specific order. These include increases in the following:

1. The size and space of the population being organized

2. The complexity of the task being performed

3. The use of markets and the money economy

4. Communication and transportation technologies

5. The use of mass democracy

6. The volume of complicated and rationalized culture

All these factors create needs for more objective and rational social relations and culture, thus driving a society to use rational-legal authority and bureaucracy almost exclusively. As society became larger and spread out over vast spans of geographic space, it became increasingly difficult to use personal relationships as a method of organization. However, increases in population size and geographic space aren't sufficient to make bureaucratic organization necessary. For example, China was able to primarily use an elaborate kinship system rather than bureaucracy to organize most of the population's behaviors for many years. Other factors are necessary to push a society toward bureaucratic organization, such as increasing complexity of the task being performed. In England and elsewhere, this occurred through the Industrial Revolution, urbanization, and high division of labor. The increasing use of money and open markets for exchange also increased the need for rationalized and speedy calculations, thus adding to rational culture.

Increasing levels of communication and transportation technologies created the demand for faster and more predictable reactions from the governing state. Insightfully, Weber (1922/1968) says that with "traditional authority it is impossible for law or administrative rule to be deliberately created by legislation" (p. 227). Remember that traditional authority is based on history and time, so the only way a new rule or law can be legitimized through tradition is to say that it is according to the wisdom of the ages. Rational-legal authority, on the other hand, can create new rules simply because it is expedient to do so. Thus, as states came into more frequent contact with one another and dealt with more complex problems, it became necessary to respond quickly to new situations with new rules, thus pushing forward a rational-legal authority and bureaucracy.

There are two additional forces that created demands for rational and objective standards. The first was democracy for the masses, rather than simply the elite, which created demands for equal treatment before the law. Initially, the idea of democracy was limited to the educated and powerful. Including the masses created the demand for equal treatment regardless of power or prestige. This, of course, led to the idea that rules and laws should be blind to individual differences. The other additional social factor that pushed for bureaucratic organization was the increase in complex and rational culture. Due to the increases in knowledge that came with science and technology, a new social identity came into existence and rose to prominence—the expert, bureaucracy's manpower.

Ideal-Type Bureaucracies

Weber gives us an ideal type for bureaucracy. Remember that this ideal type is not intended to tell us what the perfect bureaucracy should look like; rather, Weber uses the ideal type as an objective yardstick against which we may measure different subjective and cultural states. Weber talks about bureaucracy in a number of different places. But if we combine his lists, we come up with six important features to the ideal type of bureaucracy: an explicit division of labor with delineated lines of authority, the presence of an office hierarchy, written rules and communication, accredited training and technical competence, management by rules that is emotionally neutral, and ownership of both the career ladder and position by the organization rather than the individual.

Each of these characteristics is a variable; organizations will thus be more or less bureaucratized. For example, the first job I ever had was with Pharmaseal Laboratories. I worked as a lead man, which meant I kept the production lines stocked with raw materials and moved the finished product to the warehouse. After about 6 months of working there, I was promoted to management. During my week-long orientation, I was told about AVOs (Avoid Verbal Orders) that documented in triplicate any communication or disciplinary action. I was introduced to the human resources department and was shown the employee files and where each of those AVO copies went. I was told of my span of control, given an organizational chart of the company, and received a complete job description for my position and the that of every person with whom I would have business. I was also informed about the company's internal promotion policy (the career ladder)—to get past the level I was at would require additional schooling and credentialing.

On the other hand, my wife's experience at a local firm indicates a more patrimonial and less bureaucratized organization. When she first arrived at this company, she asked for the organizational chart and job descriptions. She also wanted to know exactly what her span of control was and to whom she reported. But what she found was a 70-person company that was run on personal relationships. Most of the communication was not documented, there was no organizational chart or policy defining the span of control or communication, and there were no job descriptions. Thus, there are differences in the level of bureaucratization, even in a society such as ours. However, that company has changed over the years and has become more bureaucratized. These changes are due to increasing pressures for rationalization and objectification, coming from much the same preconditions of which Weber initially spoke.

Effects of Bureaucratic Organization

Contemporary theorists point out that living in a society that organizes through bureaucracy can produce the *bureaucratic personality*. There are at least four characteristics of this kind of temperament. First, individuals (like you) tend to live more rationally due to the presence of bureaucracy, and not just at work. People generally become less and less spontaneous and less emotionally connected to others in their lives. They understand goals, the use of time and space, and even relationships through rational criteria. Second, people who work in bureaucracies also tend to identify with the goals of the organization. This varies with a person's position in the organization. Workers at levels that are less bureaucratized tend to complain about the organization; on the other hand, managers, who exist at more bureaucratized levels, tend to support and believe in the organization. Again, this isn't something that we just put on at work—we *become bureaucratized.*

Third, because bureaucracies are based on technical knowledge, people in bureaucratic societies tend to depend on expert systems for knowledge and advice. In traditionally based societies, people would trust the advice of those they loved, those who had extensive experience, or those who stood in a long lineage of mentoring and discipleship. Conversely, in societies like the United States, we look to those who have credentials to help us. The sense of mutual dependency of small village life has been uprooted and sterilized of all social connections. Further, honor in modern society is given to those with credentials; age and experience are of no consequence. And, fourth, bureaucracies lead to sequestration of experience. By that I mean that different life experiences are separated from one another, such as dying being separated from living. In traditional society, life was experienced holistically. People would see birth, sickness (emotional, mental, and physical), and death as part of their normal lives. In the same home, children were born and grandparents died. Today, most of those experiences are put away from us and occur in bureaucratic settings where we don't see them as part of our normal and daily life; such settings include hospitals, rest homes, sanatoriums, and so on. The world has thus become tidy, clean, and rational.

Society is affected as well by bureaucracy. There are two main effects. The first is the *iron cage of bureaucracy*. Once bureaucracies are in place, they are virtually inescapable and indestructible for several reasons: They are the most efficient form of organizing large-scale populations, they are value-free, and they are based upon expert knowledge. We've just seen that individuals within a rationalized society become more and more dependent upon expert knowledge; the same is true with leaders. Regardless of whether the leaders are in charge of political, religious, or economic organizations, they become increasingly dependent upon rationally trained personnel and expert knowledge in the bureaucratic information age; and experts tend to share their knowledge only with other experts or experts in training, which is what graduate school is all about. Further, one of the qualities of a professional is self-administration, which means that bureaucratic experts become a self-recruiting and self-governing class, existing apart from any other organizational control. Thus, neither the people ruled, nor the rulers, nor the experts can escape the domination of the bureaucratic form. We saw the effects of the iron cage in our earlier discussion about participatory democracy. That sort of thing is why many people speak of the irrationality of rationality. One of the forces that brought about widespread bureaucratization was the need for equal treatment and democracy—yet bureaucracy is one of the forces that now disables democracy.

Definition

The *iron cage of bureaucracy* is an important effect in Weber's theory of bureaucracy and rationalization. There are two ways the idea is used. First, once in place, bureaucracies are difficult if not impossible to get rid of: "Bureaucracy is the means of transforming social action into rationally organized action" (Weber 1922/1968, p. 987). Second, while in the beginning of modernity people strove to have a rationally controlled life, the pervasive presence of bureaucratic organization has forced people to rationally control life, to become "narrow specialists without mind" (Weber 1904–1905/2002, p. 124).

There is yet another way that bureaucracies act as an iron cage. Bureaucracies are value-free, which means they can be used for any purpose, from spreading the gospel to the eradication of ethnic minorities. This implies that bureaucracies are quite good at co-optation. To *co-opt*, in this context, means to take something in and make it part of the group, which on the surface might sound like a good thing. But because bureaucracies are value- and emotion-free, there is a tendency to downplay differences and render them impotent. For example, one of the things that our society has done with the race and gender social movements of the 1960s is to give them official status in the university. One can now get a degree in race or gender relations. Inequality is something we now study, rather than the focus of social movements. In this sense, these movements have been co-opted.

The second main social effect bureaucracies have is *credentialing*. A position within a bureaucracy, you'll remember, is achieved through diplomas and certification. Who you are and whom you know aren't supposed to determine your office or job placement in a bureaucracy; expert training as evidenced by certification determines that. That being the case, society needs a legitimated process through which credentials can be conferred. The United States has chosen to use the education system. But one of the effects of that decision is that education is no longer simply about learning, nor is it seen as part of the democracy system. The education system is used to credential technical expertise rather than cultivate an informed citizenry. Universities are thus becoming populated by professional schools—the school of business, nursing, social work, computer technology, criminal justice, and so on—and there is mounting tension between the traditional liberal arts and these professional schools. Many students express this tension (and preference for credentialing) when they ask, "How will this course help me get a job?"

The emphasis on credentials coupled with the American view of mass education and the use of education to give credentials has created *credential inflation*. Every year, more and more people are going to college to get a degree so that they can be competitive in the job market. That means that there are more college graduates competing for jobs, which in turn allows companies and organizations to increase their demands for credentials. The result is that your college degree will gain you entrance to a job that 40 years ago only required a high school diploma. These effects are expansive. Let me give you a couple of examples. My sister began her career as a social worker when many caseworkers did not have a bachelor's degree; she was hired with a degree in home economics. Today, of course, a bachelor's in social work is required at the entrance level of the profession, and to be a manager requires a master's in social work. My other example is a friend with whom I went to graduate school. He got his PhD in biology, but couldn't find a job—there were too many biology PhDs around. So, he went back to school and did a 2-year postdoctorate degree, hoping that the additional credential would help. It didn't. After another failed job search, he went back to school for his MD.

Thinking About Modernity and Postmodernity

The chief characteristic of modernity for Weber is the progression of rationalization. Due to such social processes as Protestantism, state formation, markets, money, democracy, population size, the complexity of tasks, and the level of abstract and rational knowledge, modern societies are organized more in terms of rational-legal than traditional authority. Thus, bureaucracy is the chief organizational form in modernity, increasing the levels of rationalization and routinization in society at large, as well as creating an iron cage for society and the bureaucratic personality. In many ways, then, Weber was disillusioned about modernity.

Weber would also argue that modern societies are complex entities. Class remains important, as Marx would agree, but modern societies are also noted for their differing and intricate status and power relations. Weber's emphasis on status comes, of course, from the place he gives culture in his analysis. Weber understands that humans are symbolic as well as material creatures. As such, we seek to be gratified through symbols and not just money. Weber also sees that social control is wrapped up in authority—the belief in the rightness of any system of domination.

Thus, the exercise of power is based on cultural issues as well. Together, class, status, and power create a complex network of social positions and societal control.

Weber was always interested in the relation between structure and culture and the interplay among class, status, and power. You remember that he saw culture leading change at one time and structure at another. Most often, history is the result of the dialogue between these two forces. Within that dialogue, the relative importance of and interaction among class, status, and power are played out. Social change can occur as the three are seen to correlate, or if the legitimacy of one or the other stratification system (class, status, or power) is questioned.

It is quite possible that as societies move toward postmodernity, the issues and relative importance of class, status, and power will change significantly. There are many factors that produce issues that influence these relations, but let's take a look at two of them: the increasing importance of symbolic gratification and life politics. Lash and Urry (1987), who conduct a Weberian analysis, argue that symbols and symbolic gratification become increasingly important for people (1) as the size of the working class and its ability to organize decrease, (2) as the size of the professional and service classes increases, and (3) as ready-made (as opposed to socially emergent) cultural images and their meaning constitute a significant and growing social reality. In back of these three elements are what Lash and Urry term "disorganized capitalism" and the proliferation of media images.

The media we can understand, and I'll come back to it in a moment, but disorganized capitalism needs some defining. *Organized capitalism,* the kind that Weber calls rational capitalism, is characterized by increasing concentration of the means of production, distribution, and social reproduction. This type of capitalism brought large corporate and regional headquarters, massive factories, central distribution points, and workers living in concentrated areas within urban spheres. *Disorganized capitalism,* on the other hand, entails the deconcentration of the means and administration of production, commercial capital, collective consumption, and residential concentration of labor power. Capitalism has moved from large national corporations to larger international corporations to multinational conglomerates. Along with the movement of capital comes the movement of the workforce—primary production moves from core nations to peripheral or developing nations. Countries such as the United States are becoming less based on production and more based on consumption. In such consumption economies, identities around class are increasingly difficult to achieve. And, as capitalist relations have moved and become decentralized, the significance of large collectives (such as corporations, workplaces, and cities) has a diminished role in the process of identity formation. Markets and media fill this void and provide a diversity of identities from which the individual may choose.

These shifts in the economy also create demands for greater education and extended periods of unstable identity. We can think of the period between childhood and adulthood as being indeterminate and thus unstable. Between those epochs in one's life, there are few institutional obligations. You're no longer utterly dependent, yet at the same time, you're not responsible for the institutions of society. In traditional societies, there was little time between being a child and being an adult. In fact, children were seen simply as little adults. However, as the need for cultural training and socialization increased, so did the time before one would take on the responsibilities of adulthood. During these periods of instability, people are prone to playing with symbols and identities. For example, look around your college campus at the diverse array of identities and symbols, and then consider a large corporation like IBM. What you will see is that the kinds of identities and the diversity of cultural symbols that IBM tolerates are very, very low when compared to the university.

The extended period between childhood and being an adult is, of course, due to the increasing demand for education and credentials (effects of bureaucracy) in a postmodern society. The professional and service sectors are based on knowledge and credentials, not on work skills. Credentialing is seen by some postmodernists as increasing the power of culture—because differences between people and access to jobs, social standing, status groups, and other valued social niches are determined symbolically, not materially. Moreover, credentialing expands the educational system, which in turn produces new sets of symbols that elevate the overall level of cultural production in a society.

Along with these changes, there has been a shift from emancipatory politics to life politics (see Giddens, 1991). *Emancipatory politics* is concerned with liberating individuals and groups from the constraints that adversely affect their lives. In some ways, this was the theme of modernity—it was the hope that democratic nation-states could bring equality and justice for all. On the other hand, *life politics* is the politics of choice and lifestyle. It is not based on group membership and characteristics, as is emancipatory politics; rather, it is based on personal lifestyle choices.

We have come to think of choice as a freedom we have in the United States. But it is more than that—it has become an obligation. Choice is a fundamental element in contemporary living, due in part to the media (all print and visual forms of communication). Think of it this way: The basic reality for people is found in social events—face-to-face interactions. If I have an experience with Duke Power Company—let's say they cut down the trees in my yard without my permission—and I tell you about it, the telling of it is once removed from the actual event. And if I were to tell the story, I would emphasize those elements that back up my concerns, rather than those of the power company. Now let's think about what the media does. If what is presented actually occurred in real life, by the time we see it in the media it has been removed many times over from the incident, especially if we are talking about global rather than local events. Each one of these levels is subject to its own politics and interpretations. That's why we can talk about the conservative or liberal media. But much of what we are presented with in the media is pure fiction or advertising, which further removes it from embedded, social reality.

In lifting out experience from its social embeddedness in face-to-face interaction, the media present people with a vast array of diverse lifestyles and cultures. The plurality of lifestyles presented to an individual not only allows for choice, but also necessitates choice. In other words, what becomes important is not the hard issue of group equality, but rather, the insistence on personal choice. In addition to mediated experience, the plurality of choice is also a function of a number of other influences, such as the deinstitutionalization of expert oversight (think of the commercials telling us not only how to diagnose our illness, but also what kind of medicine we need), the pluralization of life-worlds (there are many potential realities within our grasp), and the legitimation of doubt (science and education tell us nothing is certain).

What is at issue in this milieu is not so much political equality (as with emancipatory politics) as inner authenticity. In a world that is perceived as constantly changing and uprooted, it becomes important to be grounded in oneself. Life politics creates such grounding. It creates "a framework of basic trust by means of which the lifespan can be understood as a unity against the backdrop of shifting social events" (Giddens, 1991, p. 215). In some ways, we can see life politics as an extension of the feminist idea that "the personal is political."

A good example of life politics is veganism—the practice of not eating any meat or meat by-products. Not only is eating flesh avoided, but also any products with dairy or eggs. In addition, fur, leather, feathers, or any goods whose production involves animal testing are eschewed. In the words of one website for vegans (www.vegan.org/about_veganism/index.html), veganism "is an integral component of a cruelty-free lifestyle." It is a political statement against the exploitation of animals, and for some it is clearly a condemnation of capitalism—capitalism is particularly responsible for the unnatural mass production of animal flesh as well as commercial animal testing. Yet for most vegans, it is simply a lifestyle, one that brings harmony between the outside world and inner beliefs, and not a collective movement to bring about social change.

What we can see through this discussion is that in postmodernity, economic class becomes less important, while cultural issues become more important. Further, status and party are not single, focused social identities around which groups can form. Rather, status identities for more and more of the population have become like playful masks that we can put on and take off at will. Yet, the importance of status in identity formation has exceeded that of class, at least in terms of this way of viewing the world. Cultural signs, symbols, and images, as well as symbolic gratification, have become paramount. For many, politics has moved from group considerations of equality and political participation to individual practices designed to create a sense of personal authenticity in a shifting landscape. In terms of the bipolarization of conflict and the probability of social change, we can see that the effects of bureaucracy and Weber's notion of crosscutting influences become more and more prominent as we move deeper into postmodernity.

Summary

- To think like Weber is to take seriously the ramifications of culture. Weberians focus on the historical, cultural, and social contexts wherein the subjective orientation of the actor takes place. To think like Weber, then, means to use ideal types and *verstehen* to explain how these contexts came to exist rather than others. To think like Weber also means paying attention to the process of rationalization and the need for legitimation.
- Religion began with the movement from naturalistic to symbolic ways of seeing the world. The movement toward symbolism and religion was influenced

by increases in economic technologies and professionalization. Religion was initially practiced in kinship-based groups with local deities. These local gods became hierarchically organized into pantheons due to the political organization of kinship groups into larger collectives, and the abstracting and rationalizing effects of the professional priesthood. Eventually, these same forces produced the idea of a monotheistic god under which all the other gods were subsumed and finally disappeared. Monotheism became ethical monotheism in response to the need of polity to control behavior on a large scale.

• The cultural foundations of rational capitalism were built by Protestantism. This religious movement (through the doctrines of predestination and asceticism, and the idea of a calling) indirectly created a rationalizing, individuating culture wherein money could be made for the purpose of making more money, rather than for immediate enjoyment. The establishment of nation-states structurally paved the way for rational capitalism by creating a free labor force, controlling large territories, standardizing money, and protecting free global markets.

• Social stratification is a complex of three scarce resources: class, status, and power. These three systems produce crosscutting interests that make social change difficult and multifaceted. Large-scale social changes become increasingly likely only as class, status, and power are seen to correlate; the legitimacy of the system is questioned; and the technical conditions of organization are met. Since social change is led by charismatic authority, each change will need to be routinized through traditional or rational-legal authority, which, in the long run, will once again set up conditions for conflict and social change.

• Bureaucratic forms of organization became prominent as societies became larger and more democratic, as tasks and knowledge became more complex, as communication and transportation technologies increased, and as markets became more widespread through the use of money. The extent of bureaucratic organization can be measured through an ideal type consisting of six variables: explicit division of labor, office hierarchy, written rules and communication, accreditation for position, affectless (without emotion or emotional connection) management by rule, and the ownership of career ladders and position by the organization. The use of bureaucracy as the chief organizing technology of a society results in the bureaucratic personality, the iron cage of bureaucracy, and social emphasis on credentials.

• A Weberian understanding of postmodern society focuses on the complex relationships among class, status, and party, with particular emphasis on the rise of the professional or credentialed class. Weber explicitly argues against a simple Marxian understanding of class and social change. For Marx, class in modern society would eventually devolve to only two classes that would face off over the means of production. Weber argues that social change is more complex, with crosscutting influences from multiple class levels along with diverse status groupings and political parties. Postmodern theorists, such as

Lash and Urry, argue that the class and status structures have become even more complex than Weber imagined, due to disorganized capitalism. Lash and Urry posit that postmodern culture takes on increasing importance due to disorganized capitalism, particularly for middle-class youth and for those in the service class. This shift in class structure toward symbolic gratification is coupled with increasing differentiation of cultural images presented through the mass media to create an emphasis on life politics rather than emancipatory politics. Increasing differentiation of the class and status structures, along with the proliferation of media images, makes it increasingly difficult for political groups to effectively form. Thus, personal choice has become more important than political change.

BUILDING YOUR THEORY TOOLBOX

Learning More—Primary and Secondary Sources

For primary sources for Weber's work, I suggest starting off with *The Protestant Ethic and the Spirit of Capitalism*. It's one of his most accessible books. After that, move on to the following:

- Weber, M. (1948). *From Max Weber: Essays in sociology* (H. H. Gerth & C. Wright Mills, Trans. & Eds.). London: Routledge and Kegan Paul.

- Weber, M. (1968). *Economy and society* (G. Roth & C. Wittich, Eds.). Berkeley: University of California Press. (Originally published 1922)

- Weber, M. (1988). *Max Weber: A biography* (H. Zohn, Trans.). New Brunswick: Transaction Books. (Definitive biography, written by Weber's wife)

There are a number of very good secondary sources for Weber's work; I suggest the following:

- Bendix, R. (1977). *Max Weber: An intellectual portrait.* Berkeley: University of California Press. (The standard Weber reference)

- Collins, R. (1986). *Weberian sociological theory.* New York: Cambridge University Press. (Systemization of Weber's theory by a prominent contemporary theorist)

- Turner, B. S. (1992). *Max Weber: From history to modernity.* New York: Routledge. (Explanation of Weber's theories of modernity)

- Turner, S. P. (2000). *The Cambridge companion to Max Weber.* New York: Cambridge University Press. (Excellent resource concerning Weber's works and influence)

Seeing the Social World (knowing the theory)

- Write a 500-word synopsis of Weber's sociological imagination, making certain to include not only how he sees the world (perspective), but also how that perspective came about (biography, historical period, intellectual influences).

- After reading and understanding this chapter, you should be able to define the following terms theoretically and explain their theoretical importance to Weber's theory: ideal types, *verstehen,* action, rationalization, magic and religion, professionalization, symbolism, traditional and rational capitalism, spirit of capitalism, bureaucracy, bureaucratic personality, instrumental-rational action, value-rational action, traditional action, affective action, credentialing, class, status, power, crosscutting stratification, legitimation, authority (charismatic, traditional, rational-legal), routinization, bureaucracy.

- After reading and understanding this chapter, you should be able to answer the following questions (remember to answer them theoretically):

   1. Explain how religion evolved to ethical monotheism.

   2. Explain how capitalism came to exist.

   3. Describe how bureaucracies came to be the central organizing feature of modernity. What are the elements of Weber's ideal type? How do bureaucracies impact you and society?

   4. Explain Weber's theory of stratification and social change.

Engaging the Social World (using the theory)

- One of the central themes in Weberian theory is rationalization (as a contemporary example, see George Ritzer's *The McDonaldization of Society*). Take a look at your life: In what ways has rationality influenced you? Do you think your life is more or less rationalized than your parents' was at your age? Let's take this one step further. Go to a place of business, like a fast-food restaurant or mall, and observe behaviors for at least 2 hours. How rationalized were the actions you observed? Overall, do you think that life is becoming increasingly rationalized? Why or why not? What are the benefits and drawbacks to rationalization? How do you think Weber felt about this process?

- Get a sense of the kinds of jobs you can get today with a college education and the jobs available with the same education 50 years ago. You can do this by using your Internet search engine, going to http://nces.ed.gov/ and searching the data, or by asking your parents and grandparents. Explain your findings using Weber's theory. What do you think society can or should do in response to these changes? Using Weber's theory, do you think this trend will continue or abate? Do you think the purpose of education has changed in the United States? To what level is education completely funded by government (that is, to what level is education free)? Why to that level, do you think?

(Continued)

(Continued)

- Think about Weber's ideal type of the spirit of capitalism. Are those traits more or less present in the United States today? Does this imply anything about capitalism in this country? Are we perhaps practicing a different kind of capitalism? If so, what would you call it?

- Weber gives us a robust theory concerning the rise of ethical monotheism. One of the interesting things that Weber's theory of religion tells us is that religion is clearly related to a political regime's interest in social control. Do you think there have been any changes in the structure or kind of religion practiced since Weber's time? If so, what kinds of changes? What kinds of social factors do you think are responsible for these changes (think about how Weber associated professionalism and economic changes with shifts in religion)? What, if anything, do these changes indicate about the state and social control?

- In addition to being spiritual centers, churches are social organizations. As such, Weber would argue that the type of authority and the concurrent organizational type that a church uses will influence the church and its parishioners. Either by using your own experience or by calling various churches in your area, what kind of authority and organization do you find to be most prevalent in churches today? Using Weber's theory, how do you think the church is being affected?

Weaving the Threads (building theory)

- Compare and contrast Marx and Weber on the origins of capitalism. Can these theories be reconciled in any way? Do you think one is more correct than the other? Why or why not?

- Compare and contrast Marx and Weber on structures of inequality and social change. What information does Weber give us about inequality that Marx doesn't? According to Weberian theory, what class did Marx miss entirely? Do you think that negates Marx's theory? How are Marx's and Weber's theories of social change the same and different? Can we bring them together to form a more powerful theoretical understanding of the conditions under which social change or conflict are likely to occur?

- Compare and contrast Spencer and Weber on the foundation of state power. How does Marx's theory critique both Spencer and Weber?

Introduction

Another Sociological Core

The way I've sectioned off the next three chapters may seem a little odd, for two reasons. First, Mead and Simmel don't appear with the other core theorists. Second, two of the chapters cover theorists of race and gender. My first reason for not including Mead and Simmel with the other theorists is precisely that it is odd. Our thinking will usually truck along happily in its own knowledge until we find something out of place or strange. I want to use the strangeness of pairing Mead and Simmel with theorists of race and gender to capture your attention. Here's what I hope you will be able to see:

Most classical theory books don't deal directly with the issue of modernity. It's obviously there implicitly, but I've chosen to make the issue explicit and use it as a central organizing feature. Even among those books that do give priority to modernity, I'm not aware of any that make the *individual* in modernity a central issue. This emphasis of course is in back of the sociological imagination sections in this book. I want you to see how biography, history, social structures, and intellectual/cultural contexts impact the thoughts and consciousnesses of the people we've studied (and by extension, I want you to see your own consciousness in the same way). This emphasis is also why I talked about the modern person the way I did in Chapter 1. Many modern institutions, such as the democratic state, education, science, and even the economy, are based upon and only possible because of a certain type of person who did not generally exist before the fifteenth or sixteenth centuries. Of course, there were individual and historical period exceptions, such as ancient Greece, but the seminal ideas and hopes behind the modern person are, by and large, unique to modern time.

There are two important features of this person I want to emphasize. The first feature comes from the work of Nicolas de Condorcet (1743–1797), who, you will remember, was an inspiration to Comte. Condorcet was an advocate for free public education and equal rights for women and all races. Condorcet wrote what some consider the last great work of the Enlightenment: *Outlines of an Historical View of the Progress of the Human Mind* (1795). He wrote the book while in hiding, in fear for his life during the French Revolution. In it, Condorcet traces the intertwined histories of the progress of science and the human mind. Condorcet concludes

> that no bounds have been fixed to the improvement of the human faculties; that the perfectibility of man is absolutely indefinite; that the progress of this perfectibility, henceforth above the control of every power that would impede it, has no other limit than the duration of the globe upon which nature has placed us. (p. 11)

For the second characteristic of the modern person, I'll reference the work of Sir William Blackstone (1723–1780). Blackstone was an English lawyer and judge who wrote a series of legal commentaries. These commentaries helped reform British law and had a major impact on law in the United States. The American law school system was based on Blackstone's ideas; a copy of Blackstone's commentaries is also in the John Adams Library, with copious notes from the second president of the United States. Concerning citizenship, Blackstone (1765–1769/1893) wrote,

> The true reason of requiring any qualification, with regard to property, in voters, is to exclude such persons as are in so mean a situation that they are esteemed to *have no will of their own.* If these persons had votes, they would be tempted to dispose of them under some undue influence or other. (p. 172, emphasis added)

This explains historically why property ownership was a condition of voting rights, but the more important aspect for us is the assumption that the fully modern person possesses the freedom to be, *to will,* and to act.

The modern person, then, was understood as having the innate ability to learn and expand his or her mind. Progress was seen not simply in political or economic terms, but also in terms of the ever-expanding mental powers of the individual. Of course, political and economic progress is dependent upon the progress of the mind, and this was Condorcet's point. In Condorcet's view, there is no limit to the improvement or progress of the mind. It's not a stretch to link this capability to Blackstone's point that the person must be free from all external dictates. These two features are the twin towers of democracy: Democracy depends on a mind that is open, a mind that continually considers new ideas and possibilities; and democracy is equally dependent upon that mind being free to act, *to will,* according to its own dictates.

We know from the history of democratic nations that those characteristics are ideal. While Condorcet may have favored equal rights for women and blacks, it appears the U.S. Constitution as originally written did not. But I'd like to submit to you that there is something deeper here. It isn't simply that the framers of the U.S. Constitution blew it. Rather, many of the ideas and tenets of modernity are dialectic (and some of them intentionally so). You remember from Hegel and Marx that dialectics implies intrinsic contractions or tensions. Democracy and the ideas of the person that undergird it are by their very nature incomplete and filled with contradictions. Historically, the tension was between "all men are created equal" on one side and women and blacks on the other. A more current example of this tension is the global debate about the right of a same-sex couple to marry. In 2001, the Netherlands was the first to grant that right nationally; since then, only nine other counties have granted that equal right, and as of February 2012, only seven U.S. states and the District of Columbia.

What do Mead and Simmel have to do with these issues? First, all of the theorists we've thus far considered assumed the modern person but didn't give us an explicit theory. They talked about the social structures surrounding the individual and how they impact the possibilities of the modern person. So, Marx called our attention to class inequality, false consciousness, and ideology; Durkheim wondered about the pathologies a fragmented collective consciousness would bring; and Weber saw how the processes of rationalization and bureaucratization could end in irrationality and the defeat of democracy. But none of them gave us a theory about how the individual could possibly have the potential of an expanding mind and the will to act in the first place.

George Herbert Mead gives us such a theory. His theory is based on pragmatism, which as you'll see is a philosophy that shows us how and why democratic consensus is possible. Mead sees that meaning and truth aren't set in stone by large-scale institutions; rather, they emerge in the daily course of interacting with people around us. He also gives us a theory of the mind that explains how and why it could ever be expansive. In addition, he theoretically explains how the inner self is formed through the language we learn and the people with whom we role-take. More important, the self in Mead's scheme is never fully set; it changes and grows as it moves through diverse social situations. In short, Mead gives us an explicit theory of the modern person, something our other theorists assumed.

What Simmel does for us here is show us the dialectic of selfhood in modernity. Modernity proclaims that the self is free and infinitely capable, and the institutions of modernity—state democracy, mass education, capitalism, voluntary religion, and so forth—are set up to facilitate that infinite progress and freedom. But in doing all this, modernity creates a type of culture that disables that free and capable individual. Simmel shows us that the institutions and social processes that grant us freedom simultaneously take that freedom away and make us disengage.

I've thus set Mead and Simmel apart to emphasize the modern self, how it is formed in potential and how it is disabled in actuality. This dialectic of the self is true for every one of us generally. Now, let me say that this isn't a death sentence or reason to be discouraged. There are dilemmas to the modern self, and knowing how these things work theoretically gives us the ability to make changes. I'll explain a bit of that in the section on postmodernism in Mead and Simmel's chapter.

But, why include Mead and Simmel in a section with race and gender? Hopefully, a glimmer of an answer has presented itself to you. I see Mead and Simmel addressing modern personhood—the issues of race and gender are about the same thing. While Mead and Simmel can be used generally to talk about the modern dialectic of the person, race and gender speak to the issue specifically. As I've noted, it's easy to see this historically: Women and blacks were denied full citizenship at the founding of the United States. But the issue is both more insidious and diffuse than that.

However, before I go on, I need to make something clear: It's the idea of *personhood* that connects these three chapters in my mind, not any theoretical similarities or connections. I'm not, for example, going to carry Mead's and Simmel's theories through to Wollstonecraft or Du Bois. It's possible to do so, but it would take another book to do it right.

We can understand the things our race or gender theorists say in light of Mead or Simmel, but that's not my point here. Rather, I've placed them together in order to emphasize the broader issue of personhood in modernity. Nowhere is that more clearly seen than in the issue of race. It is there, in the complete denial of personhood through slavery, that the insidious nature of the modern subject shows itself most clearly. The issue is also there in gender. Though not in the same way as race, the construction of gender in modern society is based on the idea of the Other, which denies the authenticity of women's experience.

In thinking about modernity, which any classical theory text asks us to do, it is imperative that we think specifically about the modern person or subject. This is important first in terms of understanding the projects of modernity and the kind of knowledge upon which those projects are based. That was the purpose behind Chapter 1 and the first section of the book. Second, it is equally important to grasp the complications that face the modern person in terms of the culture in which we live (Simmel), and, more significantly, in terms of our primary categories of distinction: gender and race.

CHAPTER 6

The Modern Person

George Herbert Mead and Georg Simmel

The topics of each of the previous chapters are central to the modern project. We've talked about social structures and complex social systems, religion and the state, capitalism and bureaucracy, culture and social integration, and so on. All of these are fairly large (macro level) institutional issues that help define modernity as a historical period. In this chapter, we're bringing our analysis down to the micro level and the person. In a significant way, this chapter is about you—the theoretical you. However, it is vital for you to see that the topic of this chapter is just as central to the project of modernity as those institutional forces.

Modernity was founded on a specific kind of person, a particular notion of human nature. Prior to the social and philosophical changes leading up to modernity, a person wasn't seen as an individual in the way we mean the term today. To help you see this, let me explain a little bit about how the modern person came about. One of the major forces in pushing modernity forward was the Protestant Reformation; it had broad-based effects beyond its religious implications. For example, in Chapter 5 we saw how Luther's notion of the calling provided a cultural base for the work ethic of early capitalism. The Protestant Reformation also helped reconceptualize the person. In traditional Catholicism, a person had a relationship with God based on his or her *being part of the Catholic Church*. Salvation wasn't seen individually, but rather, collectively—a person went to heaven because he or she was part of the Church, part of the Bride of Christ.

Membership in the Church was obtained through the sacraments, such as baptism (generally at birth) and the Holy Eucharist (Communion). Further, people didn't have a direct, individual relation or experience with God; a person's connection with God was continually mediated by priests, saints, the Virgin Mary, and various other intermediaries—people weren't even deemed capable of reading the Bible on their own. This type of person also existed in the political realm: Subjects of the monarchy were to be guided and cared for. The important thing I want you to see here is that none of this implied a person individually capable of making important decisions and guiding his or her life. In fact, it implies just the opposite. But with the advent of Protestantism, the person was singled out, made to stand before God on his or her own confession of faith. This is why people converting from Catholicism would be re-baptized: Being baptized as an infant did not involve personal choice. As I said, Protestantism is just one force in the redefinition of the person, but it gives us a sense of how people were viewed prior to modernity.

The Enlightenment and modernity brought with it a new kind of person, "a fully centered, unified individual, endowed with the capacities of reason, consciousness, and action" (S. Hall, 1996a, p. 597). Both science and citizenship are based on this idea of a new kind of person—the supreme individual with the power to use his or her own mind to determine truth and to use reason to discover the world as it exists and make rational decisions. This belief gave the Enlightenment its other name: the Age of Reason. This new idea, this reasoning person, obviously formed the basis of scientific inquiry; more important for our purposes, it also formed the basis for the social project. Democracy is not only possible because of belief in the rational individual; this new person also necessitates democracy. The only way of governing a group of individuals, each of whom is capable of rational inquiry and reasonable action, is through their consent.

Further, Alexis de Tocqueville (1835–1840/2002) in his study of *Democracy in America* asserts that equality in democracy suggests the "indefinite perfectibility of man [sic]" (p. 426). When citizens are ranked according to birth, "none seeks any longer to struggle against an inevitable destiny" (p. 427). But when castes disappear, and people gather in diverse groups, "as old opinions disappear and others take their place, the image of an ideal and always fugitive perfection is presented to the human mind" (p. 427).

Up to this point, all our theorists assumed this potential without explaining theoretically how it could exist. George Herbert Mead gives us just such a theory. With Mead, we'll see theoretically how an individual is free to make decisions and to guide his or her life toward higher goals. Georg Simmel, however, assumes the individual, much like our previous theorists. Yet Simmel is important here because he gives us an explicit theory about how different social factors in modernity shape and form such a person. The person you are is not incidental to modernity; it is tied up with the Enlightenment and such ideas as equality and freedom. Perhaps more important, *the type of person that it is possible for you to be* is historically specific and socially created—the possibilities of personhood alter with changes in social practices, structure, and culture. This issue of personhood, of human potential, is also central to our next chapters on race and gender. Whereas Mead and Simmel talk about the person in general, our race and gender theorists expose the fact that the person assumed by Mead and Simmel is the white male, and the experiences and potential of selfhood are restricted and impacted by race and gender.

George Herbert Mead—Symbolic Interaction

George Herbert Mead was born on February 27, 1863, in South Hadley, Massachusetts. Mead began his college education at Oberlin College when he was 16 years old and graduated in 1883. After short stints as a schoolteacher and surveyor, Mead did his graduate studies in philosophy at Harvard. In 1893, John Dewey (a prominent pragmatist) asked Mead to join him to form the Department of Philosophy at the University of Chicago—the university was also the site of the first department of sociology in the United States. Mead's major influence on sociologists came through his graduate course in social psychology, which he started teaching in 1900. Those lectures formed the basis for Mead's most famous work, *Mind, Self, and Society* (1934), published posthumously by his students. Mead died on April 26, 1931.

Of Mead, the American philosopher John Dewey (1931) said, "I attribute to him the chief force in this country in turning psychology away from mere intro-spection and aligning it with biological and social facts and conceptions" (p. 311). Dewey's quote indicates that George Herbert Mead is at least partially responsible for aligning psychology with the facts of biology and sociology. In doing so, Mead helped found an approach known as social psychology. The uniqueness of Mead's approach, when compared to psychology, is that Mead *begins in the social rather than the individual,* as compared to psychology, which begins in the individual:

> Social psychology has, as a rule, dealt with various phases of social experience from the psychological standpoint of individual experiences. The point of approach which I wish to suggest is that of dealing with experience from the standpoint of society. (Mead, 1934, p. 1)

Mead refers to his own method as *social behaviorism.* In using such a term, Mead is comparing himself to the kind of behaviorism that psychologist John B. Watson advocated. Watson focused on outward human behavior exclusively, arguing that investigating the internal workings and states of the human mind was unnecessary. Mead wants to maintain the emphasis on human action, but argues that understanding the mind is vital for understanding human behavior. In fact, it is the mind that makes human behavior unique. According to Mead, the mind itself is behavior, a kind of pre-behavior before actual action can take place. In other words, before people act, they think about possible personal and social consequences.

Mead's emphasis on social behaviorism also indicates that he is interested in the place social communication has in behavior. In other words, Mead sees social interaction as the chief motivating or organizing principle behind human behavior, rather than as operant conditioning. In a brilliantly creative move, Mead argues that the mind is internalized social communication—that the internal thinking that precedes and makes unique human behavior is socially constructed.

As a result of this particular approach, most sociological social psychology begins with George Herbert Mead. Of course, there were other theories, such as those of Marx, Durkheim, and Weber, that contained some social psychological elements, and some theorists, such as Simmel and Cooley, who were explicitly concerned with the individual or self, but it is Mead who formed the solid foundation for a sociological perspective of what goes on inside a person. Simmel's work doesn't help us much with the internalization process, as he is mostly concerned with the intersections among the individual, subjective culture, and objective culture. Simmel's theory assumes an intrinsic individual psyche; Mead's theory assumes that the self is utterly social.

Mead gives us an entirely social way of understanding consciousness and thought. According to Dewey (1931), this is the issue that Mead was constantly thinking about: "the problem of individual mind and consciousness in relation to the world and society" (p. 311). In the end, Mead's theory implies that mind and consciousness are social entities; they are formed through symbolic interaction and role taking. More than that, Mead's theory also implies that the self is historically situated, rather than a universal property of human nature as Enlightenment philosophers generally assumed. Thus, you not only think and feel different things from your parents and grandparents, but your inner self is also probably qualitatively different. The very essence of your self is potentially different because it was formed in connection with different social factors. The differences between you and your grandparents may not be extreme, but using Mead's perspective, we would say that the self of twenty-first century America is fundamentally different from that of first century Greece.

Hint

This theoretical hint is more in the way of a reminder. In our chapter covering Marx, we talked a bit about the enduring issue of consciousness. I encourage you to go back and review that section because Mead is giving us a sociological understanding of consciousness, one that is different from that of Marx. Marx's approach is based on his version of materialism: The world of created products acts as a reflective mirror through which human consciousness is created. Mead's approach is quite different and is based on the reflexiveness that language and role taking give us. For a solid expansion of this approach, see Tom Burns (1998), *The Social Construction of Consciousness*, Parts 1 and 2.

THEORIST'S DIGEST

Central Sociological Questions

Society as we know it came to exist in modernity. But, for Mead, society doesn't exist as macro-level structures and systems. What we mean by society is actually made up of various sets of attitudes, ways of seeing and being in the world. Society is enacted and comes to exist as individuals interact with one another and take on these attitudes. Central to this social interaction, and thus to society, is a certain kind of self, one capable of taking his or her own actions as social objects and deciding to act. Mead's (1934) quest, then, is to explain how this modern person exists: "How can an individual get outside himself (experientially) in such a way as to become an object to himself? This is the essential psychological problem of selfhood or of self-consciousness" (p. 138).

Simply Stated

At the core of Mead's question is the issue of separation: The watching perspective of an individual is separate from the actions of the self. Mead wants to explain how that separation takes place. There are two chief mechanisms through which people become separated from their own actions: language acquisition and role taking. Role taking progresses through three stages as the individual develops a self that is increasingly separated from his or her own actions.

Key Concepts

Pragmatism, emergence, action, meaning, natural signs and significant gestures, social objects, emergence, role taking, perspective, self, mind, play stage, game stage, generalized other stage, the I and the Me, interaction, society.

Concepts and Theory: Truth, Meaning, and Action

There were many influences on Mead's thinking. In fact, his work is an early example of theoretical synthesis, bringing together several different strands of thought to create something new (for Mead's influences, see Morris, 1962). But for our purposes, we will concentrate on Mead's debt to the philosophy of pragmatism.

Pragmatism is the only indigenous and distinctively American form of philosophy, and its birth is linked to the American Civil War (Menand, 2001). The Civil War was costly in the extreme: The number of dead and wounded exceeds that of any other war that the United States has fought, and the dead on both sides were family members and fellow Americans. This extreme cost left people disillusioned and doubtful about the ideas and beliefs that provoked the war. It wasn't so much the content of the ideas that was the problem, but rather, the fact that ideas that appeared so right, moral, and legitimate could cause such devastation. It took the United States almost 50 years to culturally recover and find a way of thinking and seeing the world that it could embrace. That philosophy was pragmatism.

Pragmatic Truth

Pragmatism rejects the notion that there are any fundamental truths and instead proposes that truth is relative to time, place, and purpose. In other words, the "truth" of any idea or moral is not found in what people believe or in any ultimate reality. Truth can only be found in the actions of people; specifically, people find ideas to be true if they result in practical benefits. Pragmatism is thus "an idea about ideas" and a way of relativizing ideology (Menand, 2001, p. xi), but this relativizing doesn't result in relativism. Pragmatism is based on common sense and the belief that the search for "truth and knowledge shifts to the social and communal circumstances under which persons can communicate and cooperate in the process of acquiring knowledge" (West, 1999, p. 151).

Definition

Pragmatism is a school of philosophy that argues that the only values, meanings, and truths that humans hold onto are the ones that have practical benefits. These values, meanings, and truths shift and change in response to different concrete experiences. Ultimate truth, then, isn't a real issue for people; practical truth is what matters, and it emerges as humans solve issues of day-to-day living in group life. Pragmatism forms the base for many American social theories, most specifically symbolic interactionism.

Understanding pragmatism helps us see the basis of Mead's concern for meaning, self, and society. As we will see, Mead argues that the self is a social entity that is a practical necessity of every interaction. We need a self to act deliberately and to interact socially; it allows us to consider alternative lines of behavior and thus enables us to act rather than react. In pragmatism, human action and decisions aren't determined or forced by society, ideology, or preexisting truths. Rather, decisions and ethics emerge out of a consensus that develops through interaction—a consensus that is based on a free and knowing subject: the self.

Pragmatism also helps us understand another important idea of Mead's: emergence. In general, the word *emergence* refers to the process through which new entities are created from different particulars. For Mead, then, meaning emerges out of different elements of interaction coming together. Let's take a hammer as an example. People create social objects such as hammers in order to survive, and hammers only exist as such for humans (hammers don't exist for tigers, though they might sense the physical object). But the meaning of objects isn't set in stone, once and for all. While the hammer exists in its tool context, its meaning can vary by its use, and its use is determined in specific interactions. It can be an instrument of construction or destruction depending upon how it is used. It can also symbolize an individual's occupation or hobby. Or it could be used as a weapon to kill, and it could be an instrument of murder or mercy, depending upon the circumstances under which the killing takes place. Thus, the "true" meaning of an object cannot be unconditionally known; it is negotiated in interaction. The meaning pragmatically emerges.

Definition

Emergence is a theoretical idea used to understand meaning and inter-action in Mead's theory of symbolic interaction. To emerge means to rise from or come out into view, like steam from boiling water. Emergent meaning, then, implies that meaning is not intrinsic to any sign or object. Meaning, rather, is a function of social interaction: The meaning of a sign, symbol, nonverbal cue, social object, and so on comes out of (emerges from) social interaction. Because the self is meaningful, it shares this quality of emergence. In this perspective, meaning—and, by implication, the self—is thus a very supple thing that is controlled by people in face-to-face interactions and not by social structure.

Human Action

Humans act—they don't react. As Mead characterizes it, the distinctly human *act* contains four distinct elements: impulse, perception, manipulation, and consumption. For most animals, the route from impulse to behavior is rather direct—they react to a stimulus using instincts or behavioristically imprinted patterns. But for humans, it is a circuitous route. The philosopher Ernst Cassirer (1944) puts it this way:

Man has . . . discovered a new method of adapting himself to his environment. Between the receptor system and the effector systems, which are found in all animal species, we find in man a third link which we may describe as the symbolic system[, which is] the way to civilization. (pp. 24, 26).

After we feel the initial impulse to act, we perceive our environment. This perception entails the recognition of the pertinent symbolic elements—other people, absent reference groups (what Mead calls generalized others), and so on—as well as alternatives to satisfying the impulse. After we symbolically take in our environment, we manipulate the different elements in our imagination. This is the all-important pause before action; *this is where society becomes possible.* This manipulation takes place in the mind and considers the possible ramifications of using different behaviors to satisfy the impulse. We think about how others would judge our behaviors, and we consider the elements available to complete the task. After we manipulate the situation symbolically in our minds, we are in a position to consummate the act. I want you to notice something about human, social behavior: *Action requires the presence of a mind capable of symbolic, abstract thought and a self able to be the object of thought and action.* Both the mind and self, then, are intrinsically linked to society. Before considering Mead's theory of mind and self, we have to place it within the more general context of symbolic meaning.

Concepts and Theory: Meaning and Interaction

According to Mead, *language* came about as the chief survival mechanism for humans. We use it to pragmatically control our environment, and this sign system comes to stand in the place of physical reality. As we've seen, animals relate directly to the environment; they receive sensory input and react. Humans, on the other hand, generally need to decide what the input *means* before acting. Distinctly human action, then, is based on the world existing symbolically rather than physically. In order to talk about this issue, Mead uses the ideas of natural signs and significant gestures.

Hint

This hint is about language and could just as easily be placed under enduring issues. Yet, I've found that the issue of language is increasingly becoming an issue for students. The problem is human versus animal language. If we take *language* in its most general definition, then many species have "language." But that approach doesn't get us very far in understanding humankind. When Mead or any sociologist uses the term, he or she is referring to the unique characteristics of human language, such as the arbitrariness and reflexivity of signs, the abstraction of symbolic meaning, and the complexity of linguistic structures. It's also important to know that recognizing human language as unique from animal communication doesn't imply superiority. Humans aren't superior, but we are different, just as a lion is different from a whale. For an informative review of communication in other species, see Winfried Nöth's treatment in the *Handbook of Semiotics* (1995), pp. 147–167.

A sign is something that stands for something else, such as your GPA that can represent your cumulative work at the university. It appears that many animals can use signs as well. My dog Gypsy, for example, gets very excited and begins to salivate at the sound of her treat box being opened or the tone of my voice when I ask, "Gypsy, want a treat?" But the ability of animals to use signs varies. For instance, a dog and a chicken will respond differently to the presence of a feed bowl on the other side of a fence. The chicken will simply pace back and forth in front of the fence in aggravation, but the dog will seek a break in the fence, go through the break, and run to the bowl and eat. The chicken appears to only be able to respond directly to one stimulus, while the dog is able to hold her response to the food at bay while seeking an alternative. This ability to hold responses at bay is important for higher-level thinking animals.

These signs that we've been talking about may be called *natural signs.* They are private and learned through the individual experience of each animal. So, if your dog also gets excited at the sound of the treat box, it is because of its individual experience with it—Gypsy didn't tell your dog about the treat box. There also tends to be a natural relationship between the sign and its object (sound/treat), and these signs occur apart from the agency of the animal. In other words, Gypsy did not make the association between the sound of the box and her treats; I did. So, in the absolute sense, the relationship between the sound and the treat isn't a true natural sign. Natural signs come out of the natural experiences of the animal, and the meaning of these signs is determined by a structured relationship between the sign and its object, like smoke and fire.

Humans, on the other hand, have the ability to use what Mead calls *significant gestures* or symbols. According to Mead, other animals besides humans have gestures, but none has significant gestures. A gesture becomes significant when the idea behind the gesture arouses the same response (same idea or emotional attitude) in the self as in others. For example, if I asked about your weekend, you would use a variety of significant gestures (language) to tell me about it. So, even though I wasn't with you over the weekend, I could experience and know about your weekend because the words call out the same response in me as in you. Thus, human language is intrinsically reflexive: The meaning of any significant gesture always calls back to the individual making the gesture.

Definition

According to Mead, *natural signs* are signs that have an intrinsic relationship to that which they signify. The sign and its object naturally go together. In comparison to natural signs, *significant gestures* are abstract and arbitrary in their relationship to an object. Significant gestures are also reflexive in that they call out the same response in the sender and receiver. Significant gestures are the central element in Mead's theory of the mind, self, and society, due to their reflexive nature.

In contrast to natural signs, *symbols* (significant gestures) are abstract and arbitrary. With signs, the relationship between the sign and its referent is natural (as with smoke and fire). But the meaning of symbols can be quite abstract and completely arbitrary (in terms of naturally given relations). For example, the concept of "Sunday" is completely arbitrary and is an abstract human creation. What day of the week it is depends upon what calendar is used, and the different calendars are associated with political and religious power issues, not nature. Because symbolic meaning is not tied to any object, the meaning can change over time. For example, in the United States there have been several meanings associated with the category of "people with dark skin."

According to Mead (1934), the meaning of a significant gesture, or symbol, is its "set of organized sets of responses" (p. 71)—notice the influence of pragmatism. Symbolic meaning is not the image of a thing seen at a distance, nor does it exactly correspond to the dictionary definition; rather, the meaning of a word is the action that it calls out or elicits. For example, the meaning of a chair is the different kinds of things we can do with it. Picture a wooden object with four legs, a seat, and a slatted back. If I sit down on this object, then the meaning of it is "chair." On the other hand, if I take that same object and break it into small pieces and use it to start a fire, it's no longer a chair—it's firewood. So the meaning of an object is defined in terms of its uses, or legitimated lines of behavior.

Because the meaning—legitimated actions—and objective availability (they are objects because we can point them out as foci for interaction) of symbols are produced in social interactions, they are *social objects*. Any idea or thing can be a social object. A piece of string can be a social object, as can the self or the idea of equality. There is nothing about the thing itself that makes it a social object; an entity becomes an object to us through our interactions around it. Through interaction, we call attention to it, name it, and attach legitimate lines of behavior to it.

Definition

The idea of *social objects* is a theoretical concept in symbolic interactionist theory. Social objects are anything in an interaction that we call attention to, attach legitimate lines of behavior to, and name. In this sense, the self and one's own feelings and thoughts can become social objects, as well as the more obvious "objects" in the environment.

For example, because of certain kinds of interactions, a Coke bottle here in the United States is a specific kind of social object. But to Xi, a bushman from the Kalahari Desert (in the film *The Gods Must Be Crazy*), the Coke bottle becomes something utterly different as a result of his interactions around it. For Xi, the Coke bottle dropped from the sky—an obvious gift from the gods. But when he brought it to his village, this playful gift from the gods became a curse, because there was only one and everybody wanted it. It became a scarce resource that brought conflict. Eventually, Xi has to go on a religious quest because of this gift from the gods (to us, a Coke bottle).

Symbolic Interaction

Notice in the preceding illustration that the meaning of the Coke bottle changed as different kinds of interactions took place. In this sense, meaning is emergent and arises out of interaction. As Mead (1934) says, "the logical structure of meaning . . . is to be found in the threefold relationship of gesture to adjustive response and to the resultant of the given social act" (p. 80). *Interaction* is defined as the ongoing negotiation and melding together of individual actions and meanings through three distinct steps. First, there is an initial cue given. Notice that the cue itself doesn't carry any specific meaning. Let's say you see a friend crying in the halls at school. What does it mean? It could mean lots of things. In order to determine (or more properly, create or achieve) the meaning, you have to respond to that cue: "Is everything alright?" But we still don't have meaning yet. There must be a response to your response. After the three phases (cue–response–response to response), a meaning emerges: "Nothing's wrong," your friend responds, "my boyfriend just asked me to marry him."

But we probably still aren't done, because her response will become yet another cue. Imagine walking away from your friend without saying a word after she tells you she's getting married. That would be impolite (which would actually be a response to her statement). So, what does her second cue mean? We can't tell until you respond to her cue and she responds to your response.

Notice that interactions are rarely terminal or closed off. Let's suppose you told your friend who was crying in the hall that marrying this guy was a bad idea. You saw him out with another woman last Friday night. At this point, the social object—marriage—which was a cue that caused her to cry in happiness, has become an object of anger. So the meaning that emerges is now betrayal and anger. She then takes that meaning and interacts with her fiancé. In that interaction, she presents a cue (maybe she's crying again, but it has a different meaning), and he responds, and she responds to his response, and so on. Maybe she finds out that you misread the cues that Friday night, and the "other woman" was just a friend. So she comes back to you and presents a cue, ad infinitum. Keep this idea of emergent meaning in mind as we see what Mead says about the self (it, too, emerges).

Concepts and Theory: Making Yourself

Have you ever watched someone doing something? Of course you have. Maybe you watched a worker planting a tree on campus, or maybe you watched a band play last Saturday night. And while you watched, you understood people and their behaviors in terms of the identity they claimed and the roles they played. In short, when you watch someone, you understand the person as a social object. After watching someone, have you ever called someone else's attention to that actor? Of course you have, and it's easy to do. All you have to say is something like, "Whoa, check him/her/it out." And the other person will look and usually understand immediately what it is you are pointing out, because we understand one another in terms of being social objects.

People-watching is a pretty common experience, and we all do it. We can do it because we understand the other in terms of being a social object. But let me ask you something. Have you ever watched yourself? Have you ever felt embarrassed or laughed at yourself? How is that different from watching other people? Actually, it isn't. But there is something decidedly odd about this idea of watching our self. It's easy to watch someone else, and it is easy to understand *how* we watch someone else. If I am watching a band play, I can watch the band because they are onstage and I am in the audience. We can observe the other because we are standing outside of them. We can point to them because they are *there* in the world around us. But how can we point to our self, call our own attention to our self, and understand our self as a social object? Do you see the problem? We must somehow *divorce our self from our self* so that we can call attention to our self, so that we can understand our self as meaningfully relevant as a social object. So, how is that done?

Role taking is the key mechanism through which people develop a self and the capacity to be social, and it has a very specific definition: Role taking is the process through which we place our self in the position (or role) of another in order to see our own self. Students often confuse role taking with what might be called role *making*. In every social situation, we make a role for ourselves. Role taking is a precursor to effective role making—we put ourselves in the position of the other in order to see how they want us to act. For example, when going to a job interview, you put yourself in the position or role of the interviewer in order to see how he or she will view you—you then dress or act in the "appropriate" manner. But role taking is distinct from impression management, and it is the major mechanism through which we are able to form a perspective outside of ourselves.

Definition

Role taking is the central mechanism in Mead's theory of the self, through which an individual is able to get outside of her or his own actions and take them as a social object. Specifically, role taking is the process through which an individual puts her- or himself in the position (role) of another for the express purpose of viewing her- or himself from that other person's role.

A *perspective* is always a meaning-creating position. We stand in a particular point of view and attribute meaning to something. Let's take the flag of the United States, for example. To some, it means freedom and pride; to others, oppression and shame; and to still others, it signifies the devil incarnate (as for some fundamentalist sects). I want you to notice something very important here: The flag itself has no meaning. Its meaning comes from the perspective an individual takes when he or she views it. That's why something like the flag (or gender or skin color or ethnic heritage) can mean so many different things. Meaning isn't in the object; *meaning arises from the perspective we take and from our interactions.*

Here's the important point: The self is just such a perspective. It is a viewpoint from which to consider our behaviors and give them meaning, and by definition, a perspective is something other than the object. In this case, the self is the perspective and the object is our actions, feelings, or thoughts. Taking this perspective, the self is how all these personal qualities and behaviors become meaningful social objects. Precisely how we can get outside ourselves in this way is Mead's driving question.

The Mind

Before the self begins to form, there is a *preparatory stage*. While Mead does not explicitly name this stage, he implies it in several writings and talks about it more generally in this theory of the mind. For Mead, the mind is not something that resides in the physical brain or in the nervous system, nor is it something that is unavailable for sociological investigation. The mind is a kind of behavior, according to Mead, that involves at least five different abilities:

1. To use symbols to denote objects

2. To use symbols as its own stimulus (it can talk to itself)

3. To read and interpret another's gestures and use them as further stimuli

4. To suspend response (not act out of impulse)

5. To imaginatively rehearse one's own behaviors before actually behaving

Let me give you an example that encompasses all these behaviors. A few years ago, our school paper ran a cartoon with a picture of three people: a man and a woman arm-in-arm, and another man. The woman was introducing the men to one another. Both men were reaching out to shake one another's hands, but above the single man was a balloon of his thoughts. In it, he was picturing himself violently punching the other man. He wanted to hit the man, but he shook his hand instead and said, "Glad to meet you."

There are a lot of things we can pull out of this cartoon, but the issue we want to focus on is the disparity between what the man felt and what the man did. He had an impulse to hit the other man, perhaps because he was jealous. But he didn't. Why didn't he? Actually, that isn't as good a question as, *how* didn't he? He was able to not hit the other man because of his mind. His mind was able to block his initial impulse, to understand the situation symbolically, to point out to his self the symbols and possible meanings, to entertain alternative lines of behavior, and choose the behavior that best fit the situation. The man used symbols to stimulate his own behavior rather than going with his impulse or the actual world.

Mead (1934) argues that the "mind arises in the social process only when that process as a whole enters into, or is present in, the experience of any one of the given individuals involved in that process" (p. 134). Notice that Mead is arguing that the mind evolves as the social process—or, more precisely, the social interaction—comes to live inside the individual. The mind, then, is a social entity that begins to form because of infant dependency and forced interaction.

When babies are hungry or tired or wet, they can't take care of themselves. Instead, they send out what Mead would call "unconventional gestures," gestures that do not mean the same to the sender and hearer. In other words, they cry. The caregivers must figure out what the baby needs. When they do, parents tend to vocalize their behaviors ("Oh, did Susie need a ba-ba?"). Babies eventually discover that if they mimic the parents and send out a significant gesture ("ba-ba"), they will get their needs met sooner. This is the beginning of language acquisition; babies begin to understand that their environment is symbolic—the object that satisfies hunger is "ba-ba," and the object that brings it is "da-da." Eventually, a baby will understand that she has a symbol as well: "Susie." Thus, language acquisition allows the child to symbolize and eventually to symbolically manipulate her environment, including the self and others. The use of reflexive language also allows the child to begin to role-take, which is the primary mechanism through which the mind and self are formed.

Stages of Role Taking

After a child begins to use language, he or she is ready to begin creating a symbolic self. This happens through *three stages of role taking*. The first stage is the *play stage,* and here the child can take the role, or assume the perspective, of certain significant others. Significant others are those upon whom we depend for emotional and often material support. These are the people with whom we have long-term relations and intimate (self-revealing) ties. Mead calls this stage the play stage because children must literally play at being some significant other in order to see themselves. At this point, they haven't progressed much in terms of being able to think abstractly, so they must act out the role to get the perspective. This is important: *A child literally gets outside of himself or herself in order to see the self.*

Definition

The important thing to see in Mead's *stages of role taking* is how the child progressively creates a perspective that is separate from his or her actions. The stages move from concrete to abstract. The play stage is the initial stage wherein the child first sees herself as a social object separate from her behaviors. In the play stage, the child literally takes on the role of one significant other and acts and feels toward herself as the other does. The game stage corresponds to the time when children are able to take the role of multiple others separately and can understand rule-based behavior. The generalized other refers to sets of perspectives and attitudes indicative of a specific group or social type with which the individual can role-take. In the formation of the self, taking the role of the generalized other allows the child to form a cohesive self and to place herself in a collective role from which to view her own behaviors.

Children play at being Mommy or being Teacher. The child will hold a doll or stuffed bear and talk to it as if she were the parent. Ask any parent; it's a frightening experience because what you are faced with is an almost exact imitation of your own behaviors, words, and even tone of voice. But remember the purpose of role taking (notice this isn't role *playing*): It is to see one's own self. So, as the child is playing Mommy or Daddy with a teddy bear, who is the bear? The child herself. She is seeing herself from the point of view of the parent, literally. This is the genesis of the self-perspective: being able to get outside of the self so that we can watch the self as if on stage. As the child acts toward herself as others act, the child begins to understand the self as a set of organized responses and becomes a social object to herself.

The next stage in the development of self is the *game stage*. During this stage, the child can take the perspective of several others and can take into account the rules (sets of responses that different attitudes bring out) of society. But the role taking at this stage is still not very abstract. In the play stage, the child could only take the perspective of a single significant other; in the game stage, the child can take on the role of several others, but they all remain individuals. Mead's example is that of a baseball game. The batter can role-take with each individual player in the field and determine how to bat based on their behaviors. The batter is also aware of all the rules of the game. Children at this stage can role-take with several people and are very concerned with social rules. But they still don't have a fully formed self. That doesn't happen until they can take the perspective of the generalized other, in the third developmental stage.

The *generalized other* refers to sets of attitudes that an individual may take toward himself or herself—it is the general attitude or perspective of a community. The generalized other allows the individual to have a less segmented self as the perspectives of many others are generalized into a single view. It is through the generalized other that the community exercises control over the conduct of its individual members. Up until this point, the child has only been able to role-take with specific others. As the individual progresses in the ability to use abstract language and concepts, he or she is also able to think about general or abstract others. So, for example, a woman may look in the mirror and judge the reflection by the general image that has been given to her by the media about how a woman should look.

Another insightful example of how the generalized other works is given to us by George Orwell (1946) in his essay "Shooting an Elephant." Orwell was at the time a British Imperial Police officer in Burma. He was at odds with the job and felt that imperialism was an evil thing. At the same time, the local populace despised him precisely because he represented imperialistic control; he tells tales of being tripped and ridiculed by people in the town. One day, an elephant was reported stomping through a village. Orwell was called to attend to it. On the way there, he obtained a rifle, only for scaring the animal or defending himself if need be. When he found the animal, he knew immediately that there was no longer any danger. The elephant was calmly eating grass in a field, and Orwell knew that the right thing to do was to wait until the elephant simply wandered off—but at the same time, he knew he had to shoot the animal.

As Orwell (1946) relates, "I realized that I should have to shoot the elephant after all. The people expected it of me and I had got to do it; I could feel their two thousand wills press me forward, irresistibly" (p. 152). He felt the expectations of a generalized other. Though contrary to his own will, and after much personal anguish, he shot the elephant. As Orwell puts it, at that time and in that place, the white man "wears a mask, and his face grows to fit it. . . . A sahib [master] has got to act like a sahib" (pp. 152–153). Not shooting was impossible, for "the crowd would laugh at me. And my whole life, every white man's life in the East, was one long struggle not to be laughed at" (p. 153). We may criticize Orwell for his decision (it's always easy from a distance); still, each one of us has felt the pressure of a generalized other.

Self and Society

For Mead (1934), there could be no society without individual selves: "Human society as we know it could not exist without minds and selves, since all its most characteristic features presuppose the possession of minds and selves by its individual members" (p. 227). We don't have a self because there is a psychological drive or need for one. We have a self because society demands it. To emphasize this point, Mead says that "the self is not something that exists first and then enters into relationship with others, but it is, so to speak, an eddy in the social current and so still a part of the current" (p. 26). Eddies are currents of air or water that run contrary to the stream. It isn't so much the contrariness that Mead wants us to see, but the fact that an eddy only exists in and because of its surrounding current. The same is true for selves: They only exist in and because of social interaction. *The self* doesn't have a continuous existence; it isn't something that we carry around inside of us. It's a mechanism that allows conversations to happen, whether that conversation occurs in the interaction or within the individual. So, the self isn't something that has an essential existence or meaning. Like all social objects, it must be symbolically denoted and then given meaning within interactions. And like all social-symbolic objects, the meaning of the self is flexible and emergent.

Definition

According to Mead, *the self* is a perspective, a conversation, and a story. The self is a perspective in the sense that it is the place from which we view our own behaviors, thoughts, and feelings. It's an internal conversation or symbolic interaction between the I and the Me through which we arrive at the meaning and evaluation we will give to our own behaviors, thoughts, and feelings, which we then thread together into a story we tell ourselves and others about who we are (the meaning of this particular social object). The self is initially created through successive stages of role taking and the internalization of language.

This emphasis of Mead's (1934) leads him to see *society* and *social institutions* as "nothing but an organization of attitudes which we all carry in us"; they are "organized forms of group or social activity—forms so organized that the individual members of society can act adequately and socially by taking the attitudes of others toward these activities" (pp. 211, 261–262). Society, then, doesn't exist objectively outside the concrete interactions of people, as Durkheim or Marx would have it. Rather, society exists only as sets of attitudes, symbols, and imaginations that people may or may not use and modify in an interaction. In other words, society exists only as sets of potential generalized others with which we can role-take.

The I and the Me

Thus far, it would appear that the self is to be conceived of as a simple reflection of the society around it. But for Mead, the self isn't merely this social robot; the self is an active process. Part of what we mean by the self is an internalized conversation, and by necessity interactions require more than one person. Mead thus postulates the existence of two interactive facets of the self: *the I and the Me*. The Me is the self that results from the progressive stages of role taking and is the perspective that we assume to view and analyze our own behaviors. The I is that part of the self that is unsocialized and spontaneous: "The self is essentially a social process going on with these two distinguishable phases. If it did not have these two phases, there could not be conscious responsibility and there would be nothing novel in experience" (Mead, 1934, p. 178).

Definition	The *I* and Me are two distinct phases of the self. The "I" phase is the initial response people have to the actions and attitudes of others—the "I" is the impulse to act. The "Me" is the product of role taking and is made up of organized sets of socially legitimated responses. In a way, the "I" is the actor, and the "Me" is the observer; the self is the conversation between both phases. Both the I and Me are necessary for a self to exist: They mutually constitute one another.

We have all experienced the internal conversation between the I and the Me. We may want to jump for joy or shout in anger or punch someone we're angry at or kiss a stranger or run naked. But the Me opposes such behavior and points out the social ramifications of these actions. The I presents our impulses and drives; the Me presents to us the perspectives of society, the meanings and repercussions of our actions. These two elements of our self converse until we decide on a course of action. But here is the important part: The I can always act before the conversation begins or even in the middle of it. The I can thus take action that the Me would never think of; it can act differently from the community.

Summary

- There are basic elements that go into making us human. Among the most important of these are symbolic meaning and the mind. We use symbols and social objects to denote and manipulate the environment. Each symbol or social object is understood in terms of legitimated behaviors and pragmatic motives. The mind uses symbolic social objects in order to block initial responses and consider alternative lines of behavior. It is thus necessary in order for society to exist. The mind is formed in childhood through necessary social interaction.

- The self is a perspective from which to view our own behaviors. This perspective is formed through successive stages of role taking and becomes a social object for our own thoughts. The self has a dynamic quality as well—it is the internalized conversation between the I and the Me. The Me is the social object, and the I is the seat of the impulses. When the self is able to role-take with generalized others, society can exist as well as an integrated self. Role taking with generalized others also allows us to think in abstract terms.

- Society emerges through social interaction; it is not a determinative structure. In general, humans act more than react. Action is predicated on the ability of the mind to delay response and consider alternative lines of behavior with respect to the social environment and a pertinent self. Thus, mind, self, and society mutually constitute one another. Interaction is the process of knitting together different lines of action. Meaning is produced in interaction through the triadic relation of cue, response, and response to response. What we mean by society emerges from this negotiated meaning as interactants role-take within specific definitions of the situation and organized attitudes (institutions).

Enduring
Issues

Mead's Perspective—Symbolic Interaction

Symbolic interaction (SI) is a primary perspective in sociology, and there are several types. The one explained in this chapter is sometimes referred to as the Chicago School, because the University of Chicago was its first institutional home. This branch of SI gained force in the 1960s as a critique of Talcott Parsons's structural functionalism (Chapter 9) and was systematized by Herbert Blumer. We'll talk about Parsons in Chapter 9, but it's also important to see that this critique is one that also could be leveled with the structuralists we've covered in our book: Spencer, Marx, and Durkheim. Blumer (1969) specifically takes on Parsons's notion of the unit act as "the quaint notion that social interaction is a process of developing 'complimentary expectations'" (p. 53). Approaches like Parsons's cast theoretical concepts at a high level of abstraction and rely heavily on quantitative data, both of which may lead to the problem of reification that Blumer points out. Symbolic interactionism, then, as Blumer conceptualized it, is an approach that points to the necessity of interpretation and moving theory from the ground up (often referred to as "grounded theory"). More generally and historically, Blumer's work follows that of George Herbert Mead and pragmatism in emphasizing the political responsibility for choices and the emergent property of democratic ethics.

In brief, Blumer's school of symbolic interaction is a theoretical perspective in sociology that assumes that human beings are fundamentally oriented toward meaning, that meaning is not a characteristic of a word or object, and that meaning emerges from the interaction. Central issues and questions for symbolic interaction include the production and use of the self in interaction, how meaning is achieved, and how actions and interactions are woven together.

Work developed from this perspective has taken a variety of paths, such as Howard Becker's labeling theory, *Outsiders: Studies in the Sociology of Deviance* (1963); Norman Denzin's cultural studies, *Symbolic Interactionism and Cultural Studies: The Politics of Interpretation* (1992); and R. S. Perinbanayagam's theory of dialogic acts, *The Presence of Self* (2000). An outstanding resource containing primary works is Ken Plummer's *Symbolic Interactionism,* volumes 1 and 2 (1991).

Another school of SI that gained force in the 1970s is often called the Iowa School, best exemplified by Manford Kuhn's work (see Kuhn & McPartland, "An Empirical Investigation of Self-Attitudes," *American Sociological Review,* Vol. 19 [1951], pp. 68–76; and Kuhn, "Major Trends in Symbolic Interaction Theory in the Past Twenty-Five Years," *Sociological Quarterly*, vol. 5 [1964], pp. 61–84). Simply put, the major difference between the two approaches is that Kuhn's work emphasizes structure, whereas Blumer stresses emergence. Kuhn argues that people develop a core self, one that is stable across situations and interactions, and that social structures, which include norms and expectations, are relatively stable and heavily influence the interaction, as does one's sense of self. In contrast to Blumer's situational approach, sociological methods in this school use structured measurements. The one Kuhn created that has had the greatest impact on the field is the Twenty Statements Test, which asks people to write 20 responses to the question, "Who am I?" Kuhn took the responses as indicators of a person's internalized objective social statuses, that is, the parts of a social structure to which a person most relates. Kuhn argues that these form personal expectations, which, in turn, structure an individual's behavior in any situation.

Georg Simmel—Formal Sociology

Georg Simmel was born in the heart of Berlin on March 1, 1858, the youngest of seven children. His father was a Jewish businessman who had converted to Catholicism before Georg was born. In 1876, Simmel began his studies (history, philosophy, psychology) at the University of Berlin, taking some of the same courses and professors as Max Weber would a few years later. In 1885, Simmel became an unpaid lecturer at the University of Berlin, where he was dependent upon student fees. At Berlin, he taught philosophy and ethics, as well as some of the first courses ever offered in sociology. In all probability, George Herbert Mead was one of the foreign students in attendance. Though Simmel wrote many sociological essays and articles, his most important work of sociology was published in 1900, *The Philosophy of Money*. All together, Simmel published 31 books and several hundred essays and articles. In 1910, he, along with Max Weber and Ferdinand Tönnies, founded the German Society for Sociology. In 1914, Simmel was offered a full-time academic position at the University of Strasbourg. However, as World War I broke out, the school buildings were given over to military uses and Simmel had little lecturing to do. On September 28, 1918, Simmel died of liver cancer.

In some ways, Simmel is the most underappreciated and least cited theorist of the classical era, and given the time in which we live, this neglect is our loss. As we will see, Simmel focuses on the relationship between the individual and culture, which is the heart of late- or postmodern theory. He is particularly concerned with the potential effects of what he calls "objective culture," culture that stands outside of the individual and faces her or him as an alien force. In evaluating Simmel's work, Guy Oakes (1984) argues that Simmel's "discovery of objectivity—the independence of things from the conditions of their subjective or psychological genesis—was the greatest achievement in the cultural history of the West" (p. 3). This observation might be truer today than when Oakes made it. As we have seen, the postmodern age is one in which the individual and culture have become the foci of social theory and our daily lives. Simmel's theory lies right at the intersection of these two issues. As such, his insights and perspective are extremely important for us.

Simmel's thought also forms the basis of what has become known as exchange or rational choice theory. Simmel and exchange theorists argue that exchange is the basis of society: People seeking to satisfy individual drives are motivated in encounters to exchange valued assets to meet those needs. Simmel also has had direct influence on contemporary conflict theory through Lewis Coser (1956). Before Simmel, conflict had been understood as a source of social change and disintegration. Simmel was the first to acknowledge that conflict is a natural and necessary part of society. Coser brought Simmel's idea to mainstream sociology, at least in America. From that point on, sociologists have had to acknowledge that "groups require disharmony as well as harmony" and that "a certain degree of conflict is an essential element in group formation and the persistence of group life" (Coser, 1956, p. 31).

Simmel is particularly significant for us because of what he adds to Mead's theory of the self. As we've seen, Mead's theory sees the self as a perspective that comes out of interactions, and he sees the meanings of symbols, social objects, and the self as emerging from negotiated interactions. Mead didn't have a well-developed concept of culture; the closest is the generalized other. Simmel, on the other hand, argues that cultural entities—such as social forms, symbols, and selves—can exist subjectively and under the influence of people in interaction, as Mead's theory would imply. However, Simmel also entertains the possibility that culture can exist objectively and independent of the person and interaction. This possibility sets up different kinds of outcomes for the individual's self-consciousness.

THEORIST'S DIGEST

Central Sociological Questions

Many sociologists argue that people are formed through social interaction and experiences with social groups and structures. Simmel is somewhat unusual because he assumes a kind of natural state for people. In other words, people are born with certain dispositions and tendencies, and interactions and institutions affect that natural state. Simmel, then, was primarily concerned with how society and objective culture influence this more natural existence. Simmel felt that modernity brought new pressures to bear on the natural state of people; thus, theory explains how modern society creates objective rather than subjective culture and how that shift influences the individual.

Simply Stated

Simmel argues that people meet their individual needs through social encounters and that these encounters can only be carried out through identifiable forms. These forms, along with cultural generally, can become alienating (objective) and thus create difficulties for the person. Culture becomes objective and alienating in modernity through such general processes as urbanization, money, and complex social networks, as well as institutionalized religion and gender.

Key Concepts

Social forms, sociability, exchange, conflict, objective culture, urbanization, the division of labor, money, web of group affiliations, normative specificity, anomie, role conflict, blasé attitude.

Concepts and Theory: The Individual in Society

At the core of Simmel's thought is the individual. In contrast to Mead, Simmel assumes there is something called human nature with which we are born. For example, Simmel feels that we naturally have a religious impulse and that gender differences are intrinsic. He also assumes that in back of most of our social interactions are individual motivations. This emphasis sets up an interesting problem and perspective for Simmel: If the individual and his or her motivations and actions are paramount, then how is society possible?

In formulating his answer, Simmel follows one of his favorite philosophers, Immanuel Kant. Kant did not ask about the possibility of society. Instead, he wondered how nature could exist as the *object* "nature" to science. Basically, Kant argued that the universe could exist as "nature" to scientists only because of the *category* of nature. Objects in the universe can only *exist* as objects because the human mind orders sense perception in a particular way. But Kant didn't argue that it is all in our heads; rather, he argued for a kind of synthesis: The human mind organizes our perceptions of the world to form objects of experience. So, nature can only exist as the object "nature" because scientists are observing the world through the a priori (preexisting) category of nature. In other words, a scientist can see weather as a natural phenomenon, produced through processes that we can discover, only because she assumes beforehand (a priori) that weather does *not* exist as a result of the whim of a god.

Simmel wanted to discover the a priori conditions for society. This was a new way of trying to understand society, rather than using a mechanistic or organismic analogy, as did Marx, Spencer, and Durkheim. In understanding society, however, Simmel wanted to maintain the integrity of the individual but at the same time recognize society as a true force. What Simmel (1971) argues is that society exists as *social forms* that come about through human interaction, and society continues to exist and to exert influence over the individual through these forms of interaction:

> Strictly speaking, neither hunger nor love, work nor religiosity, technology nor the functions and results of intelligence, are social. They are factors in sociation only when they transform the mere aggregation of isolated individuals into specific forms of being with and for one another, forms that are subsumed under the general concept of interaction. (pp. 23–24)

These forms or categories of behavior, Simmel argues, are the a priori conditions of society.

Definition

Social forms are Simmel's basic perspective of social life. A form is a patterned mode of interaction through which people meet personal and group goals; forms exist prior to the interaction and provide rules and values that guide the interaction and contribute to the subjective experience of the individual; and forms also imply social types—types of people that occupy positions within a social form (examples: stranger, adventurer, competitor, miser).

Now, think about this: If Simmel is primarily concerned about the individual, what are the implications for the person if social forms take on objective existence? There is a sense in which Mead doesn't see symbols as objective. If the meaning of symbols emerges through social interaction, then they are always subjective, at least to some degree. What Simmel wants us to consider is the possibility that signs, symbols, ideas, social forms, and so forth can exist independently of the person and exert independent effects. The question then becomes, how does objective culture impact the subjectivity of the person?

Subjective and Objective Cultures

Simmel was the first social thinker to make the distinction between subjective and objective culture the focus of his research. Individual or *subjective culture* refers to the ability to embrace, use, and feel culture. Collectives can form group-specific cultures, such as the spiked Mohawk haircut of early punk culture. To wear such an item of culture immediately links the individual to certain social forms and types, and a group member would subjectively feel those links. Individuals and dyads are able to produce such culture as well. An individual could have special incense that he or she blends just for extraordinary, ritual occasions; or a couple could create a picture that would symbolize their relationship. This culture is very close to the individual and his or her psychological experience of the world.

Objective culture is made up of elements that become separated from the individual's or group's control and reified as separate objects. Think about tie-dye T-shirts, for example. You can now go to any department store and buy such a shirt. You do not have to be a hippie to wear it, nor are you necessarily identified as a hippie, nor do you necessarily feel the connection to the values and norms of the hippie culture. It exists as an object separate from the individuals who produced it in the first place. Once formed, objective culture can take on a life of its own, and it can exert a coercive force over individuals. For example, many of us growing up in the United States believe in the ideology and morality of democracy, though in truth we are far removed from its crucial issues, ideals, and practices.

Definition

The issue of *objective culture* is central to Simmel's theorizing. Objective culture must be understood in reference to *subjective culture,* which is culture that is essentially and wholly meaningful and understood by an individual. Culture becomes objective as its size, diversity of components, and complexity increase. Among the effects of increasing objective culture are anomie and the blasé attitude.

The diagram in Figure 6.1 illustrates the relationship between subjective and objective culture. What we see on the far left is probably what Simmel has in mind under ideal conditions. People need culture to interact with others, and in small, traditional communities the culture can be kept graspable and thus subjective. The double-headed arrows indicate reciprocal relations—subjective culture influences and is affected by people and interactions. Objective culture, on the other hand, stands apart from the individual psychology. Notice that culture becomes more objective as interactions are extended to distant others and as certain features of modernity become more prominent. When this happens, a lag is produced between the individual and objective cultures. As the size and complexity of the objective culture increase, it becomes more and more difficult for individuals to embrace it as a whole. Individuals come to experience culture sporadically and in fragments. How individuals respond to this tension between subjective experience and culture is of utmost concern to Simmel.

Figure 6.1 Subjective and Objective Culture

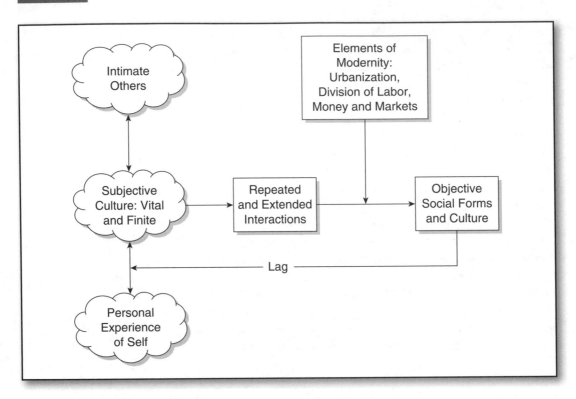

Simmel identifies three general variables of objective culture. As any of these variables increases, culture becomes more objective and less subjectively available to the individual. First, culture can vary in its *absolute size*. The pure bulk of cultural material can increase or decrease. In modernity, the amount of objective culture increases continuously. For example, in the year 2000, the world produced approximately 1,200 terabytes of scanned printed material. A terabyte contains over 1 trillion bytes. If we were to make a single book that contained just 1 billion characters, it would be almost 32 miles thick. A trillion is 1,000 billion. It has been estimated that to count to 1 trillion would take over 190,000 years, if we counted 24/7/365. The human race created over 1,200 *trillion* bytes of printed information in 2000. That's not counting the Internet. And that figure increases by 2 percent to 10 percent each year.

Culture can also vary by its *diversity of components*. Let's take fashion, for example. Not only are there simply more fashion items available (absolute size), but there are also more fashion types or styles available—there are fashions for hip-hop, grunge, skater, hardcore, preppy, glam, raver, piercer, and so on, ad infinitum. Finally, culture can vary by its *complexity*. Different cultural elements can either be linked or unlinked. If different elements become linked, then the overall complexity of the objective culture increases. For example, when this nation was first founded, there were only a few different kinds of religions (a couple of different Protestant denominations and Catholicism). Due to various social factors, the objective culture of religion has increased in its size and diversity, resulting in any number of different kinds of religion in America today. The culture of religion has also become more complex, especially in the last few years. Today, we find people who are joining together in what was previously thought to be antithetical forms of religion. So, for example, we can find Christian-Pagans in North Carolina. As these different forms become linked together, the religious culture becomes increasingly more complex.

For Simmel, cultural forms are necessary to achieve goals in a social setting. However, if these forms become detached from the lived life of the individual, they present a potential problem for the subjective experience of that individual. In an ideal world, there is an intimate connection between the personal experience of the individual and the culture that he or she uses. However, as the gap between the individual and culture increases, and as culture becomes more objective, culture begins to attain an autonomy that is set against the creative forces of the individual.

Concepts and Theory: The Self in the City

In the following subsections, we will be exploring the ideas expressed as "Elements of Modernity" in Figure 6.1. These are the specific characteristics of the modern age that tend to decrease subjective and increase objective cultures. Simmel's primary concern in this shift is the effects it has on the individual's personal experience of the self. There are three interrelated forces, or variables, in modernity that tend to increase objective culture: urbanization, the division of labor, and the use of money and markets. Urbanization is simply the process through which a larger proportion of a society's population comes to live in cities rather than rural settings. While simple, urbanization appears to be the primary dynamic of Simmel's theory, as it increases the level of the division of labor and the extent that money and markets are used. In addition to these, urbanization also changes one's *web of affiliations,* or social network, from a dense, primary network to a loose, secondary one, In the following subsections, we'll explore these three outcomes of urbanization—division of labor, money and markets, and social network—and how they influence the individual's experience of his or her self. As you'll see, Simmel's argument is complex, so read the next sections carefully *by thinking through and keeping track of the theoretical connections and effects.* At the end of the discussion, I'll give you a fuller model that captures Simmel's theory.

Definition

The *web of group affiliations* is a concept developed by Georg Simmel that describes social networks in modernity. In traditional societies, an individual's web of social connections was determined by birth and tradition (organic motivation). In modern societies, group membership is based on free choice (rational motivation). The more modern a person's web of group affiliations is, the more likely it is that she or he will develop a unique personality as well as experience anomie, role conflict, and a blasé attitude.

Simmel's (1950) concern with objective culture is nowhere clearer than in his short paper "The Metropolis and Mental Life." In it, he says,

> The most profound reason . . . why the metropolis conduces to the urge of the most individual personal existence . . . appears to me to be the following: the development of modern culture is characterized by the preponderance of what one may call the "objective spirit" over the "subjective spirit." (p. 421)

The initial factor in back of this objective spirit is urbanization—the process that moves people from country to city living. This move was a major factor in creating the modern era, and in that sense, urbanization is historically specific. But it's extremely important for us to keep in mind that this process didn't stop at some point. The dynamics that brought about urbanization are ongoing; thus, contemporary societies still have varying degrees of urbanization, with some areas being more or less urbanized. Moreover, the levels of the effects from urbanization vary as well. Urbanization created three other factors: increasing division of labor, expanding use of money and markets, and greater ratio of rational to organic group membership.

The Division of Labor

Historically, people generally moved from the country to the city because of industrialization. As a result of the Industrial Revolution, the economic base of society changed and with it, the means through which people made a living. As populations became increasingly concentrated in one place, more efficient means of providing for the necessities of life and for organizing labor were needed. This increase in the *division of labor* happened so that products could be made more quickly and the workforce could be more readily controlled. Simmel (1950) argues that the division of labor also increases because of worker-entrepreneur innovation: "The concentration of individuals and their struggle for customers compel the individual to specialize in a function from which he cannot be readily displaced by another" (p. 420).

The division of labor demands an "ever more one-sided accomplishment," and we thus become specialized and concerned with smaller and smaller elements of the production process. This one-sidedness creates objective culture: We are unable to grasp the whole of the product and the production process because we are only working on a small part. The worker in a highly specialized division of labor becomes "a mere cog in an enormous organization of things and powers which tear from his hands all progress, spirituality, and value in order to transform them from their subjective form into the form of purely objective life" (Simmel, 1950, p. 422).

Simmel claims that the consumption of products thus produced has a trivializing effect. This is basically the same issue with which Marx was concerned: Commodities produced in modern economies have little if any intrinsic meaning. Before modernity, most products were handmade; had very clear meanings to groups and individuals; and were usually embedded in socially reciprocal relations of gift giving, support, and lineage. While we can still accomplish this today by making a gift by hand (from start to finish) or passing down heirlooms, it isn't what characterizes our world. Both Marx and Simmel are saying that using such trivialized or empty commodities has an alienating effect on the person. Both the level of specialization and commodification that come from high divisions of labor create higher levels of objective culture.

Money and Markets

Urbanization also increases the level of exchange in a society and thus the use of money-facilitated markets. When thinking about the presence of money in society, it is important to understand it in comparison to barter. *Barter* is the exchange of goods or services for other goods or services. In a barter system, there is no universal value scheme. Everything is equated on an item-by-item basis, and every item is equally real. Thus, the bushel of corn that I grew may be exchanged for the two chairs that you made or for the 10 loaves of bread that Francis baked. The value of a product or service can vary tremendously, depending on local conditions or the subjective state of the trader.

Definition

Urbanization is an important factor in modernity and is defined as the process through which more and more of a given population moves from rural settings to the city. Urbanization increases in response to capitalism generally and industrialization specifically. It results in such things as higher levels in the division of labor, increases in the use of money and the size and velocity of exchanges and markets, rational versus organic group memberships, overstimulation, increasing social diversity, and so on.

Money creates a universal value system wherein every commodity can be understood. Of necessity, this value system is abstract; that is, it has no intrinsic worth. In order for it to stand for everything, it must have no value in itself. The universal and abstract nature of money frees it from constraint and facilitates exchanges. But it also has other effects for both the individual and society at large; we will talk about four of these effects next. These effects increase the level that money (or even more abstract systems such as credit) is used.

One effect of money is that it increases individual freedom by allowing people to pursue diverse activities (paying to join a dance club or a pyramid sales organization) and by increasing the options for self-expression (we can buy the clothes and makeup to pass as a raver this week and a business professional the next; we can even buy hormones and surgery to become a different sex). Second, even though we are able to buy more things with which to express and experience our self, we are less attached to those things because of money.

We tend to understand and experience our possessions less in terms of their intrinsic qualities and more in terms of their objective and abstract worth. So, I understand the value of my guitar amplifier in terms of the money it cost me and how difficult it would be to replace (in terms of money). The more money I have, the less valuable my Sunn amplifier will be, because I could then afford to buy a finer, hand-wired, boutique amp. Thus, our connection to things becomes more tenuous and objective (rather than emotional) due to the use of money.

Third, money also discourages intimate ties with people. Part of this is due to the universal nature of money. Because of its all-inclusive character, money comes to stand in the place of almost everything, and this effect spreads. When money was first introduced, only certain goods and services were seen as equivalent to it. Today in the United States, we would be hard-pressed to think of many things that cannot be purchased with or made equivalent to money, and that includes relationships. Much of this outcome is due to indirect consequences of money: The relationships we have are in large part determined by the school or neighborhood we can afford. Some of money's consequences are more direct: We buy our way into country clubs and exclusive organizations. Money further discourages intimate ties by encouraging a culture of calculation. The increasing presence of calculative and objectifying culture, even though spawned in economic exchange, tends to make us calculating and objectifying in our relationships. All exchanges require a degree of calculation, even barter, but the use of money increases the number and speed of exchanges. As we participate in an increasing number of exchanges, we calculate more and we begin to understand the world more in terms of numbers and rational calculations. Money is the universal value system in modernity, and as it is used more and more to assess the world, the world becomes increasingly quantified ("time is money," and we shouldn't "waste time just 'hanging out'").

Fourth, money also decreases moral constraints and increases anomie. Money is an amoral value system. What that means is that there are no morals implied in money. Money is simply a means of exchange, a way of making exchanges go easier. Money knows no good or evil: It can be equally used to buy a gun to kill schoolchildren as to buy food to feed the poor. So, as more and more of our lives are understood in terms of money, less and less of our lives have a moral basis. In addition, because moral constraints are produced only through group interactions, when money is used to facilitate group membership, it decreases the true social nature of the group and thus its ability to produce morality.

Thus, *money* has both positive and negative consequences for the individual. Money increases our options for self-expression and allows us to pursue diverse activities, but it also distances us from objects and people and it increases the possibility of anomie. In the same way, there are both positive and negative consequences for society. However, while it may seem that the negative consequences outweigh the positive for the individual, the consequences for society are mostly positive.

Definition

Money as a theoretical concept is a generalized media of economic exchange that can vary in its abstraction. Initially, precious metals were used as money—this form of money was valued in and of itself. Later, paper was used as a symbol for gold and silver; this money was more abstract but nevertheless had an objective base. In time, the gold and silver in back of paper money was dropped and paper money became pure symbol. Credit and debit cards then came to function in the place of paper money. Simmel sees the increasing use and abstraction of money as having both positive and negative effects: It increases personal freedom, rationalization, calculability, the number and extent of exchange relations, continuity between groups and individuals, and trust in the national system. At the same time, the increased use of money decreases the level of emotional attachment that individuals can have to things and other people, and it reduces the level of moral constraint.

We will look at a total of *four effects for society* of the increasing use of money. First, the use of money creates exchange relationships that cover greater distances and last longer periods of time than would otherwise be possible. Let's think of the employment relationship as an illustration. If a person holds a regular job, he or she has entered into a kind of contract. The worker agrees to work for the employer for a given number of hours per week at a certain pay rate. This agreement covers an extended period of time, which should only be terminated by a 2-week notice or severance pay. This relationship may cover a great deal of geographic space, as when the workplace is located on the West Coast of the United States and the corporate headquarters is on the East Coast. This kind of relationship was extremely difficult before the use of money as a generalized medium of exchange, which is why many long-term work relationships were conceptualized in terms of familial obligations, such as the serf or apprentice. And, with more generalized forms of the money principle, as with credit and debit cards, social relations can span even larger geographic expanses and longer periods of time (for example, I am obligated to my mortgage company for the next 25 years, and I just completed an eBay transaction with a man living in Japan). What this extension of relations through space and time means is that the number of social ties increases. While we may not be connected as deeply or emotionally today, we are connected to more diverse people more often. Think of society as a fabric: The greater the number and diversity of ties, the stronger is the weave.

Second, money also increases continuity among groups (level of cultural and social homogeneity). Money flattens, or generalizes, the value system by making everything equivalent to itself. It also creates more objective culture, which overshadows or colonizes subjective culture. Together, these forces tend to make group-specific culture more alike than different. What differences exist are trivial and based on shifting styles of such things as clothing, cars, music, home décor, and so on. Thus, while the weave of society is more dense due to the effects of money, it is also less colorful, which tends to mitigate group conflicts because there are fewer differences.

Third, money strengthens the level of trust in a society. What is money, really? In the United States, it is nothing but green ink and nice paper. Yet we would do and give almost anything in exchange for enough of these green pieces of paper. This exchange for relatively worthless paper occurs every day without anyone so much as blinking an eye. How can this be? The answer is that in back of money is the U.S. government, and we have a certain level of trust in its stability. Without that trust, the money really would be worthless paper. A barter system requires some level of trust, but generally speaking, you know the person you are trading with and you can inspect the goods. Money, on the other hand, demands a trust in a very abstract social form—the state—and that trust helps bind us together as a collective.

Fourth, behind this trust is the existence of a centralized state. In order for us to trust in money, we must trust in the authority of a single nation-state. When money was first introduced in Greek culture, its use was rather precarious. Different wealthy landowners or city-states would imprint their image on lumps or rods of metal. Because there was no central governing authority, deceit and counterfeiting were rampant. The images were easily mimicked and weights easily manipulated. Even though money helped facilitate exchanges, the lack of oversight dampened the effect. It wasn't until there was a centralized government that people could completely trust money.

Thus, when markets started using money for exchanges, they were inadvertently pushing for the existence of a strong nation-state. Centralized authority is the structural component to a society's trust, and it binds us together. This factor has an effect on individual freedom of expression. While the increasing use of money and markets along with decreases in the level of normative regulation spur individual freedom of expression, there's a counterforce for regulation coming from a centralized state. Modern states have increasing interest in controlling your behaviors. The list of direct and indirect regulation is almost endless, but an especially clear example is that of cigarette smoking. While it may still be legal to smoke tobacco, it's highly regulated in terms of where and when a person may smoke; at some firms, not smoking is a condition of employment.

Social Networks: Rational Versus Organic Group Membership

As a further result of urbanization, Simmel argues that social networks (what he calls the web of group affiliations) have changed. When we talk about social networks today, what we have in mind are the number and type of people with whom we associate, and the connections among and between those people. Network theory is an established part of contemporary sociology, and Simmel was one of the first to think in such terms. For his part, Simmel characterizes two types of social networks, organic and rational, which correspond to the motivations behind group membership. We'll first consider *organic social networks,* which are typical of small, more rural towns and settings. Simmel uses the term *organic* to imply that these sorts of group networks come about and develop in a way that resembles the growth of a plant or animal—they occur naturally.

In small rural settings, there are relatively few groups for people to join, and most of those memberships are strongly influenced by family. We are motivated to join the same groups as members of our family do. In these social settings, the family is a primary structure for social organization, and families tend not to move around much. So, there are likely to be multiple generations present. As a result, the associations of the family become the associations of the child. A child reared in such surroundings will generally attend the same church and school, and work in the same place as his or her parents, grandparents, cousins, and so on. Further, most of these groups will overlap. For example, it would be very likely that a worker and his or her boss attend the same church and that they will have gone to the same school.

Simmel notes that people in these settings tend to join groups because of *organic motivations*—because they are naturally or organically connected to the group. Many of the groups with which a person affiliates in this setting are primary groups. *Primary groups* are noteworthy because they are based on ties of affection and personal loyalty, endure over long periods of time, and involve multiple aspects of a person's life. Under organic conditions, a person will usually be involved with mostly primary groups, and these groups have some association with one another. This kind of community will thus contain people who are very much alike. They will draw from the same basic group influences and culture, and the groups will possess a compelling ability to sanction behavior and bring about conformity.

On the other hand, people join groups in modern, urban settings out of *rational motivations*—group membership due to freedom of choice. The interesting thing to note about this freedom is that it is forced on the individual—in other words, there are few organic connections. In large cities, people usually do not have much family around and the personal connections tend to be rather tenuous. In Southern California, for example, people move on average every 5 to 7 years. Most only know their neighbors by sight, and the majority of interactions are work related (and people change jobs about as often as they change houses). What this means is that people join social groups out of choice (rational reasons) rather than out of some emotional and organic connectedness. These *rational networks* or groups tend to have the characteristics of *secondary groups* (goal and utilitarian oriented, with a narrow range of activities, over limited time spans).

As a result of rational group affiliations, it is far more likely that individuals will develop unique personalities. A person in a more complex or rational web of group affiliations has multiple and diverse influences, and the groups' capacity to sanction is diminished. From Simmel's point of view, a group's ability to sanction is based on the individual's dependency upon the group. If there are few groups from which to choose, then individuals in a collective are more dependent upon those groups, and the groups will be able to demand conformity. This power is crystal clear in traditional societies where being ostracized meant death. Of course, the inverse is also true: The greater the number of groups from which to choose and the more diverse the groups, the weaker the moral boundaries and normative specificity (the level of behaviors that are guided by norms). In turn, this decrease in sanctioning power leads to greater individual freedom of expression.

Many students, for example, are able to express themselves more freely after moving away from home to the university. This is especially true of students who move from rural to urban settings. Not only is the influence of the student's childhood groups diminished (family, peers, church), but there are also many, many more groups from which to choose. These groups often have little to do with one another. Thus, if one group becomes too demanding of time or emotion or behavior, you can simply switch groups. So, it may be the case that you experience yourself as a unique individual having choices, but it has little to do with you per se: It is a function of the structure of your network.

While decreased moral boundaries and normative specificity lead to greater freedom of expression, they can also produce *anomie*—also a concern of Durkheim's, if you remember. For the individual, it speaks of a condition of confusion and meaninglessness. Unlike animals, humans are not instinctually driven or regulated. We can choose our behaviors. That also means that our emotions, thoughts, and behaviors must be ordered by group culture and social structure or they will be in chaos and will have little meaning. When group regulation is diminished or gone, it is easy for people to become confused and chaotic in their thoughts and emotions. Things in our life and life itself can become meaningless. So, while we may think that personal freedom is a great idea, too *much* freedom can be disastrous.

Complex webs of group affiliations can have two more consequences: They can increase the level of role conflict a person experiences, and they contribute to the blasé attitude. *Role conflict* describes a situation in which the demands of two or more of the roles a person occupies clash with one another (such as when your friends want to go out on Thursday night but you have a test the next morning). The greater the number of groups with which one affiliates, the greater is the number of divergent roles and the possibility of role conflict. However, the tendency to keep groups spatially and temporally separate mitigates this potential. In other words, modern groups tend not to have the same members and they tend to gather at different times and locations. So we see the roles as separate and thus not in conflict.

Complex group structures also contribute to the *blasé attitude*, an attitude of absolute boredom and lack of concern. Every social group we belong to demands emotional work or commitment, but we only have limited emotional resources, and we can only give so much and care so much. There is, then, a kind of inverse relationship between our capacity to emotionally invest in our groups and the number of different groups of which we are members. As the number and diversity of social groups in our lives goes up, our ability to emotionally invest goes down. This contributes to a blasé attitude, but it also makes conflict among groups less likely because the members care less about the groups' goals and standards.

Definition

The *blasé attitude* is a concept from Simmel's theory of urbanization. The word *blasé* means "to be uninterested in pleasure or life." According to Simmel, the blasé attitude is the typical emotional state of the modern city dweller. It is the direct result of the increased emotional work and anomie associated with diverse group memberships and increased cognitive stimulation due to the rapidness of social change and flow of information.

This blasé attitude is also produced by all that we have talked about so far, as well as overstimulation and rapid change. The city itself provides multiple stimuli. As we walk down the street, we are faced with diverse people and circumstances that we must take in and evaluate and react to. In our pursuit of individuality, we also increase the level of stimulation in our lives. As we go from one group to another, from one concert or movie to another, from one mall to another, from one style of dress to another, or as we simply watch TV or listen to music, we are bombarding ourselves with emotional and intellectual stimulation. In the final analysis, all this stimulation proves to be too much for us and we emotionally withdraw. Further, this stimulation is in constant flux. Knowledge and culture are constantly changing. Modern knowledge constantly changes because of the basic assumptions in back of science (the modern way of knowing): Scientific knowledge is based on skepticism, testing, and the defining value of progress. The general culture of modernity is affected by changes in knowledge, yet cultural change is also fueled by ever-expanding capitalist markets and commodities and mass media.

We have covered a great deal of conceptual ground in this section. It's been made all the more complicated because each of the things we have talked about brings both functional and dysfunctional effects, and the effects overlap and mutually reinforce one another. I've diagramed Simmel's theory in Figure 6.2. Note that I've concentrated on the effects of urbanization on the individual, which is Simmel's focus. At first glance, the complexity of the model may seem overwhelming. However, if you follow each of the paths, you'll find that it simply expresses what we've been talking about over the past few pages. For example, starting at the level of division of labor, we can see that increases in the division of labor create specialized cultures, which in turn increases the level of objective culture (relative to subjective), which then increases the likelihood that people will experience a blasé attitude. You can also work backward in the model. If you want to know when it's more likely that people will have a sense of freedom in personal expression, start at that box and work backward from the arrows pointing to the box. Working through a model such as this is a great study aid in understanding how the theory works. Be certain you know how each concept or variable influences the others, not only in what direction but also what happens substantively.

You'll notice that there is an additional concept at the end: the level of exaggerated differences and displays. These displays of difference come about because of three issues (as you can see, there are three arrows going into the concept). Notice the long arrow coming from the level of commodification. You'll remember that modern production creates commodities that are devoid of any substantial or stable meaning. In order to sell them in mass markets, all truly personal or group-specific qualities have been bleached out of the products. Simmel argues that in the face of this, we feel compelled to exaggerate any differences that do exist in the things we use in order to stand out and experience our personal selves. Two other issues contribute to this felt need: an increasing sense of confusion and meaninglessness (blasé attitude) and a feeling of lawlessness (anomie). These, too, compel individuals to reach out, to be noticed, to establish themselves as a center when the modern objective culture doesn't provide it for them. As Zygmunt Bauman (1992) notes, "To catch the attention, displays must be ever more bizarre, condensed and (yes!) disturbing; perhaps ever more brutal, gory and threatening" (p. xx). Thus, modernity increases our freedom of expression, but it also forces us to express it more dramatically with trivialized culture.

Summary

- There are two central ideas that form Simmel's perspective: social forms and the relationship between the subjective experience of the individual and objective culture. Simmel always begins and ends with the individual. He assumes that the individual is born with certain ways of thinking and feeling, and most social interactions are motivated by individual needs and desires. Encounters with others are molded to social forms in order to facilitate reciprocal exchanges. These forms constitute society for Simmel. Objective culture is one that is universal yet not entirely available to the individual's subjective

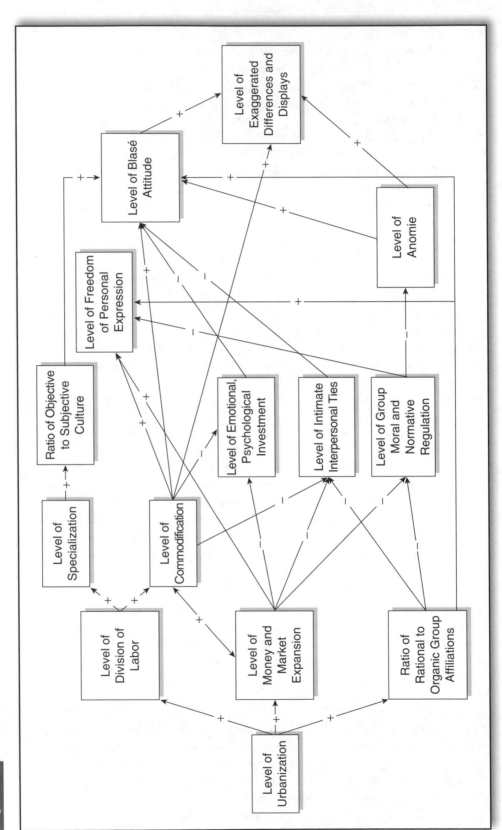

Figure 6.2 Extended Effects of Urbanization

experience. Thus, the person is unable to fully grasp, comprehend, or intimately know objective culture. The tension between the individual on the one hand and social forms and objective culture on the other is Simmel's focus of study.

- Urbanization increases the division of labor and the use of money, and it changes the configuration of social networks. All of these have both direct and indirect influence on the level of objective culture and its effects on the individual. The use of money increases personal freedom for the individual, yet at the same time it intensifies the possibility of anomie, diminishes the individual's attachment to objects, and increases goal displacement. People join groups based on either organic or rational motivations. Organic motivations imply less personal freedom and greater social conformity coupled with increased personal and social certainty; rational motivations are prevalent in urban settings and imply greater personal freedom coupled with less emotional investment and possible anomie and role conflict.

Enduring
Issues

Simmel's Perspective—Formal Sociology

Like Weber, but for different reasons, Simmel's theories don't form a cohesive whole. He theorizes about a number of different issues, many in response to needing to make a living as a lecturer rather than having a professorship. There is, however, a specific approach that is unique to Simmel: *formal sociology*. Today, we usually understand the word *formal* as meaning "proper" or "official." However, this isn't how Simmel intends the idea. Formal sociologists use the word to refer to the shape, frame, or structure that people use to create predictable patterns of social interaction. In doing so, formal sociologists make a *distinction between the form and content* of an interaction. The content is what the interaction is about: the interests and purposes of the participants, such as education, marriage, business, and so forth.

Formal sociologists, on the other hand, study the structures of social interaction that cut across such content areas. For example, a conflict in education will take the same general form as a battle in war, even though the content varies from situation to situation. Formal sociology, then, is the study of the central organizing configurations of interaction, and its intention is to create a geometry of social life.

Simmel has also influenced contemporary theory in a variety ways. His ideas concerning culture are becoming increasingly important in the work of some postmodernists (see Weinstein & Weinstein, 1993). Simmel also influenced exchange and conflict theories. He was one of the first to explicate the implications of exchange on social encounters. Rather than theorizing about the structure of the economy per se, like Marx and Weber, Simmel is instead fascinated by the influence of the social form of exchange on human experience—specifically, how value is established and how it affects the use of power in social encounters.

In addition, Simmel has had direct influence on contemporary conflict theory. Before Simmel, conflict had been understood as a source of social change and disintegration. Simmel was the first to acknowledge that conflict is a natural and necessary part of society. Lewis Coser (1956) was principally responsible for bringing Simmel's ideas about conflict to mainstream sociology.

Thinking About Modernity and Postmodernity

Postmodern theory is usually concerned with arguing that things are decidedly different today from how they were at the beginning of modernity. The unique features that are identified are diverse for different theorists. Some focus on the social hope of modernity—equality and social justice; others pay attention to the drastic changes in the economy between industrial and postindustrial or speculative capitalism; and many take issue with the assumptions of positivism, that there is an empirical world that we can discover and use for the betterment of humankind. Still others focus on what's happened to culture in an age of mass media, advertising, and the Internet. I've found that no matter what the beginning point, postmodernists generally say something about the modern subject, the idea of the person intrinsic to the Enlightenment. Of course, this isn't true of all postmodernists, but in my reading it is a recurring theme. To the extent that that's true, we now come to a central postmodern point: your experience living in postmodernity. Simmel and Mead give us inroads to consider how the experience and construction of the self is specific to our time and social context. We'll be coming at this issue from a cultural point of view.

Simmel argues that culture became increasingly objective (became greater in size, diversity, and complexity) in response to social factors specific to modernity (urbanization, the division of labor, the use of money, and rationally motivated group memberships). As we've seen, there are complex and incongruous results from these forces: for example, greater individuality and freedom coupled with increased anomie and blasé attitude. Postmodernists suggest that factors specific to this age—mass media and advertising—have pushed Simmel's concerns to another level.

I'm going to give you a simplified explanation of one of the more important points of postmodern theory. In doing so, I will undoubtedly overstate and gloss over things, but I think it will help us to get a handle on the idea of postmodernity. One of the important issues of postmodern theory concerns the way in which culture comes about. Durkheim is worried about the moral base of culture, and Simmel is concerned with the proportion of subjective versus objective culture. Both Durkheim's moral base and Simmel's subjective culture are created and firmly established in real social groups. For most of human history, culture was subjective and moral because the same social groups that created it used it. Most of us think that is still the case. I mean, Americans use American culture and Chinese use Chinese culture, right? Yes, but according to postmodernists, it isn't the same at all. Today in the United States, we have and use American culture, but our social group doesn't create a good portion of the culture we have. In fact, much of the culture we are surrounded by isn't created by an actual social group at all; it is generated by capitalist concerns in mass media and advertising.

Postmodernists argue that a major shift has occurred between modern society and postmodern society. Simmel was caught in the middle of this historical transition, which is one reason why his theory is so unique. In actual social groups, the kind where Durkheim's moral base and Simmel's subjective culture prevail, culture is created and used in the same social context. For the sake of conversation, let's call this "grounded culture." In our society, however, much of our culture is created or modified by capitalists, advertising agencies, and mass media. We'll call this "commodified culture." Members of traditional social groups were surrounded by grounded culture; members of postmodern social groups are surrounded by commodified culture. There are vast differences in the reasons why grounded versus commodified culture is created. Grounded culture emerges out of face-to-face interaction and is intended to create meaning, moral boundaries, norms, values, beliefs, and so forth. Commodified culture is produced according to capitalist and mass media considerations and is *intended to seduce the viewer* to buy products. With grounded culture, people are moral actors; with commodified culture, people are consumers. Postmodernists argue that there are some pretty dramatic consequences for people in the shift to commodified culture.

Simulacrum and Hyperreality

Jean Baudrillard (1981/1994) argues that this kind of commodification and thus trivialization can lead to some serious problems in postmodern culture. He argues that many of the things we do today in advanced capitalist societies are based largely on images completely separated from real life. For example, most people in traditional and early industrial societies worked with their bodies. Today, people in postindustrial societies don't work with their bodies; rather, they work *out* their bodies. In other words, the body has become a cultural object (pure sign) rather than a means to an end. In postindustrial society, the body no longer serves the purpose of production; rather, it has become the subject of image creation: We work out in order to alter our bodies to meet some cultural representation.

This shift obviously influences our clothing as well. Humans have always had and used clothing, and there has always been a bond between the use of the body and the function of clothing. In previous eras, there was an explicit link between the real function of the body and the clothing worn—the clothing served the work being performed or was indicative of social status. For example, a farmer would wear sturdy clothing because of the labor performed; his clothing indicated his work. (If you saw him away from the field, you would still know what he did for a living by his clothing.) However, in the postmodern society, clothing has become the creator of image rather than something serviceable or directly linked to social status and function.

Now here is where it gets interesting: A fit body represented hard work, and clothing in the past signed or symbolized the body's work, but what do our body and clothing symbolize today? Today, as noted, we work out the body rather than work *with* the body. We work out so we can meet a cultural image—but what does this cultural image represent? In times past, if you saw someone with a lean, hard body, it meant the person lived a menial, hard life. There was a real connection between the sign and what it referenced. But what do our spa-conditioned bodies reference? Baudrillard's point is that *there is no real objective or social reference* for what we are doing to our bodies. What does this imply about the clothes we wear? The clothes themselves aren't connected to any social or actual reality either.

For example, you probably wear jeans. Levi Strauss created jeans in 1853 for miners working the California Gold Rush. The story goes that after Levi Strauss arrived in California, a miner asked what he was selling. Strauss told the prospector that he had canvas for tents and wagon covers. The miner responded, "You should have brought pants. What we have don't last!" Being a smart entrepreneur, Strauss sewed the canvas into pants and later switched to denim for durability and comfort. Most of us today, however, don't wear jeans because we work hard and need something that is comfortable and will last. Then, why will we spend upwards of $100 for a pair of jeans? The simple answer is fashion. We see an advertisement that projects an image of what we think we want to look like. The prospector's jeans referenced, or were a sign of, the reality of hard labor. Our jeans, on the other hand, are an image of an image. Baudrillard calls this *simulacrum:* a sign of an image that never existed in reality.

Our culture, then, has been reduced from use-value to pure sign-value. Using Marx's ideas (Chapter 3), Baudrillard argues that the movement from use-value to exchange-value entails an abstraction of the former; in other words, the exchange-value of a commodity is based on a representation of its possible use. Use-value and exchange-value come together when I buy a new truck based on the use it has. However, advertising and mass media push this abstraction further. In advertising, the use-value of a commodity is replaced with a sign-value. What this means is that when I see an ad for the truck I want to get, chances are that what I observe are *images of masculinity:* rugged individualism, pretty girls, bottles of beer, and rock and roll. Advertising does not seek to convey information about a product's use-value; rather, advertising places a product in a field with unrelated signs in order to enhance its cultural appearance. Thus, there are certain signs and meaning contexts that get attached to commodities. As a result, we tend to relate to the sign rather than the use-value. People purchase commodities now more for the image than for the function they perform. Getting back to my truck, while I may plan to occasionally use the truck to haul stuff, that is a minor consideration in buying it. I buy it—and care for it and accessorize it—because of the image it can project. But, as we've seen, it is an image of an image that exists only in the mass media of advertising. The majority of our culture is like this, which prompted Baudrillard to argue that we live in *hyperreality* rather than actual reality. Our cultural world is no longer real; it is at best virtual. We can see this abstraction and pure sign-value even clearer in such things as reality television and video games.

Definition

Hyperreality is a theoretical concept from the work of Jean Baudrillard and refers to the preference in postmodern societies for simulated experiences rather than real ones. In Baudrillard's scheme, the real has been stripped of all meaning, authenticity, and subjective experience through the proliferation of simulacrum. People thus seek emotional experiences in images of reality that have been exaggerated to the grotesque and fantastic (breast implants and the digital manipulation of photographs are good examples).

A great deal of theoretical ink has been spilled in the debate concerning the relationship between the subject and culture in postmodernity. It is important to note that Simmel was aware of the link between culture and the individual's subjective experience long before postmodernism was even a word. It is ironic to think that the forces that produce individuality and personal freedom are the same forces that endanger the individual and her or his particular experience. Yet that is exactly Simmel's argument. Individual freedom is increasing as cultural cohesion and meaning are decreasing. This creates the basic postmodern question: Is the individual on the cusp of experiencing a core sense of individuality that sits at a yawning abyss of meaninglessness where emotional depression (blasé attitude) is normal?

Reflexivity and the Fragmenting of the Self

What would Mead's theory add to this understanding of the postmodern self? Recall that Mead argues that the self becomes integrated through role taking with a generalized other. When role taking with only significant others, the self will seem segmented, divided as it is among the different points of view. The generalized other is able to link all those individual perspectives into one abstract whole, thus giving the self a sense of integration. But what happens if the generalized other is fragmented or constructed from vacuous images? This is precisely the problem that postmodernists give us.

Rather than a self that is created through role taking with actual social others, the postmodern self is created through role taking with media images. These images are not connected to one another in any real way. Through the media, individuals are inundated with disconnected images of different groups, different people, different values, and different modes of expression. Mead's theory implies that these images have increased the amount of reflexivity and self-monitoring in which individuals engage. The problem is that this increased reflexivity is directed toward a cultural self that is socially saturated and filled with constant change and doubt. The result is what Gergen (1991, pp. 73, 150) refers to as "multiphrenia" (vertigo of self-values filled with expressions of inadequacy) and the "pastiche personality" (a social chameleon that uses bits and pieces of available identities to fit in). Jameson (1984b) goes so far as to argue that the culture of the postmodern era "stands as something like an imperative to grow new organs, to expand our sensorium and our body to some new, as yet unimaginable, perhaps ultimately impossible, dimensions" (p. 80).

As a result, a core or authentic self is increasingly difficult to create and experience in our time. But are we left with that difficulty? Does Mead perhaps give us a way of understanding how we experience the self that might counter these issues? I've been concerned with these social psychological problems for a good while, and published a book addressing these concerns in 1998: *The Meaning of Culture: Moving the Postmodern Critique Forward*. I'm always leery of theorists who include their own work in a book such as this, and I hesitate in doing so, but I think Simmel and postmodernism have raised concerns that directly impact you and thus think that you might find something interesting here. The theory in *The Meaning of Culture* focuses on the issue of *reflexivity*, thought turned back on itself.

As we've seen, Mead is certainly aware of the reflexive dimension of creating a self. In fact, according to Mead, both the mind and the self are created through reflexive thought. The self is found in the social object (the Me) as well as in the reflexive conversation an individual has with her- or himself from the perspectives of specific and generalized others. Though Mead does not go into much detail, he is aware that this reflexiveness, although necessary for social life, is not altogether comfortable for the individual. It is perhaps most problematic in postmodernity.

Being too reflexive about things can actually deconstruct them (Allan, 1998). Sometimes this can be good. Racial and sexual inequalities can only continue if people take for granted that the social categories are real. When we realize, for example, that race is not an essential quality but rather a symbolic meaning that has been created through historical, social processes and defined in interactions, then we can deconstruct race and it will no longer influence our behaviors. There is, however, a limit. If subjective reflexivity pushes deconstruction too far, we are then faced with nothing but a yawning abyss of meaninglessness—a placeless surface that is incapable of holding personal identity, self, or society. Meaning is set loose from its social and material moorings and becomes completely free floating.

This is part of what postmodernists are getting at. Reflexivity has increased in postmodernity because of a number of social factors, like the legitimation of doubt (scientific inquiry), constantly changing knowledge, disconnected media images, rapidly shifting markets, and so on. This increased reflexivity is particularly important for the self. Too much reflexivity can destroy the meaning that we give the self. However, Mead gives us a theory through which we can understand how the destructive tendencies of postmodern reflexivity can be lessened.

Fusing the I and the Me

Let me give you a practical example of how reflexivity works. Let's pretend you are at a bar and see someone that you think is attractive. What do you do? Well, if you're like most of us, you think about it and you role-take. You put yourself in that person's shoes and see how you would look to her or him. You might try out different scenarios in your mind. Depending on how these reflexive thoughts come out, and depending on who that person is with and what that other person may think of you, you may or may not approach the person. If you are like most of us, that process can be emotionally draining.

But let me add one small detail to the story. In this scenario, you are at the bar, you see someone attractive, and you are drunk. Now what do you do? Chances are good that you will simply walk right up to that person without so as much as a second thought. Why? Because you're drunk, right? Yes, but let's think like theorists and go below the surface. What is it that being drunk does? Being drunk short-circuits the reflexive loop inside you. And because of that, you exist more in the immediate moment than you generally do. In other words, the impulses of the I aren't subject to the scrutiny of the Me (at least, we can't hear it talking). This is a shadow example because this dampening of the reflexive loop is in some ways artificial, since it is due to alcohol, but it helps us see a bit of what Mead is talking about.

Mead recognizes an important phase in the interaction between the I and the Me. There are times in which the I and the Me can be fused and the reflexive loop stopped. During those times, things feel more real and we sense the "true" meaning of life. We experience those times when we aren't reflexive as essentially *real*. In other words, we feel that we are not attributing meaning; rather, it simply is what it is (direct sensual experience).

Mead argues that those times when the I and the Me are fused are intense emotional experiences. This emotional response is all the more powerful when experienced with a group. "The wider the social process in which this is involved, the greater is the exaltation, the emotional response, which results" (Mead, 1934, p. 274). This description is reminiscent of Durkheim's explanation of ritual. Mead supplies what is missing from Durkheim: the internal effects of increased emotional energy in response to ritual performance.

Mead says that fusing takes place in circumstances where there is complete identification of the person with the expectations of the situation. This can most easily happen in group activities. Mead gives the example of saving someone from drowning. Everybody who sees the person drowning will respond in the same manner. There is a sense of common effort and identification among all people, although some may call for help and others may dive in. Other situations that give rise to this phenomenon are religious and patriotic events. Durkheim, of course, sensitizes us to another kind of practice that brings about reduced reflexivity, that being ritual. There is yet another source for this kind of experience. It's called flow.

Mihaly Csikszentmihalyi (1990) uses the term *flow* to capture what he calls "optimal experience": "the state in which people are so involved in an activity that nothing else seems to matter" (p. 4); it is due to being so involved in one's activity that "[s]elf-consciousness disappears, and the sense of time becomes distorted" (p. 71). In a state of flow, people are not reflexive; they are completely immersed in the situation. Based on a psychology of optimal experience, Csikszentmihalyi argues that flow comes about in rule-bound situations in which there is a clearly known goal, the individual is pushed to the functional limits of her or his skill level, and there are unambiguous signals about how well she or he is doing. In other words, the situation must enable people to concentrate all their attention on the task at hand. Moreover, the author asserts, "whenever one does stop to think about oneself, the evidence [of good performance] is encouraging . . . and more attention is freed to deal with the outer and the inner environment" (p. 39).

If the postmodernists are correct, Mead's theory would predict that people who participate in more activities in which the reflexive loop is dampened and the I and the Me fused will have more core-self experiences. In particular, from just this discussion, we can see how increasing participation in emotion-producing rituals (like rock concerts) and activities that focus one's cognitive energy (like extreme sports) could create an array of experiences that feel essentially real and produce a firm sense of self, in contrast to what many postmodernists claim. I think that Simmel and postmodern theory are right: Contemporary culture is in itself meaningless and free floating. But that isn't the end of the story. What matters is how you and I use and interact around this culture. We can, through certain kinds of interactions and practices, make this culture subjectively real.

BUILDING YOUR THEORY TOOLBOX

Learning More—Primary and Secondary Sources

The chief primary source of George Mead's theory is found in a compilation of student notes:

- Mead, G. H. (1934). *Mind, self, and society: From the standpoint of a social behaviorist* (C. W. Morris, Ed.). Chicago: University of Chicago Press.

Unlike Mead, Georg Simmel published quite a bit, so there are a good number of primary sources. I suggest that you start off with the first two readers, and then move to his substantial work on money:

- Simmel, G. (1959). *Essays on sociology, philosophy, and aesthetics [by] Georg Simmel [and others]: Georg Simmel, 1858–1918* (K. H. Wolfe, Ed.). New York: Harper & Row.

- Simmel, G. (1971). *Georg Simmel: On individuality and social forms* (D. N. Levine, Ed.). Chicago: University of Chicago Press.

- Simmel, G. (1978). *The philosophy of money* (T. Bottomore & D. Frisby, Trans.). London: Routledge and Kegan Paul.

To read more about Mead from secondary sources, I would recommend the following:

- Baldwin, J. D. (1986). *George Herbert Mead: A unifying theory for sociology.* Beverly Hills, CA: Sage.

- Blumer, H. (1969). *Symbolic interactionism: Perspective and method.* Englewood Cliffs, NJ: Prentice Hall.

- Blumer, H. (2004). *George Herbert Mead and human conduct* (T. J. Morrione, Ed.). Walnut Creek, CA: AltaMira Press.

- Cook, G. A. (1993). *George Herbert Mead: The making of a social pragmatist.* Urbana, IL: University of Chicago Press.

For Simmel, the following are excellent secondary resources:

- Featherstone, M. (Ed.). (1991). A special issue on George Simmel. *Theory, Culture & Society, 8*(3).

- Frisby, D. (1984). *Georg Simmel.* New York: Tavistock.

Seeing the Social World (knowing the theory)

- Write a 250-word synopsis of the theoretical perspective of symbolic interaction.

- Write a 250-word synopsis of the theoretical perspective of formal sociology (feel free to use outside sources).

- After reading and understanding this chapter, you should be able to define the following terms theoretically and explain their theoretical importance to symbolic interaction: pragmatism, emergence, action, meaning, natural signs and significant gestures, social objects, emergence, role taking, perspective, self, mind, play stage, game stage, generalized other stage, the I and the Me, interaction, society.

- After reading and understanding this chapter, you should be able to define the following terms theoretically and explain their theoretical importance to Simmel's theory of the individual in modern society: social forms, sociability, exchange, conflict, subjective and objective cultures, urbanization, the division of labor, money, web of group affiliations, normative specificity, anomie, role conflict, blasé attitude.

- After reading and understanding this chapter, you should be able to answer the following questions (remember to answer them theoretically):

  1. Define pragmatism and apply the idea to meaning, truth, and self.

  2. Explain how the mind and self are effects of social interaction.

  3. Demonstrate how the mind and self are necessary for the existence of society.

  4. Define objective culture, and be able to explain how urbanization and the use of money increase the level of objective culture.

  5. Identity and describe the effects of urbanization and rational group formation on the individual.

Engaging the Social World (using the theory)

- More and more people are going to counselors or psychotherapists. Most counseling is done from a psychological point of view. Knowing what you know now about how the self is constructed, how do you think sociological counseling would be different? What things might a clinical sociologist emphasize?

- Using Google or your favorite search engine, do a search for "clinical sociology." What is clinical sociology? What is the current state of clinical sociology?

(Continued)

(Continued)

- Mead very clearly claims that our self is dependent upon the social groups with which we affiliate. Using Mead's theory, explain how the self of a person in a disenfranchised group might be different from one associated with a majority position. Think about the different kinds of generalized others and the relationship between interactions with generalized others and internalized Me's. (Remember, Mead himself doesn't talk about how we feel about the self.)

- How would Mead talk about and understand race and gender? According to Mead's theory, where does racial or gender inequality exist? From a Meadian point of view, where does responsibility lie for inequality? How could we understand class using Mead's theory? From Mead's perspective, explain how and why things like race, class, gender, and heterosexism are perpetuated (contrasting Mead's point of view with that of a structuralist).

- Remembering Simmel's definition and variables of objective culture, do you think we have been experiencing more or less objective culture in the last 25 years? In what ways? In other words, which of Simmel's concepts have higher or lower rates of variation? If there has been change, how do you think it is affecting you? Theoretically explain what the proportion of subjective to objective culture will be like for the generation following yours. Theoretically explain the effects you would expect.

- Perform a kind of network analysis on your web of group affiliations. How many of the groups of which you are a member are based on organic and rational motivations? In what kinds of groups do you spend most of your time? Over the next 5 years, how do you see your web of affiliations changing? Based on Simmel's theory, what effects can you expect from these changes?

Weaving the Threads (building theory)

Mead and Simmel offer an opportunity to synthesize their theories to get a better understanding of how the self is impacted by modern factors. Think of Mead telling us about how the person is socialized, and use Simmel to explicate the social and cultural context of modern socialization. Mead talks about language acquisition and role taking. Think of language acquisition more generally as culture, and understand role taking occurring within urban settings, with all that implies. After you've begun to get a sense of what the modern self/person would be, think about how the self of someone growing up and living in Los Angeles, California, might be different from that of someone growing up and living in a small, rural town.

CHAPTER 7

Seeing Gender

Harriett Martineau and Charlotte Perkins Gilman

As we've seen, the social project of modernity is based on the idea, value, and belief in equality, and this belief comes from a more fundamental assumption about human beings: Every individual is capable of discerning truth, is born with the capacity to reason, and is not only able but required to make decisions that guide his or her life and contribute to the welfare of society at large. However, you also know that the social project fell well short of the mark—the belief in equality was in truth founded upon practices of inequality. Western capitalism, for example, was built on black slave labor as well as upon the free labor that women provided and continue to provide. Women were also excluded from owning property in Western societies until the middle of the nineteenth century, and they didn't have the right to vote in democratic elections until well into the twentieth century. Gender continues to be the basis for the unequal distribution of income: In 2008, the median income for women in the United States with a college degree was 26 percent less than for men (U.S. Bureau of Labor Statistics, 2009). Gender is also a factor in violence. In 2008, about 552,000 women above the age of 12 were victims of intimate violence compared to 101,000 men (Bureau of Justice Statistics Selected Findings, 2009). In addition, it's estimated that 91 percent of rape victims are female versus 9 percent that are male (UC Santa Cruz Student Health Center, n.d.). In short, gender matters in terms of these issues of democracy: social justice, equality, and happiness.

Generally speaking, the social disciplines distinguish between sex (the biological assignment of male and female) and gender (the social constructions of meaning, norms, values, and so on that are built upon the biological categories). When compared to other systems of inequality, gender has unique characteristics. It is the oldest system of economic and political discrimination, and gender is ubiquitous: It's found in virtually every situation and thus crosscuts every other system of stratification. But gender is deeper still: It involves every facet of the body—from health to image to sex and sexuality. Gender is perceived to impact the way we think and feel and relate to our self, and gender is the foundation of marriage and family and thus an intrinsic part of how we become human through socialization. Further, gender is structured in such a way as to keep group members from connecting, sharing grief, and building solidarity—there is no neighborhood or ghetto for gender. Moreover, gender is required of every single person; thus, children are intentionally and thoroughly trained in gender, sometimes even before birth. Finally, gender is the social category that is most powerfully linked by most people to biological, genetic, and religious causes and legitimations.

Waves of Feminism

In Western European societies, these inequalities of gender have been addressed in three distinct waves of feminism. Ideas about the equal rights of women first began to emerge during the Enlightenment. The first significant expression of these issues was Mary Wollstonecraft's book *A Vindication of the Rights of Woman* (1792/1993). In it, she clearly linked gender to the discourse of human nature found in the

Enlightenment and thus "brought the issue of women's rights for a short moment in the 1790s into the general debate about civil rights" (Todd, 1993, p. xv). Wollstonecraft acknowledged the physical strength differences between men and women but argued that the differences in personality, mind, and action were due to education—education in the thorough socializing sense. Women, she argued, are socialized to be alluring, simply objects of beauty fit only for the pedestal. Women's "strength and usefulness are sacrificed for beauty" in that "they are treated as a kind of subordinate being, and not as a part of the human species" (Wollstonecraft, 1792/1993, p. 71).

The first wave of feminism became organized as the result of the 1848 Seneca Falls Convention, which called for equal rights for women to vote and own property, to full access to educational opportunities, and equal compensation. The Convention adopted a statement written by Elizabeth Cady Stanton (1848/2007) that was patterned after the Declaration of Independence: "We hold these truths to be self-evident: that all men and women are created equal; that they are endowed by their Creator with certain inalienable rights; that among these are life, liberty, and the pursuit of happiness" (p. 58). After a 72-year fight, women finally gained the right to vote in the United States in 1920; Great Britain followed in 1928.

The second wave of feminism grew out of the civil rights movements of the 1960s. Important for the second wave of feminists were the publication of Simone de Beauvoir's *The Second Sex* (1949/1989) and Betty Friedan's *The Feminine Mystique* (1963). Beauvoir affirmed that gender distinctions involve issues pertaining to human nature, but she used an existentialist philosophy as her base rather than the Enlightenment—in many ways, what she said is more profound.

Beauvoir (1949/1989) pointed out that part of what makes us human is our consciousness (see Chapter 3 of this volume); and intrinsic to our consciousness—one of the ways it is built—is the category of the Other. Otherness is built into most world religions, and it is a fundamental aspect of *self*-consciousness: *Self*-consciousness can only exist if there is an Other (otherwise, everything and everyone is the same, which is what some religions tell us). For men, the Other category is woman, and it's men who have defined it as such. The profound problem with this is that the Other doesn't have its own independent existence: "Thus *humanity is male* and man defines woman not in herself but as relative to him" (Beauvoir, 1949/1989, p. xxii, emphasis added). In their Otherness, women "do not authentically assume a subjective attitude. . . . [T]hey have taken nothing, they have only received" (p. xxv).

Friedan (1963) pointed to the same problem but came at it from a different perspective. The "feminine mystique" was the picture of a suburban housewife. Beginning in the latter part of the 1940s and coming into its own in the 1950s, being a "housewife" took on increased cultural value. Books were written, classes were organized, and degrees were offered in home economics to enable the American women to be the best she could be, which, of course, meant being a housewife caring for her children and husband. This was the mystique of feminine fulfillment.

The problem was that women weren't happy, but the *real* problem was that women couldn't name the problem. The mystique had so engulfed the hearts of Americans that it was unthinkable that being a housewife could be a source of unhappiness. Friedan (1963) argued that two things needed to happen. One, housework had to be seen for what it is and not as a career in itself. Friedan urged women to say no to the voices of marketers and advertisers who tell them they require a specific product brand to be the perfect, happy housewife. Two, like men, women will only find happiness in creative work of their own. She thus advocated a national education system for women that would be supported well enough that the demands of children and husbands wouldn't require the unpaid work of a woman.

Central to the organization of the second wave was the founding in 1966 of the National Organization for Women. Important issues for these feminists were pay equity; equal access to jobs and higher education; and the control over women's bodies, including but not limited to sexuality and reproduction, and the reduction of violence and rape.

The third wave of feminism is much harder to define, both in terms of its characteristics and its history. Rory Dicker (2008) argues that an important historical moment was the Clarence Thomas Supreme Court confirmation hearings (p. 103). Thomas, an African American, was appointed to replace Thurgood Marshall, the first black Supreme Court justice. During the hearings, African American lawyer Anita Hill accused Thomas of sexual harassment. It came down to he said/she said, and Thomas was confirmed in spite of the accusations, but it set off a firestorm of controversy.

Writing in response to the hearings and cultural upheaval, Rebecca Walker (1992/2007)—an African American woman (and the daughter of author Alice Walker)—said that the hearings were about "checking and redefining the extent of women's credibility and power" (p. 398). The hearings were also about one woman's *experience* in the face of a *structure* of prejudice. The night after the hearings, Rebecca Walker tells us, she was talking with a black man who saw the confirmation as a step backward for racial equality. She was outraged because this man didn't see the issue of gender at all, just race. A week later, she was on a train and overheard two men talking about sex and women: "'Man, I fucked that bitch all night and then I never called her again'" (p. 399). The man then hit on Walker, she told him in no uncertain terms what she thought, and then she moved to another part of the train. As Walker says of the incident, "I am livid, unrelenting in my anger at those who invade my space . . . who refuse to hear my voice" (p. 400). To her readers, she writes, "Let Thomas' confirmation serve to remind you . . . the fight is far over. Let this *dismissal of a woman's experience* move you to anger" (p. 400, emphasis added). Walker tells us that being third wave means demanding the validity of a woman's experience. It is in many ways a counter to the effects of being Other.

Yet third-wave feminism isn't just one thing. In her chapter on the third wave, Dicker (2008) touches on a number of manifestations, such as womanism, ecofeminism, queer theory, power feminism, and the grassroots cultural work of the radical feminist group Riot Grrrls. Kathleen Hanna (1992/2007), of the punk band Bikini Kill, writes in the "Riot Grrrl Manifesto,"

> WHY RIOT? BECAUSE us girls crave records and books and fanzines that speak to US that WE feel included in and can understand in our own ways . . . BECAUSE we must take over the means of production in order to create our own meanings.

BECAUSE viewing our work as being connected to our girlfriends-politics-real lives is essential if we are gonna figure out how [what] we are doing impacts, reflects, perpetuates, or DISRUPTS the status quo. (p. 395)

The time period of the two women I've chosen for this chapter coincides with the first wave of feminism, Martineau toward the beginning and Gilman toward the end. Charlotte Perkins Gilman was politically active and wrote theory as well as several works of fiction. Harriet Martineau, on the other hand, isn't someone we can classify as being directly involved in the feminist movement. Martineau was clearly concerned with gender and saw the condition of women to be the chief measurement of any society's democracy. For Martineau, the quality of a good society is measured by the happiness of its disenfranchised. But Martineau was a generalist in a way that Gilman was not, and her primary focus was on democracy.

Harriet Martineau—Gender and Democracy

Harriet Martineau was born in Norwich, England, on June 12, 1802, the sixth of eight children. Martineau was a sickly child, and by the age of 12, she had lost most of her hearing. Many of the adult pictures of Martineau show her with her hearing trumpet (an early hearing aid), which she acquired when she was 18. Martineau grew up in a Unitarian household and was generally educated at home (most women at this time were barred from the university). Unitarians believe in one God, in contrast to the doctrine of the trinity, and are committed to the use of reason in religion, universal brotherhood, and a creedless church. Unitarians were part of the religious dissenters' movement in England, which fought against the dominance of the Anglican Church. Though she eventually became agnostic, Martineau's religious background clearly informs her thinking, as we will see when we get to her typology of religion. Her father was a modest capitalist, whose business eventually collapsed due to competition from larger, more industrialized companies. This business failure forced Harriet to seek gainful employment, which she found, writing for the *Monthly Repository,* a Unitarian journal. Martineau began by writing a series of short stories with ethical and political points. She saw writing as a way of teaching and illustrating the ideas of political economy for the general public. Her talent for writing soon became obvious: In only her second year of full-time writing, she published a novel, a book on religious history, a variety of essays, and 52 articles for the *Monthly Repository.*

Among her best-known works are *How to Observe Morals and Manners* (a book of sociological methods), *Society in America* (a three-volume sociological study of America), *Deerbrook* (a novel about English domestic life), and the translation of Comte's *Positive Philosophy* into English. She wrote *Morals and Manners* on the ship traveling to America and spent 2 years in the United States gathering data for *Society in America.* Martineau was a prolific writer, traveled extensively, and was well respected by the intellectuals of her day. Martineau's first serialized piece, for example, outsold Charles Dickens, who was a contemporary of hers. By the time of her death, Martineau had published over 70 books and around 1,500 newspaper articles. Harriet Martineau died of natural causes at 74 years of age. She passed away at The Knoll, a home she designed and had built in the Ambleside Lake District in England, and she left her own obituary.

THEORIST'S DIGEST

Central Sociological Questions

Martineau is what we would call today a general theorist. She's obviously concerned with gender, but sees it as part of the bigger project of modernity and nation building. Thus, Martineau's work is focused on understanding how democracy is progressing. She answers this question by first explaining how certain institutions and issues ought to be in a modern democracy. Notice that for Martineau, and most early social thinkers, democracy could not be understood apart from ethics and morals: Democracy is built upon a specific ethical, moral base. After laying out what democracy should be, Martineau empirically compares those standards against how the United States was doing at the time.

Simply Stated

Martineau's theoretical work isn't concerned with explaining how something works or came into existence; she is focused, rather, on being a critical observer of the progress of democracy. Martineau's methodology goes something like this: In preparing to observe society, there are three issues a person must first resolve. She or he must have a clear notion of what defines human nature, the categories of distinction upon which observation will be made must be clear, and the observer must divest her- or himself of all ethnocentrism. In her work, Martineau considered many social institutions as measures of democracy; among them are gender and family, religion, and education.

Key Ideas

Natural law and happiness, morals and manners, preparing the observer, observational safeguards, things, gender, marriage, religion, education.

Concepts and Theory: Observing Society

Harriet Martineau is one of the earliest sociologists. She is best known for her translation of Auguste Comte's *Positive Philosophy*, one of the major foundation stones of sociology. In praise of Martineau, Comte wrote,

> And looking at it from the point of view of future generations, I feel sure that your name will be linked to mine, for you have executed the only one of those works that will survive amongst all those which my fundamental treatise has called forth. (as cited in Harrison, 1913, p. xviii)

Like Comte, Martineau felt that the social sciences could be used to improve the human condition. Martineau took Comte's positivism as a tool to measure how well the societies born out of the Enlightenment were faring. She looked specifically at the United States, as the nation most likely to embody the hopes of enlightened thinkers. In the words of Seymour Martin Lipset (1962), "Americans today have forgotten, if they ever knew, the meaning the United States once held for the world." To the poor, it was a country where "they might live as free and equal men." To the intellectuals, political activists and rulers "America was both a reality and a symbol" (p. 5).

The standard against which she evaluated the United States is one that was deeply embedded in the Enlightenment as well as the Declaration of Independence. It was the Declaration's line concerning the pursuit of happiness that Martineau (1838/2003) saw as the core of this democratic experiment: "every element of social life derives its importance from this great consideration—the relative amount of human happiness" (p. 25). Her method and purpose was "to compare the existing state of society in America with the principles on which it is professedly founded; thus testing Institutions, Morals, and Manners by an indisputable, instead of an arbitrary standard" (Martineau, 1837/1966, p. viii). Martineau's statements on methods are important because they constitute one of the first accounts of social methodology, which predates both Weber's and Durkheim's books. More important, however, is the reason why Martineau wrote her book on methodology: *She felt that every responsible citizen ought to make informed, scientific observations of society.*

Morals and Manners

Martineau's approach to observing society is to make a distinction between morals (stated beliefs) on the one hand and manners (observable practices) on the other. In her comparison of morals and manners, Martineau prefigures the contemporary work of Robert Wuthnow (1987). Wuthnow argues that the moral foundation of any society is its cultural code. The content *per se* isn't really a concern for Wuthnow; rather, he is interested in the distinctions between the code and the way people relate to the code. He sees a dynamic aspect to the distance between morals and manners. Individuals must maintain a close but distant relationship between the ideal and the real. If the distance gets too close, people will tend to become either discouraged, because no one can uphold the ideal, or self-righteous and judgmental in their supposed adherence. On the other hand, if the boundary between the two is blurred and the individual ceases to be able to distinguish between the ideal and the real, then either cynicism or loss of justification for action will result. Though Martineau doesn't include the social psychological effects, the use of comparison is the same: How do our actions measure up to the unseen beliefs we hold?

Martineau (1838/2003) advocates the observation of what she terms Things: "The great secret of wise inquiry into Morals and Manners is to begin with the study of THINGS, using the DISCOURSE OF PERSONS as a commentary upon them" (p. 73). Like Durkheim, she sees that morals are part of the "common mind" of a people and that this public consciousness can be understood through social facts. However, Martineau's approach is much broader than Durkheim's. I've listed the social facts that Martineau uses to measure morals and manners in her work in America in Table 7.1. I'm not going to be describing each element of her methodology because they are of varying degrees of theoretical importance. I do, however, want to point out the variety and objectivity of the things she has on her list. In order to understand human practices, Martineau uses such things as typologies, buildings, the physical environment, social practices such as suicide, cultural artifacts such as cemeteries, social networks such as attachment to birthplace and family, land ownership, health, marriage, education, police, urbanization, newspapers, songs, literature, and so on. One is almost overwhelmed at the breadth by simply reading the list. The really interesting point is that almost all of these items are actual things and facts that can be observed. One more issue I'd like to point out is that this list, along with her follow-up interviews with people, represents a very sophisticated and contemporary triangulation of methodologies. She used observational techniques, interviews, archival, and content analysis methods.

Table 7.1 Martineau's Measurements of Morals

Structure	*Observance*
Religion	General type: licentious, ascetic, or moderate Physical places of worship Condition of the clergy Superstitions Suicide
General Moral Notions	Cemetery (age, and cause of death; visions of future) Degree of attachment to birthplace and family Representative characters—two kinds: dead (deepest beliefs) and living (currents of belief) National celebrations Treatment of guilty Songs Literature
Domestic State	Physical environment Distribution of resources Method and ownership of land Extent of commerce Physical health Condition of marriage
Idea of Liberty	Amount of feudal arrangements still present Kind of police (military, private, public) Urbanization Conditions of servants Freedom of press Education Persecution of opinion

Preparing the Observer

Further, Martineau sets out requirements for observation, the first of which concern the observer. Martineau is making an important point in social research: Sometimes the observer is the instrument of measurement. To demonstrate the importance of preparation, Martineau points out that if someone asked you or me to explain the physics of the universe, or the biology of the human body, or the geographic strata of the Grand Canyon, we would respond by saying the request is silly, unless we had some basis to claim expertise. Yet the average person does not hesitate for a moment to "imagine that he can understand men at a glance" (Martineau, 1838/2003, p. 14). I wonder what Martineau would think about the prevalence today of people expressing their opinion about social issues on radio, television, and blogs—mostly without an ounce of preparation whatsoever. Martineau thought having an opinion about social things without training was just as ludicrous as my having an opinion about Einstein's theory of special relativity. This was why Martineau wrote her book on methods of observation.

For Martineau, there are two main areas of preparation: philosophical and moral. The first thing to notice about her preparations is that they pertain to the inner person. Martineau points out that the instrument of observation is the human mind. Here we can really see the difference between old-school positivism and neo-positivism (see Chapter 1). Of course, neo-positivism requires training, but the real instrument of measure is the survey or statistical analysis. For the old school, it's the inner person that must be prepared.

Philosophical preparation. The first philosophical preparation you have to make is to know what it is you want to know. For Martineau, what she wants to know is the condition of the society with regard to happiness. I'd like you to see a couple of things here. First, Martineau didn't get the idea of happiness off the top of her head; she drew it from the intellectual discourse of her time. Darrin McMahon (2006) tells us that prior to the Enlightenment, happiness was viewed as a reward for exceptional behavior or having arrived at perfection. For thinkers of the Enlightenment, happiness was "a *natural human endowment* attainable in theory by every man, woman, and child" (p.13, emphasis added). In fact, as Martineau points out, if people aren't happy, then something was wrong "with their beliefs, with their form of government, with their living conditions, with their customs" (p. 13). The second thing I'd like you to see is in the italicized portion of the preceding quote. We again see the importance of the ideas and assumptions about human nature for studying society.

Another element of philosophical preparation involves understanding that human beings *create* social institutions and their beliefs and practices. Martineau lived at the beginning of the age of modernity, the period of time when the world began to shrink. Communication and transportation technology had shown the civilized nations that there are many truths and many ways of organizing human behavior. Martineau observes that this information should give the researcher the ability to step out of her or his own culture. Martineau sees two important results from this open point of view. The prepared observer of human morals will "escape the affliction of seeing sin wherever he sees difference" and will be freed from emotional responses that can cloud observation (Martineau, 1838/2003, p. 34). Rather than experiencing shock and alarm, the observer is able to perceive and understand the morals and manners of a people as they exist for that society.

Moral preparation. In addition to philosophic preparation, the observer of the human condition must make moral readiness. In this, Martineau pulls no punches, for she tells us that to make accurate observations the observer "should be himself perfect." Here we can see the depth of relationship between the observer and the observed. Martineau truly believes that the inner condition of the researcher is the prime concern regarding preparation for making social observations. Martineau also recognizes that no one is perfect, and if we should wait for perfection, no observations would be made at all. Thus, we are to take stock and understand the issues that would most easily beset our observations and make us prejudiced. Knowing what can pervert our vision, we can "carry with us restoratives of temper and spirits which may be of essential service to us in our task" (Martineau, 1838/2003, p. 52). In other words, we should know enough about the human spirit, and our own psychological and emotional development in particular, to be able to plan ahead to meet the issues that may cause us to be prejudiced.

Moreover, and more practically, the researcher must have unreserved sympathy. Martineau says that a geologist can have a heart of stone and a statistician can be as abstract as a column of numbers and still be successful. But the observer of society will be subject to deception at every turn if she cannot find her way into the hearts and minds of people. The observer of humanity must herself be most human. She must be able to feel her way through the labyrinth of social institutions and interactions and find what is truly meaningful to the subjective actor. She must be able to empathetically understand what people do and why they do it, what people believe and why they believe it, and what people feel and why they feel it.

Notice first that Martineau's emphasis echoes Weber's concern with subjectivity and *verstehen,* except Martineau predated Weber by about 80 years. I also want to point out that this priority of the person that Martineau gives us is a central feature in modern feminist epistemology. According to Patricia Hill Collins (2000), a black feminist epistemology entails the ethics of caring, which implies "that personal expressiveness, emotions, and empathy are central to the knowledge validation process" (p. 263). Dorothy E. Smith (1987) tells us, "a sociology for women preserves the presence of subjects as knowers and as actors. It does not transform subjects into the objects of study" (p. 105).

Safeguarding Observations

Martineau adds safeguards to the philosophic and moralistic preparation of the investigator. She advises that no preemptory decision be made, not only in public when she returns to her home culture but also in the private, daily journal she keeps. The researcher, in other words, must *deny her- or himself the luxury of coming to any conclusions until massive amounts of data are gathered.* This advice stands in stark contrast to the American system of academia that honors the number of publications over the profoundness and depth of thought and insight. Further, the observer must guard against making generalizations too quickly. The connections we see between different situations and beliefs may in fact be only on the surface. The data we gather must, then, be complex and nuanced (Martineau again predating Weber). Only by knowingly guarding ourselves against premature generalization can we hope to do justice to the variety of subjective worlds that we observe. Obviously, Martineau isn't telling us not to make generalizations. Rather, she is telling us to make them slowly and only after compiling enough detail to respect the social worlds of others. Of course, one way to check our observations of others is to actually ask them, a practice Martineau recommends.

The final safeguard that the traveler must keep is hope. The path that Martineau advises is long and arduous. The researcher must make philosophical and moral preparations. Her heart and mind must be ready to observe human behavior. The observations must be made in great detail and breadth. She must keep a journal of her observations and considerations, being careful to not generalize or draw conclusions too quickly. This kind of research takes time, patience, and part of our soul. With such costs, it's easy to become discouraged, particularly when we have to be so careful and take so much time. But Martineau (1838/2003) tells us to keep hope:

> Every observer and recorder is fulfilling a function; and no one observer or recorder ought to feel discouragement, as long as he desires to be useful rather than shining; to be the servant rather than the lord of science; and a friend to the home-stayers rather than their dictator. (pp. 20–21)

Concepts and Theory: Gender and Democracy

Martineau investigated many institutions in her quest to understand the progress of democracy in America. We'll consider three of them; each is important for different reasons. First, gender and family are of specific concern because home is the place of primary socialization, where morals and manners are thus bestowed and can best be observed. In addition, the elite of any society will be happy (happiness is a key point in modernity), but the true test of democracy is the happiness of those that society excludes—in this case, women. Second, religion is one of the most powerful institutions deciding the success or failure of democracy. According to Martineau, modernity and democracy can only prosper under one type of religion: moderate religion that has few rituals and is concerned primarily with education and self-actualization. Third, the single most important institution in a modern democracy is education. Every level of education must be freely available, and a university education must be based upon critical thinking.

Gender and Family

The home is important for several reasons in Martineau's scheme. It is the primary part of what she calls the domestic state. As you can see in Table 7.1, the domestic sphere of society includes economic and distributive structures (manufacturing, markets, roadways, and so forth), commerce in general (Martineau includes dealings between individuals or groups in society, and interchanges of ideas and opinions, or sentiments), the general level of health, as well as marriage. Among all these domestic institutions, family (marriage in particular) stands out. It's the place where the primary socialization of children takes place, and a place where the morals and manners of a people can best be observed. Further, the treatment of women in marriage is one of the most important keys for understanding how a society measures up to the universal law of happiness. Keep in mind that happiness is the basic human right and that marriage is the fundamental social institution. Thus, unhappy marriages are a strong indicator of the moral state of an entire society. Martineau also says that the treatment of blacks is a measuring rod for happiness. It's the position of the oppressed that is particularly important for critiquing a society's overall standing with regard to happiness. The elite in any society are going to be happy; but whether or not human happiness is generally valued can only be gauged by how universal it is in a collective, and that is determined by the experience of society's minority. The reason for this goes back to the issue of natural rights. If human beings have natural rights, then the groups it excludes from those rights and the degree of that exclusion judge the success of government.

The Cultural Logic of Gender

Cultural logic is an idea that expresses the fact that culture has implications, often unanticipated or unintended. The idea is particularly insightful regarding gender. One example of cultural logic is eating disorders. Cultures that put a high premium on women being thin tend to have higher incidences of eating disorders among women. It's no surprise, then (there's a logic to it), that the United States has the highest incidence of eating disorders in the world. The only nation that is even close is Japan. It's also no surprise that eating disorders were almost unknown in Japan prior to the 1960s, when Western culture began to make inroads; actual cases in Japan doubled between 1976 and 1981, and quadrupled between 1988 and 1992. Martineau uses the idea of cultural logic to talk about gender relations, and in this she was (again) ahead of her time.

Martineau argues that the state of marriage in particular provides insight into the moral fabric and practices of a people. She asserts there is a link between marriage and the practice of gender. While the institution of marriage and the structure of gender are two distinct things, they also have reciprocal effects on one another. Thus, the unhappiness in marriage that is prompted by gender inequality is given strength by our cultural concepts of gender. Specifically, Martineau finds clear links between the condition of marriage and the beliefs that men should have courage and women should be chaste. The connection between gender on the one hand (masculinity = courage; femininity = chastity) and marital unhappiness on the other is a cultural link.

What's wrong with men being identified with bravery? Martineau argues that as long as the chief defining feature of masculinity is courage, then men's primary obligation is to society. Bravery is a social virtue. It is played out on the stage of society for the benefit of society: Men are brave so that society can exist. They demonstrate their courage most clearly on the battlefield and are given medals, parades, and sacred burials in return. Being brave in war is the pinnacle toward which all other forms of masculine bravery is oriented, whether it is bravery in the face of pain in sports or emotional strength in the face of sad stories. Little boys must be brave and not give in to pain or emotion because someday they may be called to battle.

The cultural logic of this gets played out in gender in a couple of ways. First, and probably most insidiously, the principal male quality is expressed in obligation to society, not to the home or to individual people—most specifically, to wives. There is, then, a clear separation between public and private spheres in this culture of courage and chastity. The logic of the culture dictates that men are responsible for public society and social power, and women are responsible for the private fields of home and sexual purity.

The other significant result of this cultural logic is that women are subject to the whims of men. Martineau tells us that honor associated with courage is shallow because it comes to be "viewed as regarding only equals," which in this case would only be other men. This kind of honor is not a personal virtue. It is associated with courage, which we have seen is a public sentiment, and is therefore a public issue. A public authenticity is at stake in this kind of honor, not personal authenticity. Moreover, its honorability is dependent upon its base. What I mean is that this kind of honor is defined by courage. Therefore,

> The requisitions of honour come to be viewed as regarding only equals, or those who are hedged about with honour, and they are neglected with regard to the helpless. Men of honour use treachery with women; with those to whom they promise marriage, and with those to whom, in marrying, they promised fidelity, love, and care; and yet their honour is, in the eyes of society, unstained. (Martineau, 1838/2003, pp. 173–174)

Let me reemphasize that Martineau does not see these as *intrinsic* features of gender. Men are not naturally aggressive, and women are not naturally moral. The sentiments and values associated with gender are found in specific kinds of societies—structure and culture go together in Martineau's scheme.

Workforce Participation

Martineau also argues that the level and kind of workforce participation women have influences the moral codes of a society. In this, of course, she foreshadows a great deal of feminist theory, as well as the work of Charlotte Perkins Gilman. There is a relationship between the way we think of marriage and the workforce participation of women. Martineau gives us two ways that domestic morals are influenced. First, to the degree that women are left out of the workforce, they will be seen as "helpless." This helplessness, Martineau tells us, provides the basis for men's aggressive domination of women, oftentimes to the point of physical violence. There is a relationship between personal power or efficacy and money. This relationship makes perfect sense, of course, in a society where one's survival is linked to work and making money. In societies where a living wage is guaranteed to all, or women are given equal opportunity, this association would be very weak. Women would not be dependent upon men for their survival, and the power between the genders would be equal. In societies "where the objects of life are as various and as freely open to women as to men," there will be found "the greatest amount of domestic purity and peace" (Martineau, 1838/2003, p. 182).

The second result of unequal workforce participation is that the institution of marriage is "debased." The contemporary ideal of marriage in our culture is that it is a union freely entered into by two people in love. In a society where marriage is the chief way that women get ahead or survive, it can never be the case that marriage is secure "against the influence of some of these motives even in the simplest and purest cases of attachment" (Martineau, 1838/2003, p. 183). Martineau actually frames this issue more pointedly. She talks about it in terms of marriage being the principal or only aim of women in society. When the main goal of women is to marry and have children, then the values associated with that goal are inculcated in the very young. Young girls are socialized to value being attractive to men (yet chaste), dating, marriage, and having children. The toys, games, and images provided for girls generally have these themes. Thus, when grown, women sense that they are unfulfilled and incomplete unless they are married and have children. One's femininity becomes synonymous with marriage and childbearing. In such a society, according to Martineau, "the taint is in the mind before the [marital] attachment begins, before the objects meet; and the evil effects upon the marriage state are incalculable" (p. 183).

Martineau (1838/2003) makes an interesting observation about the condition of women and the legitimacy of the state. She says that every civilized nation claims legitimacy by its treatment of women. In modern society, the rights of women were probably an issue before racial inequality. No matter what race or society or culture in which you find yourself, there are men and women and gender relations. As I've mentioned, one of the earmarks of modernity is the idea of the natural equality of all men. Before modern democracy, people were judged to be different by birth. For example, in a monarchy, the royal family is believed to be naturally superior. At the core of the modern state is the belief that all men are *created* equal. Of course, this idea was initially understood in a limited manner, and women were thus seen as naturally different from men. Once planted, however, this idea took root and grew to eventually include women.

Because of the link between state legitimacy and equality, nation-states claiming to be modern point to their treatment of women as proof of modernity. However, as Martineau notes, different proofs may be used and in the end only serve to legitimate the exclusion of women. One society of people may claim that women are honored in their country because women are "cherished domestic companions" and afforded high respect. This respect might be shown by allowing women to go first through a door, by standing when they enter or leave a room, or by covering them from head to toe in a veil of sacredness. Another society will claim honor because religion safeguards women by not requiring them to deal with the harsh aspects of life; another will declare the honor of women because all of society knows that behind every great man is a great woman. The list of legitimations goes on, but Martineau claims that only in societies where women are guaranteed equal workforce participation is there true honoring of women and true equality.

Concepts and Theory: Religion, Education, and Democracy

Religion is another important social structure for measuring society's morals and manners. As we've seen, religion is consistently theorized as an important social institution. For example, Durkheim argued that religion forms the foundation of society. Out of it sprang the primary categories of thought and the moral basis of society. Durkheim noted that the two functions of religion eventually split, with science taking over speculation about the origins and workings of the universe and religion the issues of faith. Religion, in Durkheim's way of thinking, will evolve to a very general form that embraces all humankind under universal values.

Religious Forms

Martineau isn't concerned with the origins of religion, as Durkheim is, but she does seem to say that religion evolves toward more inclusive, less judgmental forms. The *forms* of religion are of chief interest for Martineau. She gives us three types of religion: licentious, ascetic, and moderate. *Licentious religions* are those where the gods are seen as part of nature and having human passions, as in the early Greek pantheons. During the early stages of this type of religion, licentious or immoral conduct is encouraged. This form changes when people are seen as less than the gods, and the power for doing good or harm is attributed to the gods. When this happens, the idea of propitiation (an act or gift given to appease the wrath of a god) enters and ritual worship begins. Even in its advanced form, this kind of religion is still preoccupied with licentious behavior, albeit in a negative way: The rituals performed for an almighty god counter the ill done by sin.

Ascetic religions are highly ritualized as well. Emphasis here is on rituals that provide proofs of holiness. Martineau argues that spiritual vices rather than immoral acts of the flesh are the center of this type of religion. The spiritual vices of "pride, vanity, and hypocrisy,—are as fatal to high morals under this state of religious sentiment as sensual indulgence under the other" (Martineau, 1838/2003, p. 79). Martineau sees religious ritual as misplaced emphasis. In this way of thinking, rituals are an empty form that takes our attention away from "self-perfection, sought through the free but disciplined exercise of their whole nature" (p. 80). *Moderate religions* are the least ritualized of the three, and these focus on self-actualization through broad education that involves the entire being.

Martineau doesn't argue that the different forms are replaced as religion evolves. On the contrary, each of these forms existed in her day and is still with us today. For example, traditional Christianity is based on propitiation and the sins of the flesh.

The older forms of Christianity, like Catholicism, still emphasize ritual as the vehicle through which come propitiation and forgiveness. The big question for the social researcher is which form dominates. The importance of the different religious forms is that religion plays a key role in forming the morals of any society. If licentious religious forms dominate, then the political and domestic morals will be very low or contain mostly ritual. Where asceticism governs, the natural behaviors and emotions of humanity are judged to be sin. Martineau (1838/2003) prefigures symbolic interaction and labeling theory by arguing that once a behavior or emotion is called licentiousness, "it soon becomes licentiousness in fact, according to the general rule that a bad name changes that to which it is affixed into a bad quality" (p. 81). Moderate religion encourages a culture of freedom and acceptance. It's a "temper" rather than a pursuit, a settled state of mind focused on inner peace and strength rather than behaviors.

Religious Forms and Democracy

Martineau's understanding of religion is evolutionary and intrinsically linked to modernity and democracy. She doesn't explain how religion changed, but she provides a typology of religion whereby observers can ascertain democratic progress by the type of religion most prominent. The first historical type is ascetic religions, which use abstinence for proof of holiness; a society wherein this type of religion dominates will be characterized by an ever-expanding inventory of sin and an emphasis on outward restraint rather than inward ethics. The second type, licentious religions, are so called because in their beginning phases, they lack moral constraints—gods and humans alike participate in lewd behavior. In the latter phases, this kind of religion still focuses on immoral behaviors, but on their exclusion: More and more actions are defined as sin and are easily atoned for through ritual. In society, licentious religions produce a condition with little concern for humanistic morals and strong emphasis on ritualized behaviors. The third, modern, type is that of moderate religions, which emphasize the overall well-being of humanity through self-actualization and education. A society that has a high level of this type of religion will have a culture that stresses freedom and true equality and broad-mindedness.

Conditions of Religion

The things of religion that Martineau advises us to look at include the places of worship, the condition of the clergy, generally held superstitions, holy days, and suicide. In talking about the places of worship, Martineau is giving us a *sociology of space*. Sociology of space looks at how the design and placement of buildings influences human sentiments and behaviors. Martineau tells us to look at where the religious buildings are located, how they are built, their number, their diversity, and whether or not the physical structure is conducive to ritual or spiritual service.

Martineau's sociology of space takes in much more than architecture. The first thing she recommends in studying domestic institutions is to look at a geological map. Martineau recognizes that how a society is situated geographically, that is, its physical environment, is important for how social structures develop. For example, nations whose territory is isolated but filled with natural resources will have a different set of structural arrangements than a society that has boundaries on all sides and is dependent upon other nations for raw materials. The natural environment also strongly influences a society's division of labor and the psychological makeup of the workers. Farming needs large expanses of fertile soil, and mining requires deposits of valued ore—farmers and miners are different kinds of people because of the different work they perform. Areas that are conducive to city dwelling and trade will contain artisans and craftspeople, who are both different from miners and farmers. Martineau tells us that the psychological dispositions of a society are significantly dependent upon these jobs and political realities, all of which are influenced by the environment.

In addition to architecture and building placement, the clergy are important for the study of religion because they are the first learned people in a society. Education begins with them, so it is important to ask, what kind of knowledge do they espouse and how are they esteemed in society? Are they respected because of their perceived holiness or their great learning? Suicide, of course, gives us some idea of religious sentiment as well: Why do people commit suicide? Is it from shame, from devotion, or to shrink from suffering? What kind of religious sanctions or motivations are there regarding suicide?

Martineau places her thinking about religion under a research agenda she calls the "idea of liberty." The way a society is structured around the idea of liberty is extremely important to Martineau. It is clearly related to the idea of happiness. The way in which the structures of a society are related to happiness is gauged by the level and kind of oppression. As we've seen, her key cases are those of race and gender. The opposite of oppression is liberty. Thus, liberty can be seen as the positive counterpart to oppression in measuring the structured potential of happiness in any society. Martineau mentions several factors that provide measures of liberty: the level of class divisions in society (the greater the bifurcation, the more likely the use of ideology and suppression); the kind of police force (national and military versus local and custodial); the level of urbanization and the influence city culture has on the nation (urbanized culture will tend to be more diverse and accepting of difference); the level of democratic governance; the level of direct or indirect governmental control of the press; and the degree of persecution of opinion. However, Martineau argues that in many ways the most important indicator of the idea of liberty is the condition of the education system in society.

Education and Freedom

The extent of popular education is a fact of the deepest significance. Under despotisms there will be the smallest amount of it; and in proportion to the national idea of the dignity and importance of man,—idea of liberty in short,—will be its extent, both in regard to the number it comprehends, and to the enlargement of their studies. (Martineau, 1838/2003, pp. 202–203)

As a modernist thinker, Martineau believes in the progress of humanity. According to Martineau, there are two great powers—force and knowledge—and the story of humanity is the movement away from one and toward the other. Social relations began through physical force and domination and the idea that "might makes right." In such societies, "the past is everything." The kind of knowledge that is of most value, then, is traditional knowledge. The tastes, ideas, sentiments, and ambitions in such a society are those that come from antiquity. This kind of knowledge is nonreflexive and maintains the status quo; it "falls back upon precedent, and reposes there" (Martineau, 1838/2003, p. 45).

Martineau (1838/2003) argues that society moves away from force and toward knowledge as the basis for moral power in response to increases in commerce, professionalization, communication, and urbanization (p. 46). Under these conditions, knowledge cannot stay fixed in the past. It is opened up to new horizons through increases in communication and commerce with other societies. Urbanization and commerce tend to expand the limits of knowledge as well as push for constant change. City living also brings people more in contact with diversity, which in turn opens up the boundaries of knowledge. Professionalization creates expert knowledge systems and, as more and more jobs and trades become professions, people in general begin to value expert knowledge over traditional information.

Concerning the education system itself, Martineau gives us both interesting (like the use of uniforms, which she says is "always suspicious") and obvious measures (such as the distribution of social types—gender, race, religion, class—and the general level of educational attainment). However, two of the most important indicators of education and its connection to the idea of freedom are its financial base and the position of the university. The influence of education's financial base should be apparent, but it bears mentioning. The first issue is *the extent of free education*. The level to which a society supports public education is an unmistakable measure of the support of the ideas of freedom and democracy. In civilized countries, education is the way to socially advance; it's the legitimate way to get ahead. A high level of support for free, public education indicates a high level of support for equal opportunity. The level of supported equality is linked to the kinds of jobs or careers one can get with a given level of education. In other words, all we have to do to understand the level of equality that society wants to support is look at the kinds of job opportunities free and public education provides. For example, if a society supports public education only through high school, it indicates that the level of equality that the state is interested in providing is equal to the jobs that require a high school education.

The second indicator of the place of education in society is *the status of the university*. Martineau (1838/2003) claims that "in countries where there is any popular Idea of Liberty, the universities are considered its stronghold" (p. 203). The reason for this link between liberty and universities might surprise you. Weber tells us that in societies where bureaucracy is the principal form of social organization, education will serve to credential a workforce. Martineau's function for the university is different from Weber's. Of it she says, "It would be an interesting inquiry how many revolutions warlike and bloodless, have issued from seats of learning" (p. 203). According to Martineau, the function of the university in its relationship to liberty is to inspire critical thinking. A university education should make you question authority and society. Freedom must be won anew by every generation. Almost by definition, democracy is alive and expansive. A university that doesn't challenge its students with new ideas and doesn't provide an atmosphere wherein new principles can be founded, is "sure to retain the antique notions in accordance with which they were instituted, and to fall into the rear of society in morals and manners" (p. 203).

Not only are the purpose, content, and environment of the university important, but so are its students, in particular their motivation for study. To the degree that students are motivated to obtain a university education for a job, education for freedom is dead in a society. Martineau (1838/2003) makes a comparison between students in Germany and those in America (recall that she is British and thus has no vested interest). German students are noted as having a thirst for knowledge: The student may "remain within the walls of his college till time silvers his hairs." The young American student, on the other hand, "satisfied at the end of three years that he knows as much as his neighbours . . . plunges into what alone he considers the business of life" (p. 205). These words were of course written almost 170 years ago. However, Martineau had great hope for the American system of education: "a literature will grow up within her, and study will assume its place among the chief objects of life" (p. 206). I will let the reader judge the fulfillment of her hope.

Summary

- Martineau's perspective is based on positivism and the Enlightenment. She believes that all humans have an inalienable right to happiness. This notion is the touchstone for all empirical and theoretical work. Because happiness is a natural right for humans, governments must be measured in terms of their facilitation or denial of this right. In particular, Martineau uses the way society treats gender and race as key indicators of happiness. Empirically, societies are measured by comparing their stated morals and practiced manners.

- Martineau feels that every citizen should be part of the enlightened observation of society. Society is generally moving forward, but its true progress toward providing for the natural rights of human beings must be measured, and societies must be held accountable to these measures. However, because the human mind is the principal tool of observation, quite a bit of preparation must take place: Observers must not judge the behavior of others simply because it is different from their own; the researcher must also be able to empathize with others.

- Martineau argues that there is a link between the culture and practice of marriage, and gender. Courage as a particularly masculine trait will obligate men to society, rather than to family or relationships; thus, it follows that men will give honor to heroic people (men), rather than to women and children. The feminine culture of chastity also has practical effects. Equating chastity with femininity makes women responsible for intimate relations. Together, these cultural equations help to create separate spheres of influence and concerns for men and women.

- Marriage and gender are also negatively affected as women are denied workforce participation. Not only are women seen as helpless and men enabled to be vicious, but marriage is always to some degree motivated by economic concerns rather than more enlightened incentives, when women are given limited workforce opportunities. Children are socialized into this gendered culture and economy. Thus, when grown, people feel that their gender is essentially tied to expressions of courage and work, chastity and relationships.

- In order for societies to progress toward providing happiness for their citizenry, education must be freely provided. The level of equal opportunity in a society is measured by the level of free education. Education must also be sought for its own sake, and not simply economic gain. In addition, the universities in a society must provide a critical, reflexive education. The kind of religion most prevalent in a society will also influence how far a society goes in providing for and safeguarding happiness. Overly ritualized religions, or religions that focus on holiness or redemption, will stand in the way of the self-actualization of the people. Moderate religions provide a mind-set that is focused on inner peace, strength, and authenticity. Societies will tend to move toward greater levels of free and open education and moderate religious forms due to the influences of commerce, professionalization, communication, and urbanization. However, the true level is always determined through empirical research.

Charlotte Perkins Gilman—The Evolution of Gender

Charlotte Stetson Perkins Gilman was born Charlotte Perkins on July 3, 1860, in Hartford, Connecticut. Her father was related to the Beecher family, one of the most important American families of the nineteenth century. Gilman's great-uncle was Henry Ward Beecher, a powerful abolitionist and clergyman, and her great-aunt was Harriet Beecher Stowe, who wrote *Uncle Tom's Cabin*, which focused the nation's attention on slavery. Gilman's parents divorced early, leaving Charlotte and her mother to live as poor relations to the Beecher family, moving from house to house. Gilman grew up poor and was poorly schooled.

Gilman's adult life was marked by tumultuous personal relationships. Gilman's first marriage in 1884, to Charles Stetson, ended in divorce, but it also provided the impetus for her novella, *The Yellow Wallpaper*. The autobiographical book tells the story of a depressed new mother who is told by both her doctor and husband to abandon her intellectual life and avoid any writing or stimulating conversation. The woman sinks deeper into depression and madness as she is left alone in the yellow-wallpapered nursery. Gilman's feeling of hopelessness against the tyranny of male-dominated institutions is heard in the repeated refrain of "but what is one to do?" Between her first and second marriages, a period of about 12 years, Gilman had a number of passionate affairs with women. However, her second marriage—to

George Gilman, a cousin—proved to be at least a somewhat successful and important relationship for Charlotte: Some of her best work was done during the courtship and marriage.

In her lifetime, Gilman wrote over 2,000 works, including short stories, poems, novels, political pieces, and major sociological writings. She also founded *The Forerunner*, a monthly journal on women's rights and related issues. Among her best-known works are *Concerning Children, The Home, The Man-Made World*, and *Herland* (a feminist utopian novel). However, Gilman's most important theoretical work is *Women and Economy*. In Gilman's lifetime, the book went through nine editions and was translated into seven different languages.

Gilman was also politically active. She often spoke at political rallies and was a firm supporter of the Women's Club Movement, which was an important part of the progress toward women's rights in the United States. It initially began as a place for women to share culture and friendship. Before the development of these clubs, most women's associations were either auxiliaries of men's groups or allied with a church. In contrast, many of the clubs of the Women's Club Movement became politically involved, but others were simply public and educational service organizations. One such club became the Parent-Teacher Association (PTA). On August 17, 1935, sick with cancer, Charlotte Gilman died by her own hand in Pasadena, California.

THEORIST'S DIGEST

Central Sociological Questions

Gilman was specifically concerned with gender in a way that Martineau was not. Gilman was also an evolutionist, much like Herbert Spencer. But where Spencer was concerned with the evolution of society, Gilman was focused on the evolution of gender inequality. Gilman is also different from Spencer in another important way: She was interested in how the evolution of gender inequality has affected and in turn is structured by the female body.

Simply Stated

In order for humans to evolve a productive economy and complex social relations, patriarchy and monogamy were naturally selected. As a result, women's bodies changed and developed exaggerated secondary sex characteristics in order to attract a mate. However, human beings have now dominated the natural environment utterly and have created complex social relations extending far past kinship. Thus, these early adaptive changes of patriarchy, monogamy, and exaggerated sexual dimorphism are now dysfunctional for human society.

Key Ideas

Gynaecocentric theory, self- and race preservation, sexuo-economic relations, morbid excess in sex distinctions.

Concepts and Theory: Critical Evolution Theory

We've seen that Harriet Martineau's perspective is somewhat informed by the idea of evolution. She sees society moving from organization by physical force to order through the moral force of education. Gilman, on the other hand, bases her entire perspective on a fully developed theory of evolution that in many ways is like that of Herbert Spencer (Chapter 2). Gilman generally understands the history of human society as moving from simple to complex systems, with various elements being chosen through the mechanism of survival of the fittest. Gilman doesn't simply use the evolutionary perspective to explain societies up to this point. As we will see, she uses evolution to explain what is going on currently in society and where we are headed.

Functional Evolution

Gilman begins with functionalism, but she does a couple of interesting things. First, she takes seriously the idea of environment. With Spencer and Durkheim, the environment is assumed, but it doesn't become a major player in the theories. Gilman makes the environment a major factor in social evolution. However, she does it with a twist. Anticipating the work of people like Peter Berger and Thomas Luckmann (1966), whom I've mentioned a couple of times in this book, Gilman argues that there came a point in our species' history where the sociocultural environment took precedence over the natural environment. The second thing she does with functionalism is imply that there are unintended dysfunctions. This insight didn't enter into mainstream sociology until the work of Robert K. Merton (1949). The third unique contribution is that Gilman adds a critical Marxian bent to functional theory.

Like Spencer, Gilman argues that humans began as brute animals pursuing individual gain. Survival of the fittest at that point was based on individual strength and competition. Men would hunt for food and fight one another for sexual rights to women. It was a day-by-day existence with no surplus or communal cooperation. Wealth was, of course, unknown, as it requires surplus and social organization. Competition among men was individual and often resulted in death. Females, as is the case with many other species, would choose the best fit of the males for mating. There was no family unit beyond the provision of basic necessities for the young, mostly provided by women. Life was short and brutal.

As competition between species within a natural environment continued, human ancestors gradually developed a social organization beyond their natural base. Very much like an organism, society began to form out of small cells of social organization that joined with other small cells. Gilman (1899/1975) says that "society is the fourth power of the cell" (p. 101). By that she means something similar to Spencer's notion of compounding: Once societies started to grow and structurally differentiate, the process continued in a multiplicative manner. The reason for this is simple: Complex societies, because of the level of social cooperation and division of labor, have a greater chance of survival then do individuals or simple societies. As societies became more complex, and thus more social, structures and individuals had to rely on each other for survival. Just like the way different organs in your body must depend upon other organs and processes, so must every social unit (individuals and structures) depend upon the others.

An interesting thing happened for humans at this point. All animals are influenced by their environment. If the environment turns cold, then the organism will either develop a way of regulating body temperature (polar bears are well-suited for the Arctic), move to a different environment, or die out. The environment equally affected humans at one time. But after we organized socially, our organization and culture developed, allowing us to live in a controlled manner with regard to the environment: Humans can now exist in almost any environment, on or off the planet. We thus distanced ourselves from the natural environment, but at the same time we created a new environment in which to live. Gilman argues that *human beings have become more influenced by the social environment of culture, structure, and relations than by the natural setting*—we have a new evolutionary environment.

This evolutionary approach with its organismic analogy has intuitive appeal. It certainly makes sense that differentiated structures and people with a high division of labor would be functionally dependent. Spencer named this as one of the main reasons that societies tend to integrate. However, Gilman has something specific that she wants us to see. To make it clear, let me state the principles of social evolution as a maxim: To say "society" is to say cooperation and dependency. The bottom line of this postulate is that, while at the beginning of our evolutionary path it was functional for individuals to struggle and compete for their own gain, it is no longer functional for them to do so now. As Gilman (1899/1975) says, "It is no longer of advantage to the individual to struggle for his own gain at the expense of others: his gain now requires the co-ordinate efforts of these others; yet he continues so to struggle" (p. 104). Gilman's point is in that last bit of the quote: We continue to struggle for individual gain even though it is no longer functionally advantageous. From an evolutionary standpoint, Gilman argues, there is something wrong. It is here that Gilman's critical viewpoint is most noticeable.

Adding Marx

Gilman's critical perspective has clear Marxian overtones. Like Marx, she argues that the economy is the most important social institution. All species are defined by their relationship to the natural environment. Stated in functionalist terms, structures in an organism are formed as the organism finds specific ways to survive (structure follows function), and the same is true in society. The economy is the structure that is formed in response to our particular way of surviving, and it is thus our most defining feature. We are therefore most human in our economic relations. Marx argued that we are alienated from our nature as we are removed from direct participation in the economy, as with capitalism. Gilman also argues that alienation occurs, but the source of that alienation is different. Where Marx saw class, Gilman sees gender. Gilman, in fact, is quite in favor of capitalism—in no other system has our ability to survive been so clearly manifest. Even so, there are alienating influences that originate in our gendered relationship to the economy, most specifically the limitation of women's workforce participation.

Gilman's perspective is thus similar to but different from Martineau's. Obviously, they are both concerned with gender, and we will see that more clearly as we go through Gilman's theoretical ideas. And they both have a sense of forward progress, from brute force to more peaceful, complex forms of society. Yet Gilman's evolutionary theory is far more developed than are Martineau's ideas, and that's true across the board. Gilman gives us a fully developed theory, whereas Martineau gives us more of a general perspective charged with pregnant ideas. Another way in which they are different is in their scope. In looking at society at one point in time, Martineau gives us a much fuller vision, but when it comes to long-term social factors, it's Gilman who excels. Martineau's point of view is more in keeping with the Enlightenment. She sees herself as a cultured citizen who believes in the philosophical progress of humanity. Happiness is the natural law, and education is our path to find it. Gilman, on the other hand, is a functional evolutionist who sees progress meted out over long periods of time by natural selection. Change for Gilman comes through massive movements of social structure over centuries.

Their most pronounced similarity is the idea of workforce participation for women. This is central to both women and is a key ingredient in contemporary feminist theory. Gilman, however, is much more detailed in her explication of this factor. Martineau argues that excluding women from the workforce reduces their social, political, and personal power, a concept that became a major tenet of contemporary gender theory. Martineau also gives us a subtler cultural take. She says that where livelihood is an issue in marriage (where women must marry to survive), marriage can never be free from the taint of the economy. Thus, marriage is debased from the ideal of two people in love to an economic relationship. Gilman focuses on this latter idea and details it in all its ramifications.

Gynaecocentric Theory

As we will see in the following pages, Gilman understands the exclusion of women from the economy in evolutionary terms. This perspective yields amazing results. We will see in the next section about *self- and race preservation* that removing women from economic participation threw off the balance between the individual and society. This disequilibrium from a functionalist's view means that the evolutionary path of humanity is not functional. Gilman explores the main point of this dysfunction in her use of gynaecocentric theory. *Gynaecocentric theory* posits that the female is the basic social model rather than the male (which had been assumed for most of our scientific and philosophic history). Gilman argues that men and women have natural, intrinsic energies that are different from one another. Based on this postulate of different energies, Gilman contends that excluding women from the workforce actually ended up producing evolutionary advantages, which we'll cover in a moment. Thus, in the long run, it turned out to be functional to remove women from the workforce. However, it is certainly not the case that Gilman thinks everything is right and wonderful for women. In the section on *sexuo-economic effects*, we'll see that denying women access to the natural economic environment has had horrible consequences.

Definition

Gynaecocentric theory originated with Lester F. Ward and is used by Gilman to explain gender inequality. Gynaecocentric theory argues that women, not men, are the general species type for humans: It is through women that the race is born and the first social connections created. In gynaecocentric theory, men and women have essential sex-specific energies.

Concepts and Theory: Dynamics of Social Evolution

I've diagramed Gilman's basic model of social evolution in Figure 7.1. As you can see, she, like Marx, argues that the basic driving force in humanity is economic production: Creative production is the way that we as a species survive. In all species, economic needs push the natural laws of selection that result in a functional balance between self-preservation and race preservation. Natural selection in self-preservation develops those characteristics in the individual that are needed to succeed in the struggle for self-survival. In the evolutionary model, individuals within a species fight for food, sex, and so on. Natural selection equips the individual for that fight. Race preservation, on the other hand, develops those characteristics that enable the species as a whole to succeed in the struggle for existence. Gilman is using the term *race* in the same way I'm using *species*; she isn't using it to make racial distinctions among ourselves. The most important point here is that the relationship between self- and race preservation is balanced: Individuals are selfish enough to fight for their own survival and selfless enough to fight for the good of the whole species.

Figure 7.1   Gilman's Basic Model of Social Evolution

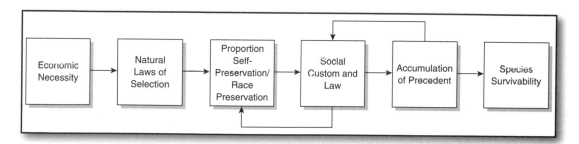

Definition

The ideas of *race preservation* and *self-preservation* are part of Gilman's evolutionary theory of gender relations. Race preservation stimulates the natural selection for skills that promote the general welfare of the collective, and self-preservation motivates the natural selection for skills that protect the individual. Self-preservation and race-preservation skills ought to balance one another. However, because women have been removed from the natural environment by male-dominated society, the self-preservation skills are overemphasized and the evolutionary system is dysfunctional.

Gilman's point here is interesting because most evolutionists assume that these two factors are one and the same. In other words, self-preservation is structured in such a way as to function as species preservation. This assumption is what allows social philosophers like Adam Smith to take for granted that severe individualism in systems like capitalism is good. Gilman insightfully perceives that these two functions can be at odds with one another. She thus creates an empirical research question where many people don't see any problem or they make a value judgment, such as the inherent goodness of capitalistic free markets.

The third box of Figure 7.1 is where human evolution becomes specific. Because of our overwhelming dependency on social organization for survival, we develop laws and customs that in turn support the equilibrium between self-preservation and race preservation, as noted by the feedback arrow. With the idea of "accumulation of precedent," Gilman has in mind art, religion, habit, and institutions other than the economy that tend to reinforce and legitimate our laws and customs. As you can see, there is a self-reinforcing cycle that encompasses the race/self-preservation proportion; social laws and customs; and the institutions, culture, and habits of a society. This social loop becomes humankind's environment. Of course, we still relate to and are affected by the natural world, but the social environment becomes much more important for the human evolutionary progress. This entire process works together to increase the likelihood of the survival of the human race.

That's the model of how things are supposed to work. Figure 7.2 depicts how things actually work under gender inequality. As you can see, the model is basically the same. The same central dynamics are present: Economic necessity pushes the laws of selection that in turn produce the proportion of self- to race-preservation characteristics, which then create laws and customs that reinforce those characteristics and produce social behaviors, culture, and institutions that strengthen and justify the laws and customs. This time, however, the social loop doesn't result in greater species survivability, but rather, in gender, sex, and economic dysfunctions.

Figure 7.2 Gilman's Model of Sexuo-Economic Evolution

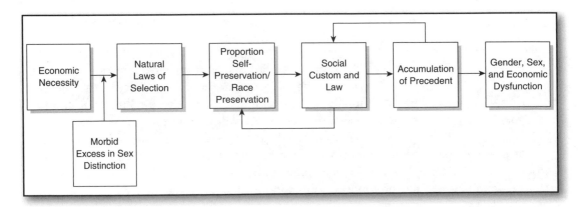

We will talk about those dysfunctions, but first we need to look at how Gilman sees the overall movement of evolution. Her conclusions are extremely interesting and perhaps startling, but in the overall way in which she makes her argument, she is years ahead of the functional arguments of her time. In fact, we can safely say that the kind of argument that she presents wasn't generally found in sociology until after Robert K. Merton's work around 1949. In general, Gilman argues that dysfunctions in some subsystems can have functional consequences for the whole.

Morbid Excess in Sex Distinction

The dynamic difference between Figure 7.1 and Figure 7.2 is found in what Gilman calls "morbid excess in sex distinction." Much like Spencer, Gilman argues that human sexuality and the kinship structure were evolutionarily selected. In the predawn history of humankind, it became advantageous for us to have two sexes that joined together in monogamy. From an evolutionary perspective, the methods of reproducing the species are endless. They run from the extremes of hermaphroditism (both sexes contained in a single organism) to multiple-partnered egg hatcheries. Because human beings need high levels of culture and social organization to survive, the two-sex model was naturally selected along with monogamy. Thus, having two sexes that are distinct from one another is natural for us. In this model, the physical distinctions between the sexes produce attraction and competition for mates, just as in other species. Eventually, however, because of the need for social organization among humans, the brutal competition for mates was mitigated and the social bond extended past mother–child through monogamy.

Patriarchy, however, tainted this process. Gilman presents a unique and fascinating argument as to why patriarchy came about. According to Gilman (1911/2001), there are distinct, natural male and female energies. The basic masculine characteristics are "desire, combat, self-expression; all legitimate and right in proper use" (p. 41). Female energy, on the other hand, is more conservative and is characterized by maternal instincts—the love and care of little ones. In this, Gilman sees us as no different than most other species—these energies are true of many male and female creatures. Yet the natural level of these energies posed a problem for human evolution. To survive as humans, we required a higher degree of social organization than simple male and female energy could provide. The male energy in particular was at odds with this need. As long as "the male savage was still a mere hunter and fighter, expressing masculine energy . . . along its essential line, expanding, scattering" (Gilman, 1899/1975, p. 126), social organization of any complexity was impossible. Men were aggressive and individualistic and thus had to be "maternalized" for social organization to grow.

In this argument, Gilman is using and expanding on Lester F. Ward's gynaecocentric theory: The female is the general race type, while male is a sex type. In fact, Gilman dedicated her book *The Man-Made World* (1911/2001) to "Lester F. Ward, Sociologist and Humanitarian, one of the world's great men . . . and to whom all women are especially bound in honor and gratitude for his Gynaecocentric Theory of Life." The word *gynaecocentric* means woman-centered. Though Herbert Spencer didn't phrase it in quite this way, the theory is very similar to his argument concerning monogamy. Family is the basis of social organization. All species, including humans, are related to one another biologically through descent. In our case, we used those biological relations to build social connections. The basic kin relationship is found between mother and child. This is the one relationship that has always been clear: It's obvious from whom the child comes. In order to build social relations on that basic tie, fathers had to be included through some mechanism. Of all the possible forms of marriage, monogamy provides the clearest connection between father and child, and thus the strongest, most basic social relationship.

Definition

Morbid excess in sex distinction is a theoretical concept in Gilman's evolutionary theory of gender. Sex distinctions are those physical and visual cues that make sexes different from one another. Gilman argues that the sex distinctions in humans have been carried to a harmful extreme due to women no longer living in the natural environment of economic pursuit, but, rather, living in the artificial environment of the home that has been established by men. Women have thus been physically and sexually changed by evolution so that they can better survive under patriarchy.

What Ward and Gilman add is the idea of intrinsic male and female energy. Male energy is at odds with family and social ties. The "giant force of masculine energy" had to be modified in order to add and extend the social tie between father and child. This was accomplished through the subjugation of women. (I told you it was interesting.) When men created patriarchy and dominated women, they also made women and children dependent upon men. In natural arrangements, both men and women produce economically; but as men dominated women, they also removed them from the world of economic production. Women could no longer provide for themselves, let alone their children. Men, then, had to take on this responsibility, and as they did, they took on "the instincts and habits of the female, to his immense improvement" (Gilman, 1899/1975, p. 128). Therefore, men became more social as a result of patriarchy.

This functional need was a two-way affair. Not only did men need to take on some of the women's traits, but also the natural environment could be more thoroughly tamed by men rather than women. By taking women out of the productive sphere, and thereby forcing men to be more productive themselves, "male energy . . . brought our industries to their present development" (Gilman, 1899/1975, pp. 132–133). Both men and human survival in general thus benefited from patriarchal arrangements.

Gilman draws two conclusions from gynaecocentric theory. First, women should not resent the past domination of men. Great harm to individuals and society has come through this system, but at the same time, necessary and great good has come as well,

[I]n the extension of female function through the male; in the blending of facilities which have resulted in the possibility of our civilization; in the superior fighting power developed in the male, and its effects in race-conquest, military and commercial; in the increased productivity developed by his assumption of maternal function; and by the sex-relation becoming mainly proportioned to his power to pay for it. (Gilman, 1899/1975, pp. 136–137)

The second thing Gilman draws from this theory is that the women's movements, and all the changes that they bring, are part of the evolutionary path for humanity. While women's economic dependence upon men was functional for a time, its usefulness is over. In fact, Gilman tells us, the women's and labor movements are misnamed; they ought more accurately to be called human movements. The reason that we've become aware of the atrocities associated with male dominance at this time in our history is that the dysfunctions are now greater than the functions. The behaviors and problems of patriarchy have always existed, but the benefits outweighed the costs and it therefore remained functional. Now, because of the level of social and economic progress, it is time to cast off this archaic form.

Sexuo-Economic Effects

However, it should be clear that Gilman's purpose is not to justify the subordination of women. Gynaecocentric theory gives her a way of understanding the effects, including the women's movements, in functional, evolutionary terms. It is, I think, to her credit that she maintains theoretical integrity in the face of the obvious suffering of women, and her own personal anguish as a woman. That said, a great deal of Gilman's work is occupied with exposing and explaining the current state of women.

Gilman characterizes the overall system and its effects as sexuo-economic. In *sexuo-economic relations,* two structures overlap, one personal and the other public. The social structure is the economy, and the private sphere is our sexual relationships. The basic issue here is that women are dependent upon men economically; as a result, two structures that ought to be somewhat separate intertwine. Historically, as women became more and more dependent upon men for their sustenance, they were removed further and further from economic participation. As we've noted, the economy is the most basic of all our institutions: It defines our humanness. A species is defined by its mode of survival; survival modes, then, determine the most basic features of a species. In our case, that basic feature is the economy. The economy is where we as a species compete for evolutionary survival. It is where we as a species come in contact with the environment and where the natural laws of selection are at work. Thus, when workforce participation is denied to women, they are denied interaction with the natural environment of the species. That natural environment, with all of its laws and effects, is replaced with another one. Man (meaning males) becomes woman's economic environment, and, just as any organism responds to its environment, woman changed, modified, and adapted to her new environment (form and function).

Definition

In Gilman's evolutionary theory of gender inequality, *sexuo-economic relations* occur when sex relations and economic relations overlap fundamentally. In a society where women's workforce participation is limited, the structures of economy and family are confounded, sex distinctions between men and women are accentuated and unbalanced, sex itself becomes pathologically important to people, women become consumers par excellence, and men are alienated from their work.

Gilman talks about several effects from living in this new environment, but the most important are those that contribute to morbid excess in sex distinction. As we've seen, all animals that have two sexes also have sex distinctions. These distinctions are called secondary sex characteristics: those traits in animals that are used to attract mates. In humans, according to Gilman, those distinctions have become accentuated so much that they are morbid or gruesome. In societies where they are denied equal workforce participation, the primary drive in women is to attract men. Rather than improving her skills in economic pursuits, she must improve her skills at attracting a man. In fact, expertise in attraction skills *constitutes* her basic economic skills (which is how prostitution can exist). Thus, the secondary sex characteristics of women become exaggerated. She devotes herself to cosmetics, clothes, primping, subtle body-language techniques, and so on. Most of these are artificial, yet there are also natural, physical effects. Because women are pulled out from interacting with the natural environment, their bodies change: They become smaller, softer, more feeble, and clumsy.

The increase in sex distinctions leads to greater emphasis on sex for both men and women. The importance given to sex far exceeds the natural function of procreation and comes to represent a threat to both self- and race preservation. The natural ordering of the sexes becomes perverted as well. In most animals, it is the male that is flamboyant and attractive; this order is reversed in humans. To orient women toward this inversion, the sexual socialization of girls begins early. In our time, young girls are taught that women should be beautiful and sexy, and they are given toys that demonstrate the accentuated sex distinctions (like Barbie), toy versions of makeup kits used to create the illusion of flamboyancy in women, and games that emphasize the role of women in dating and marriage. Another result of this kind of gender structure and socialization is that women develop an overwhelming passion for attachment to men at any cost; Gilman points to women who stay in abusive relationships as evidence. In the end, because of the excess in sex distinction brought on by the economic dependence of women, marriage becomes mercenary.

Obviously, the economic dependency of women will have economic results as well. Women are affected most profoundly as they become "non-productive consumers" (Gilman, 1899/1975, p. 116). According to Gilman, in the natural order of things, productivity and consumption go hand in hand, and production comes first. Like Marx, Gilman sees economic production as the natural expression of human energy, "not sex-energy at all but race-energy" (p. 116). It's part of what we do as a species, and there is a natural balance to it when everybody contributes and everybody consumes. The balance shifts, however, as women are denied workforce participation. The consumption process is severed from the production process, and the whole system is therefore subject to unnatural kinds of pressures. Noneconomic women are focused on noneconomic needs. In the duties of her roles of wife, homemaker, and mother, the woman creates a market that focuses on "devotion to individuals and their personal needs" and "sensuous decoration and personal ornament" (Gilman, 1899/1975, pp. 119–120). The economy, of course, responds to this market demand. Contemporary examples of this kind of "feminized" market might include such things as fashion items (like jewelry, chic clothing, alluring scents, sophisticated makeup, and so forth); body image products (such as exercise tapes, health clubs, day spas, dietary products, and the like); fashionable baby care products (like heirloom baby carriages, designer nursery décor, and special infant fashions by The Gap, Gymboree, Baby B'Gosh, Tommy, Nike, Old Navy, ad infinitum); and decorative products for the home (such as specialized paints, wallpaper, trendy furnishings, and so on). Gilman argues that the creative efforts of men, which should be directed to durable commodities for the common good, are subject to the creation and maintenance of a "false market." Women are thus alienated from the commodities they purchase because they don't participate in the production process nor are their needs naturally produced; men are alienated because they are producing goods and services driven by a false market.

As Figure 7.2 shows, the morbid excess in sex distinction disables the laws of natural selection and puts self- and race preservation out of balance; we've just explored some of these shifts. As a result, the customs and laws of society respond, as do the social institutions and practices (accumulation of precedent). Institutions and practices that properly belong to humankind become gendered. For example, the governing of society is a "race function," but in societies that limit women's workforce participation, it is seen as the duty and prerogative of men. Decorating is also distinctly human; it is a function of our species, but it is perceived as the domain of women. Religion, an obvious human function, is dominated by men and has been used to justify the unequal treatment of women, as have law, government, science, and so forth.

Yet Gilman does tell us that things are changing. She sees the women's movements as part of our evolutionary path of progress. Implied in this conclusion is the idea that equality is a luxury. Under primitive circumstances, only the strong survive. The weak, the feeble, and the old are left behind or killed. As social relationships come into existence, cooperation is possible, and as people cooperate, a surplus develops. Yes, for the most part, the powerful control the surplus and turn it into wealth. Yet, at the same time, it is then feasible to support some who had once been cut off. These people are integrated into society, and society continues to grow. It becomes more complex and technically better able to control the environment and produce surplus. In turn, other groups are brought into the fold, and increasing numbers of disenfranchised groups are able to live and prosper. It is

Gilman's position that just as technical progress is our heritage, so is ever-expanding equality. Evolution of the species not only involves economic advancement, but increasing compassion as well; ethical and technical evolutions are inexorably linked.

At the present evolutionary moment, we are only hindered by our own blindness. We can afford the luxury of complete equality, but we fail to see the problem. Gilman says that this is due to a desire for consistency. We don't notice what we are used to. Even evil can become comfortable for us, and we will miss it when it's gone. This is one of the reasons why slavery wasn't seen as evil for so many years of our history. We also have a tendency to think in individual rather than general terms. It's easier for us to attribute reasons for what we see to individuals rather than broad social factors. This, of course, in C. Wright Mills' terms, is the problem of the sociological imagination. It is difficult to achieve.

> Being used to them, we do not notice them, or, forced to notice them, we attribute the pain we feel to the evil behavior of some individual, and never think of it as being the result of a condition common to us all. (Gilman, 1899/1975, p. 84)

Summary

- Gilman is a critical evolutionist. All species fight for survival and create organic structures to help them survive in their environment. In this case, the function produces the structure. For humans, the primary evolutionary structure is the economy. It is through the economy that we as a species survive.

- When women were removed from the economy through male domination, several things happened. First, women were taken out of the natural environment of economic production and given a false environment—men, and indirectly the home and family, became women's environment. Gilman is a social evolutionist and argues that, as every species will, women changed in response to change in their environment. This change in environment made the distinctions between the sexes more pronounced. Rather than being equipped to economically produce, women became equipped to pursue a husband (in fact, their survival depended upon it). Their bodies became smaller and softer from disconnection with the natural world, and women augmented these changes through artificial means (such as makeup, clothing, etc.). Human beings now have a morbid excess in sex distinction, and as a result, the natural order has been reversed. In most animals, it is the male that is marked for attraction; in humans, it is the female. In order to produce this artificial order, girls have to be socialized from birth to want to be attractive and to value having a husband over all else. Further, in societies where women's workforce participation is limited, women become unnatural consumers. The natural relationship is production followed by consumption. Women are simply consumers without production. As such, their perceived needs are radically changed and they produce a market focused on beauty, decoration, and private relationships.

- This evolutionary path has also had functional consequences. The natural energy of men is aggressive, individualistic, and dominating. The male is, then, ill-suited for social life. However, in making women and children dependent upon him, he also obligated himself to provide for them (to be socially involved and committed). Thus, through gender oppression, the natural energy of men has been modified: They now have maternal feelings and higher social abilities. Further, large societies require highly developed systems that can control the environment and extract resources. Gilman argues that men and women together could not have achieved the needed level of economic and technological development. In Gilman's evolutionary scheme men are naturally more aggressive and objectifying. Having men solely responsible for the economic sphere meant that society produced more and achieved higher levels of technological control. Thus, pulling women out of the workforce was functional so that the economy could develop fully.

- Humans have, however, reached the point where the suppression of women is more dysfunctional than functional. The women's movements are sure signs that evolution is pushing us toward higher levels of equality.

BUILDING YOUR THEORY TOOLBOX

Learning More—Primary and Secondary Sources

For Martineau, I recommend you start with *How to Observe Moral and Manners* and then move onto Lipset's edited version of *Society in America*:

- Martineau, H. (2003). *How to observe morals and manners.* New Brunswick, NJ: Transaction Publishers.

- Martineau, H. (1962). *Society in America* (Seymour Lipset, Ed.). New Brunswick, NJ: Transaction Publishers.

For Gilman, start with *Women and Economics* and then *The Man-Made World*. Gilman also wrote a goodly number of works of feminist fiction; I suggest the edited volume by Barbara H. Solomon:

- Gilman, C. P. (1975). *Women and economics: A study of the economic relation between men and women as a factor in social evolution.* New York: Gordon Press.

- Gilman, C. P. (2001). *The man-made world.* New York: Humanity Books.

(Continued)

(Continued)

- Gilman, C. P. (1992). *Herland and selected stories* (B. H. Solomon, Ed.). New York: Signet Classic.

In terms of introductions to these and other neglected theorists, I recommend the following:

- Lengermann, P. M., & Niegrugge-Brantley, J. (1998). *The women founders: Sociology and social theory, 1830–1930.* Boston: McGraw-Hill.
- Freedman, E. B. (Ed.). (2007). *The essential feminist reader.* New York: The Modern Library.
- Dicker, R. (2008). *A history of U.S. feminisms.* Berkeley, CA: Seal Press.

Seeing the Social World (knowing the theory)

- Write a 250-word synopsis of the theoretical perspective of feminism.
- After reading and understanding this chapter, you should be able to define the following terms theoretically and explain their importance to Martineau's theory: natural law and happiness, morals and manners, preparing the observer, observational safeguards, things, gender, marriage, religion, education.
- After reading and understanding this chapter, you should be able to define the following terms theoretically and explain their importance to Gilman's theory of the origins of gender inequality: gender, social evolution, self-preservation, race preservation, gynaecocentric theory, sexuo-economic relations, morbid excess in sex distinction.
- After reading and understanding this chapter, you should be able to answer the following questions (remember to answer them theoretically):

1. Explain Martineau's method and how it involves the researcher.
2. Explain Martineau's measures of morals and manners and how they apply to analyzing democracy.
3. Explain Martineau's theory of gender. In your explanation, be certain to address the issues of cultural logic and workforce participation.
4. Explain Martineau's theory of the place religion and education have in a democracy.
5. Explain the differences between self-preservation and race preservation and discuss their implications for gender.
6. Explain the sexuo-economic theory of evolution and its effects on the economy and gender.
7. Discuss the reasons why the perspectives of oppressed groups are able to give the kinds of critical insights necessary for social change.

Engaging the Social World (using the theory)

- Using Martineau's idea of happiness, how would you evaluate the society in which you live? Has happiness dropped out of our national discourse of democracy? If so, why do you think it has?

- Writing at the beginning of the twentieth century, Gilman thought we would have "evolved" past where we were in her time. Use the ideas of morbid excess in sex distinction and sexuo-economic effects to argue for or against our evolution in the past century.

- Evaluate the idea of critical knowledge. If we accept the idea that knowledge is a function of a group's social, historical, and cultural position, then is this idea of critical knowledge correct? If so, what are the implications for the way in which we carry on the study of society? If you disagree with the idea of the critical outsider view, from what position is true or correct knowledge formed?

Weaving the Threads (building theory)

- Gilman gives us a very specific theory of gender oppression. Compare her theory with what Marx said about gender oppression. How are these thinkers similar and different on the issue of gender? Which do you think is more accurate? Why? How do the different theories create different ideas about how to bring about gender equality?

- One of Weber's factors in social stratification is status. Gender is a social status group. Compare and contrast Gilman's and Weber's theory of status and social change. What does Gilman add? How significant is her addition? Justify your answer.

CHAPTER 8

Seeing Race

Frederick Douglass and W. E. B. Du Bois

While gender is more universal, the use of race as a category of distinction has historically been more destructive. Though women have been seen as less than men, in general their essential humanity hasn't been denied. The modern category of race, however, is based on such a distinction. While people have obviously always been aware of differences of skin tone and facial features, in premodern societies race wasn't an important way that people used to mark difference. Religion, territory, and eventually being civilized versus uncivilized were much more important categories for most of human history. Race, as such, only became important with the dawn of modernity and specifically capitalism. Capitalism provided the motivation (accruing profit at the least possible expense) and the means (commodification) to make race the primary marker of difference in modern nations. While slavery had always existed, it wasn't until the advent of capitalism that chattel slavery—people seen as property—could exist. It was also modernity that brought to the forefront the issue of human nature as a political concern. You'll recall that political rights in modernity exist for the individual simply because of his or her humanity. Thus, one's standing as a human became an issue of concern and definition.

We are without a doubt at a significant moment of change in U.S. race relations. In 2008, the United States elected its first black president. But are we now in a "post-racial" society, one in which race no longer makes a difference? Current statistics appear to say no. The 2008 median income for white families was $55,530, while black families' median income was $34,281 (DeNavas-Walt, Proctor, & Smith, 2008). The U.S. Department of Justice (2008) reports that at midyear 2008, black males were incarcerated at a level almost 7 times that of white males, and that 1 in 21 black males was in prison compared to 1 in 138 white males.

The National Urban League (2009) has developed an overall statistical measure of African American equality as compared to whites: The 2009 Equality Index was 71.1 percent.

Race Literature

I've found that the writings of people who study and think about race tend to be distinct and different from that of people writing about gender. Probably the best known of classic race theorists in the United States is W. E. B. Du Bois, who wrote from the 1890s through the early 1960s. Du Bois gave us a number of significant theoretical ideas, such as double consciousness, and produced some of the first social scientific studies of black Americans. Yet, reading Du Bois is more than reading theory and data; it is an experience. In his writing, Du Bois moves back and forth among intellectual argumentation, song, prayer, poetry, irony, parable, data, riddles, analogy, and declaration. He weaves a tapestry for the reader, one that touches every part of the reader's being. He wants us to be able to understand the objective state of blackness as well as experience its soul.

This dual approach of blending theory and experience seems fairly common among African American writers. An African American colleague of mine who teaches African American Social Thought in our department said it this way:

"When [black people] write . . . they are presenting a lens of dual reality—blackness in America, America on blackness: 'How do I feel about my country and how does my country feel about me?'" However, this issue isn't limited to sociologists. Quite a few African American authors, writing about race in fiction, essay, poetry, and the like, tend to include what we see as social and sociological theory, because, as my colleague says, "challenging race legacy touches on everything we know to be sociological!" (S. Cureton, personal communication, October 9, 2009).

An important reason for this twofold nature of black writing is due to the fact that to write about blackness is to write about humanness, as we'll see clearly with Frederick Douglass. Thus, centering race in writing, either creative or scientific, can always evoke the existential cry "How does it feel to be a problem?" in response to "measuring one's soul by the tape of a world that looks on in amused contempt and pity" (Du Bois, 1903/1996d, pp. 101–102). In addition, this approach to doing sociology and writing chronicles and puts in the public record the very human experiences of blacks living under inhuman circumstances. This record serves two purposes: It first preserves a truer history of black experience for the African American community. The writings, both disciplinary and otherwise, contain generational knowledge that provides the building blocks for a black identity that "speaks to the existential issues of what it means to be a degraded African" and "involves self-respect and self-regard, realms inseparable from, yet not identical to, political power and economic status" (West, 2001, p. 97). The second purpose is that these writings offer for "public consumption the 'soul of blackness' . . . in a society that addresses blackness as deviant" (S. Cureton, personal communication, November 20, 2009). They proclaim the strength of a people proven in a cauldron of suffering.

Frederick Douglass—The American Discourse of Race

Frederick Augustus Washington Bailey (last name later changed to Douglass) was born to a black slave mother and a white father. The exact date of his birth isn't certain, though his guess was that he was born in February 1817. Most slaves didn't know their day of birth; their masters kept it from them. Questions about birthdays "were regarded by the masters as evidence of an impudent curiosity" (Douglass, 1882, p. 2). He never knew his father, and he and his mother were separated when he was an infant, most likely to keep her from telling who his father was. His only memory of his mother was of her sneaking into his room at night to sleep beside him. Young Frederick lived with his grandmother, Betty Bailey, on the plantation where she was a slave until he was around 8 years old, after which he was sent to Baltimore where he became a house slave to Hugh Auld. When Frederick was about 12, Auld's wife, Sophia, began teaching him how to read. When Auld found out, he made his wife stop, saying that reading would only make a black slave discontented. Frederick Douglass later said that it was the first antislavery speech he had ever heard.

Frederick continued his own quest for literacy by talking to and learning from white children in the street and by watching men write in the house where he worked. Soon he was reading everything he could get his hands on: newspapers, books, and journals, anything to feed his hungry mind. He implicitly knew and would later say that education was the pathway to freedom. It was in those early days of literacy that he first heard the word *abolition*. In his autobiography, he recounted a time when he was 12 and came across a copy of *The Columbian Orator*. The *Orator* was a textbook of the time that contained political essays and other works. The text was used to teach children not only nationalist sentiments, but also how to speak publically and persuasively, a skill that Frederick fully developed.

True to Thomas Auld's prediction, reading apparently did make Frederick discontented. Auld began having trouble with Frederick and took him, at the age of 16, to Edward Covey, a known "slave breaker." Frederick was beaten regularly and given little food. The experience nearly broke him. He attempted escape on three different occasions. A light finally dawned for the future abolitionist in the person of Anna Murray, a free black woman living in Baltimore. Her life and presence gave Frederick new hope, and finally, in 1838 he made a successful escape to New York City. Murray had provided him with money and a sailor's uniform, and he carried identification papers provided by a free black sailor. Of that day Douglass (1882) wrote,

> A new world had opened upon me. If life is more than breath, and the "quick round of blood," I lived more in one day than in a year of my slave life. . . . I felt as one might feel who had escaped from a den of hungry lions. (p. 170)

Within a short period of time, Frederick Bailey changed his name to Frederick Douglass, married Anne Murray, and moved to New Bedford, Massachusetts.

Douglass continued his self-directed education, subscribed to William Lloyd Garrison's *The Liberator*, and soon began attending abolitionist meetings. Garrison (n.d.) was a white abolitionist who didn't pull any punches when it came to racial inequality: "I do not wish to think, or speak, or write, with moderation. . . . I am in earnest—I will not equivocate—I will not excuse—I will not retreat a single inch—AND I WILL BE HEARD" (n.p.). Soon after beginning to read *The Liberator*, Douglass (1882) had a chance to hear him speak: "I sat away back in the hall and listened to his might words—might in truth—might in simple earnestness" (p. 181).

In the summer of 1841, Garrison invited Douglass to accompany him to an abolitionist meeting, which he gladly accepted. While the meeting was in progress, a man who had heard Douglass speak with his "colored friends" at a schoolhouse, approached him and invited him to address the meeting. Douglass said he "trembled in every limb" but spoke to the group. After the meeting, John Collins, the general agent for the Massachusetts Antislavery Association, asked Douglass to become one of its regular agents. The position would involve travel and speaking to groups, and Douglass was at first hesitant, unsure of his skill and afraid that public speaking would expose him to the slave master he had fled.

Yet he ultimately took the position and spent the next few years speaking in churches and at abolitionist meetings. Though he always drew a crowd, people in the crowd would often heckle and at times violence broke out. Because of his eloquence and intelligence, many whites were convinced Douglass had never been a slave. Eventually, this wore on him, so he decided to write out his life. He published his first autobiography: *The Life and Times of Frederick Douglass* (1882). The book was a best seller in the United States and was also published in Europe. Because of the book's success, his friends and associates encouraged Douglass to go to Ireland; they were afraid his previous owner would find him out. Douglass went and toured Ireland and Britain, giving speeches for 2 years; he returned to the United States in 1847.

After his return, Douglass again took up the cause of black people in the United States. He published an abolitionist paper, *The North Star*, and others from 1847 to 1863. Douglass's (1847) views were clearly expressed in his first issue: "It has long been our anxious wish to see, in this slave-holding, slave-trading, and Negro-hating land, a printing-press and paper, permanently established, under the complete control and direction of the immediate victims of slavery and oppression" (n.p.). Douglass supported the efforts for women's rights in print, actions, and speech. He attended the Seneca Falls Women's Rights Convention (1848) and signed The Declaration of Rights and Sentiments (declaration of women's rights), and his last speech was delivered to the National Council of Women (1895).

In his new efforts, however, Douglass parted company with his mentor William Garrison. The primary point of contention was the U.S. Constitution. Garrison felt that the Constitution was a document forged in hell and was proslavery. Garrison's approach to emancipation, then, was radical and revolutionary. Douglass, on the other hand, argued that the Constitution could be used to help free slaves; he saw it as a document that could, in the end, legitimate freedom and equality for all.

During the American Civil War (1861–1865) and after, Douglass fought to have black men allowed in the Union Army, financially supported the underground railroad, and advised Presidents Lincoln and Andrew Johnson. After Reconstruction (1865–1877), Douglass held many high-level posts in government, among them American consul-general to Haiti (1889–1891). Following his speech to the National Council of Women, Frederick Douglass died of a heart attack on February 20, 1895.

THEORIST'S DIGEST

Central Sociological Questions

Douglass was an activist. He used his theoretical insights into the institution of slavery in public discourse in order to free slaves and move forward the democratic project of life, liberty, and the pursuit of happiness.

Key Ideas

Universalism, particularism, Other, ruling species, chattel slavery, emigration, ontological wounds, Dred Scott, democracy as an ongoing transcendent project.

Concepts and Theory: The Discourse of Slavery

To talk about Douglass, I want to introduce you to the idea of *public sociology.* The idea was initially introduced by Herbert Gans (1989) in his presidential address to the American Sociological Association (ASA). Gans argued that sociology should become more involved in public issues and discourse. One of the reasons Gans identified for why sociology isn't as socially involved as he wanted is *scientism,* the belief that sociological methods should be modeled on the natural sciences. In arguing against scientism, Gans first pointed out that even "natural scientists do not operate according to the idealized conception of their method" (p. 11). More specifically, according to Gans, there is a prejudice for the natural science model in modern culture generally, and academic disciplines of various ilks feel pressured to conform in order to qualify as legitimate.

Gans's address began a movement for a "public sociology," one that was later advanced by Michael Burawoy (2005) in his own ASA presidential address: "the standpoint of sociology is civil society and the defense of the social. In times of market tyranny and state despotism, sociology—and in particular its public face—defends the interests of humanity" (p. 24). Ben Agger (2000) sees the business of public sociology as moving from social facts to literary acts. Agger's "intent is to foster a public sociology, which acknowledges that it is a literary version, confesses its animating assumptions and investments, and addresses crucial public issues" (p. 2).

As always, my purpose isn't to promote one brand of sociology over another. Our discipline is multifaceted, which I think is appropriate given the complexity and depth of our subject matter (society). I bring public sociology to your attention as a way of understanding how and where the work of someone like Frederick Douglass fits. He wasn't educated as a sociologist (he wasn't formally educated at all, but then neither was Spencer), nor was he a professional sociologist (but then Durkheim was probably the first that could be considered a professional sociologist as such). However, Douglass did use a sociological perspective and he did influence the discourse of race in the United States in a profound manner, and as such, Douglass did the work of a public sociologist.

Frederick Douglass was in many ways the most important public spokesperson of his time. Without a doubt, he had the most powerful voice in creating the discourse of race in the United States. Stuart Hall (1996b) defines *discourse* as "a group of statements which provide a language for talking about—i.e., a way of representing—a particular kind of knowledge about a topic" (p. 201). Discourses are produced through language and practices. They are ways of talking about and acting toward an idea or group of people. One of the most powerful insights concerning discourses is that "anyone deploying a discourse must position themselves *as if* they were the subject of the discourse" (p. 202). The example that Hall gives is the discourse of the West. Ever since the distinction between the East and the West was made, the West has been seen as more advanced, more modern, and so on. This is in fact one of the reasons the distinction was made—to talk about the West as superior. In this discourse, the West is the model toward which the

"Rest" must strive. This discourse also places an obligation upon the West to assist the Rest in their move up the societal ladder. While you as an individual may not believe in the supremacy of the West, in order to talk about the relationship between the West and the Rest you must adopt a position as if you did believe it. For example, anytime we use the terms *third world nation, modernization,* or *globalization,* we are positioning ourselves within the West/Rest discourse and implicit Western superiority.

Definition

Discourse is a theoretical concept that is widely used but is most specifically associated with the contemporary work of Michel Foucault. A discourse is an institutionalized way of thinking and speaking. It sets the limits of what can be spoken and, more importantly, *how* something may be spoken of. In setting these limits, discourses delineate the actors in a field, their relationships to one another, and their subjectivities. Discourses are thus an exercise of power.

There are two issues in Douglass's public discourse that I want to focus on: race as Other and democratic universalism. These two are intimately tied in Douglass's mind. They are antagonistic categories—like good and bad, or sacred and profane—that cannot exist together in the same space. There is deep wisdom here. On the one hand, Douglass sees that democracy has to be all-inclusive. If life, liberty, and the pursuit of happiness are not available to all, then they are potentially not available to any. If one group can be excluded, then it's clear that no higher ethical reason would be needed to exclude any other group. This means that the distribution of democratic freedoms is predicated on a group's power to obtain them, which in itself denies the possibility of democracy. This is one reason why Douglass (1848/2009a) exhorted, "Never refuse to act with a white society or institution because it is white, or a black one, because it is black. But act with all men without distinction of color" (p. 211).

Yet I think Douglass saw something more basic as well. There is tension inherent in democracy: *Other* in the face of *all men* creates conflict. It's easy to see that the history of the United States is the working out of that conflict. We know that blacks and women were particularized and excluded from the universal ideas of life, liberty, and the pursuit of happiness. What most of us fail to see—but Douglass saw—is that this tension is built into modern democracy and we will never get away from it. Jeffrey C. Alexander (2006), a well-known and respected contemporary theorist, built his theory of the civil sphere around this insight:

> that the very effort to speak universalism must always and everywhere take a concrete form, opens our consideration of civil society not only to the particularistic but also the repressive dimensions of modern democratic life. (p. 50)

Cornel West (1999) puts it this way: "To be American is to raise perennially the frightening democratic question: What does the public interest have to do with the most vulnerable and disadvantaged in our society?" (p. xix).

Race as Other

The central issue in Douglass's public sociology concerns the nature of racism. There are at least two factors in racism. The first interrogates racism's origin and the second its power. Douglass approaches both of these from the same perspective: human nature. There were, and still are, arguments about the separation of the races being essential or necessary. In race relations, we usually think of this argument being made by whites against blacks, but it's found in black culture as well, as with certain forms of Black Nationalism. Douglass (1882), however, saw racism as learned behavior. The mental and emotional state of both slaveholder and slave does not come by nature; "nothing but rigid training, long persisted in, can perfect the character of the one or the other" (p. 54). Of his experiences with white children, Douglass writes, "I do not remember ever to have met a *boy* while I was in slavery, who defended the system; but I do remember many times, when I was consoled by them, and by them encouraged to hope" (p. 56). For Douglass, then, racism is a social construction, one that nevertheless shapes a person's innermost being.

The power of racism and the institution of slavery are founded upon ideas concerning human nature. Throughout this text, I've emphasized the importance of beliefs concerning human nature. Modernity is founded on beliefs in natural rights and reason; the success or failure of modern democracy is in many ways tied up with those beliefs. If freedom and equality are based on human nature, then in order to deny those self-evident truths one must deny the humanity of another. The nature of slavery, then, is defined by a master "who claims and exercises a right of property in the person of a fellow-man," and a slave who has been "divested of all right—reduced to the level of a brute—a mere 'chattel' in the eye of the law—placed beyond the circle of human brotherhood—cut off from his kind" (Douglass, 1850/2009b, p. 216).

It's important to see that this denial is more profound than what occurs in gender. In Chapter 7, we considered Beauvoir's notion of the Other, with man as the standard and woman the Other. This is one reason why the second wave of feminism was concerned about the language of gender. Using such terms as *man* and *he* to represent people generally, helps establish men as the universal. Douglas Hofstadter (1985) calls this the "slippery slope of sexism": "When a generic term and a 'marked' term . . . coincide, there is a possibility of mental blurring on the part of listeners and even on the part of the speaker" (p. 151). Hofstadter gives an example of a newscast that announces that the four-man space shuttle is on schedule. There may in fact be three women and a man, but we can't know that from the report. The point is that some of the general, "men as all humanity," is transferred over to specific men in particular. This sort of Otherness in gender denies the woman's subjective experiences and authenticity.

The use of Otherness in racism is even more profound. Though women were seen as secondary, the Otherness is restricted to gender—their essential humanity wasn't denied. In religious terms, women are *of Adam,* even though required to submit. Slavery denies even that connection. Blacks were seen as property with no more rights or dignity than an animal. The word Douglass uses, *chattel,* shares the same Middle English and Medieval Latin root as the word *cattle.*

Douglass characterizes the black man in this system as the "silent dead." Ask a slave, he declares, what he or she thinks of enslavement and you will receive no answer: "There comes no *voice* from the enslaved" (Douglass, 1850/2009b, p. 219). To have an opinion and to voice that opinion presupposes a self who can be spoken to and who can respond. No such self exists in the slave. To create black slavery, *the essential Other* had to be created—if all men are created equal, then the slave is no man. This is *speciation*—the creation of a new species. In speaking about the experiences of blacks under white, colonial rule, Frantz Fanon (1961/2004) tells us Marx's analysis of the *ruling class* has little to do with understanding white/black race relations: "The *ruling species* is first and foremost the outsider from elsewhere, different from the indigenous population, 'the others'" (p. 8, emphasis added).

Though colonialism and slavery don't structurally exist today, the discourse that came out of those institutions can still spread its effects, as Cornel West tells us in his book *Race Matters* (2001). A key in his analysis of contemporary black experience is black existential angst. In general, *existential angst* refers to the profound insecurity and dread that accompanies living as a human being. For example, humans are the only animal that knows it is going to die. This knowledge drives us to search for the meaning of life; this quest, however, implies that the meaning may not be found or, worse, that life is simply meaningless, which is what brings the dread. West has this sort of thing in mind but modifies it by making it specific to African Americans: "black existential *angst* that derives from the lived experience of ontological wounds and emotional scars inflected by white supremacist beliefs and images permeating U.S. society and culture" (p. 27). The word *ontology* means the study of existence; these wounds and scars thus originate in whites denying blacks their *human existence.*

Democracy and Universalism

Howard Brotz (2009) tells us, "The first issue to arise within Negro opinion concerned emigration" (p. 1). The goal of emigration, or the Back-to-Africa movement, was to return free blacks to Africa. The movement formally began with the establishment of the American Colonization Society in 1816. In the beginning decades of the nineteenth century, there was a significant increase in the Northern black population. In the North, the movement was generally motivated by whites' fears of integration and white workers' fears of losing their jobs to a cheaper workforce. The South feared the presence of a free black population in the United States; it would, they reasoned, create unrest among blacks in slavery. The movement lost momentum in mid-century but was rekindled during and after Reconstruction (1865–1877). White-on-black violence increased dramatically after the Civil War through such paramilitary groups as the Ku Klux Klan, the White League, and the Red Shirts; and the Back-to-Africa, or emigration, movement was seen as a solution. The American Colonization Society eventually returned somewhere around 13,000 blacks to Africa, most of them to Liberia.

The Back-to-Africa movement was one of the issues that eventually split Douglass and Garrison, especially as it became intertwined with the Constitution. Though he eventually left the American Colonization Society, Garrison argued that the Constitution was racist to its core and the only way to bring equality was to dissolve the Union; Douglass argued that the Constitution was free of racism and could be used to bring about universal equality. Douglass's position is nowhere clearer than in his response to the Dred Scott Decision. *Dred Scott v. John F. A. Sandford* is a case that was brought to the U.S. Supreme Court in 1857. Scott was a slave who was born in a slave state but whose master had taken him to live in states where slavery was prohibited. The Supreme Court in a 7–2 decision ruled that Scott couldn't sue because blacks had no rights of citizenship. Further, the judges argued that the framers of the Constitution saw blacks as an inferior race having no rights that a white man was legally bound to respect, and that a decision in favor of Scott could be seen as granting the rights of freedom of speech and freedom of assembly to all men of the Negro race.

In his response to the Dred Scott Decision, Douglass addressed Garrison directly. Garrison and his crew had been telling Douglass (1857/2009d), "while in the Union, we are responsible for slavery" (p. 255). Douglass agreed, but Douglass disagreed fervently with Garrison's solution of dissolving or leaving the Union: "in telling us that we shall cease to be responsible for slavery by dissolving the Union, he and they have not told us the truth" (p. 255). Douglass likened it to family responsibility: A man may desert his family and move miles away. Though he would be out of sight of his children, "it cannot free him from responsibility. Though he should roll the waters of three oceans between him and them, he could not roll from his soul the burden of this responsibility" (p. 255). Douglass sees his and every American's responsibility as tightly bound to the ethical and moral responsibilities of democracy. Douglass finds the basis for this argument in the Constitution.

Douglass (1857/2009d) spoke out powerfully and eloquently in defense of the Constitution. His main premise is that the Constitution speaks for "'We the people'—not we, the white people—not we, the citizens, or the legal voters—not we, the privileged class . . . not we, the horses and cattle, but we the people—men and women, the human inhabitants of the United States" (p. 257). His argument is thus based on universalism. *Universalism* can best be understood in contrast to particularism: the belief that the right to social justice and equality is restricted to one particular group. This idea is key to the work of the civil sphere. "We need to understand," Alexander (2006) tells us, that civil society is a sphere that is "*morally more universalistic* vis-à-vis the state and the market and other social spheres as well" (p. 31, emphasis added).

In another place, Douglass (1863/2009c) makes his argument more pointed: "The Constitution did not hold the slave tight enough. The Declaration of Independence did not hold the slave at all" (p. 269). Notice that here Douglass actually gives more weight to the Declaration of Independence. The importance of this reference is that it echoes one of his contemporaries, Abraham Lincoln. In his Gettysburg Address, Lincoln famously said, "Four score and seven years ago our fathers brought forth upon this continent a new nation, conceived in Liberty, and dedicated to the proposition that all men are created equal." Historian Garry Wills (1978) points out that Lincoln's phrase "four score and seven years ago" "takes us back to 1776, the year of the Declaration" (p. xvi).

The reason that Douglass and Lincoln refer to the Declaration is that it embodies "the pure spirit of the nation's Idea" (Wills, 1978, p. xviii). Lincoln confirms that the phrase "all men are created equal" did nothing to separate us from Great Britain; it was put in

> for future use. Its authors meant it to be—as, thank God, it is now proving itself—a stumbling block . . . to all those who in after times might seek to turn a free people back into the hateful paths of despotism. (Lincoln, as quoted in Wills, 1978, p. xviii)

We talked about the importance of the *Idea* of democracy in Chapter 1 as well. Through Douglass, we're beginning to see the significance of the idea: It's the *idea of democracy, not the reality* that propels us forward. The idea of *all men are created equal* is a constant challenge to the way we live our lives with other people. It's that tension between fact and ideal that is the energy of democracy.

In this way, civil society is transcendent—and in this way, civil society is never complete. These two go together: Democracy is an ongoing project (never complete) precisely *because* it is transcendent. In an insight that many today have forgotten, Douglass (1863/2009c) tells us that social equality "*does not exist anywhere*" (p. 271, emphasis added). In other words, democracy wasn't finished after the founding documents were penned and democratic institutions put in place. No, equality does not *exist* anywhere as a settled state—it is the *pursuit* of equality, freedom, and happiness that marks a democratic society. The question for Douglass isn't about whether social equality exists, nor is it about blacks being "mentally equal to his white brother," nor is it about whether a black man "will be likely to reach the Presidential chair." The issue for Douglass is whether or not black and white people in this country can "be blended into a common nationality, and enjoy together . . . the inestimable blessings of life, liberty, and the pursuit of happiness" (p. 271). Democracy, then, never exists as such. It must be won anew not only with every generation, but with every possibility of extending life, liberty, and the pursuit of happiness to others.

Summary

- Our theoretical interest in Douglass's work revolves around the idea of discourse. Discourse is a specific way of talking about something. It is common—we all use discourse—but it is also powerful because the discourse of an issue determines how we can think and talk about it. Moreover, discourse can be oppressive because it sets out the possibilities and impossibilities of specific social relations and experiences.

- Discourses also contain contradictory elements that can be used by disenfranchised groups to further their cause in a democracy. Douglass is particularly adept at this, using the ideas of Other and universalism to counter the discourse of slavery and race.
- There are also elements in Douglass's work that anticipate important work in race and democracy by more than a century and a half, as exemplified by Jeffrey Alexander and Cornel West.

W. E. B. Du Bois—The Culture of Race

William Edward Burghardt Du Bois was born in Great Barrington, Massachusetts, on February 23, 1868. His mother was Dutch African and his father French Huguenot-African. Du Bois's father, who committed bigamy in marrying Du Bois's mother, left the family when the boy was just 2 years old, but Du Bois grew up without experiencing the depth of suffering blacks felt at that time. Du Bois's education in the abject realities of blackness in America came during his college years at Fisk. He taught two summers in rural Tennessee and saw firsthand the legacy of slavery in the South. As a result, Du Bois became an ardent advocate of social change through protest. His belief in confrontation and dissent put Du Bois at odds with Booker T. Washington, a prominent black leader who favored accommodation. This disagreement with Washington influenced the early part of Du Bois's political life; he founded the Niagara Movement in large part to counter Washington's arguments.

Through most of his life, Du Bois generally favored integration, but toward the end he became discouraged at the lack of progress and increasingly turned toward Black Nationalism: He encouraged blacks to work together to create their own culture, art, and literature, and to create their own group economy of black producers and consumers. The cultural stand was directed at creating black pride and identity; the formation of a black economic community was the weapon to fight discrimination and black poverty. Du Bois was also a principal force in the Pan-African movement, which was founded on the belief that all black people share a common descent and should therefore work collectively around the globe for equality. As I mentioned, in the latter part of his life, Du Bois became disheartened at the lack of change regarding the color line in the United States. In the end, he renounced his citizenship; joined the Communist Party; and moved to Ghana, Africa.

More than any other single person, Du Bois was responsible for black consciousness in America and probably the world during the twentieth century. His book, *The Souls of Black Folk* (1903/1996d), defined the problem of the color line. He was a founding member of the NAACP and its chief spokesperson during its most formative years. Du Bois also produced the first scientific studies of the black condition in America.

I have listed the major points in Du Bois's life in the timeline below. I encourage you to not skip over this information; read it and let it sink in. The timeline is but an outline of a life that spanned almost a century and helped to change the face of American society.

- February 23, 1868: Born in Great Barrington, Massachusetts
- 1888: Received his BA from Fisk University, Nashville (turned down by Harvard)—taught two summers in rural Tennessee, which introduced him to the deep poverty of the Southern black person
- 1888–1895: Completed second BA and PhD at Harvard, during which time he spent 2 years in Berlin, attending lectures by Max Weber and becoming friends with both Marianne and Max Weber
- 1894–1910: Taught at Wilberforce, University of Pennsylvania, and Atlanta University (where he was a professor of economics and history for 14 years); published 16 research monographs
- 1899: Published *The Philadelphia Negro: A Social Study* (the first ethnography of a black community in the United States)
- 1900: Led the first Pan-African conference, in London
- 1903: Published *The Souls of Black Folk,* which first set out the problem of the color line and Du Bois's opposition to Booker T. Washington's strategy
- 1905: Founded the Niagara Movement, precursor to NAACP
- 1909: Helped found the National Association for the Advancement of Colored People (NAACP); left academia and became editor of the association's journal, *The Crisis*
- 1910–1934: Worked at the NAACP, edited *The Crisis,* became the principal leader of blacks in America; advocated Black Nationalism
- 1919–1927: Engineered four Pan-African congresses
- 1934–1944: Left the NAACP in a disagreement over assimilation versus separation; returned to Atlanta University
- 1935: Published Black Reconstruction: An Essay Toward a History of the Part Which Black Folk Played in the Attempt to Reconstruct Democracy in America, 1860–1880 (Marxist interpretation of the Reconstruction)
- 1940–1944: Chaired the Department of Sociology at Atlanta University; editor of social science journal, *Phylon*
- 1940: Published *Dusk of Dawn: An Essay Toward an Autobiography of a Race Concept* (viewed his career and life as representative of black–white relations)
- 1944–1948: Returned to the NAACP; ended in disagreement; continued discouragement and movement toward political left
- 1951: Indicted as unregistered agent for a foreign power; acquitted
- 1961: Joined the Communist Party; moved to Ghana, continued work on his long-term project, *Encyclopedia Africana*
- 1962: Renounced American citizenship
- August 27, 1963: Died in Ghana at age 95

THEORIST'S DIGEST

Central Sociological Questions

Du Bois's central question was simple, straightforward, and powerful: How can black Americans be freed from oppression? His emphasis in this question is interesting, especially when viewed in comparison to someone like Marx who is a structuralist. Du Bois's thrust is cultural. So, the question is, "How is culture used to oppress blacks?"

Simply Stated

A good deal of Du Bois's writings were deeply spiritual. I don't intend that in any specifically religious way, but they were appeals to the human spirit. Some of his writings were directed at articulating what it is like to live in a society where one's human existence is made a problem; some were directed at rousing the black community to become coworkers in the cultural field along with whites—these writings also exhorted the white community and exposed ways that whites excluded black America. Du Bois's theoretical work focused on how culture was used to oppress—through ideological history, cultural representations, and double consciousness—and how capitalism and the white middle class were dependent upon the exploitation of blacks around the world.

Key Ideas

The centered subject, second sight, history as ideology, representation, stereotypes, looking glass self, double consciousness, exploitation, dark nations, personal whiteness.

Concepts and Theory: The Experience of Oppression and Critical Knowledge

In all his work, Du Bois centers the subject in his writing. One of the things most of us are taught in college English is that we should write from a de-centered point of view. We are supposed to avoid using "I" and "me" and should always write from an objective perspective. Yet Du Bois begins one of his most famous works, *The Souls of Black Folk,* with "Between me and the other world. . ." In another place he wonders, "Who and what is this I . . .?" Du Bois isn't being self-centered, nor is he unaware of the rules of composition. He is being quite deliberate in his use of personal pronouns and the centered subject. Du Bois is a social scientist and produced one of the first systematic studies of African Americans, yet he also tells us that race is not something that can be understood through the cold, disassociated stance of the researcher. Race and all marginal positions must be experienced to be understood. Du Bois uses his life as the canvas upon which to paint the struggles of the black race in America and in the world: "that men may listen to the striving *in the souls of* black folk" (Du Bois, 1903/1996d, p. 107, emphasis added).

Standpoint of the Oppressed

There are two things I think this subjective stance implies: The first is my own comment, and the other is something I think Du Bois has in mind. First, any secondary reading of Du Bois, such as the book you have in your hands, falls short of the mark. This is generally true of any of the thinkers in this book—you would be much richer reading Durkheim than reading what somebody says about Durkheim—but it is particularly true of Du Bois. Part of what you can acquire from reading Du Bois is an experience, and that experience is a piece of what Du Bois wants to communicate.

The second implication of Du Bois's multidimensional, subjective approach is theoretical. Du Bois (1903/1996d) says that

> the Negro is a sort of seventh son, born with a veil, and gifted with second-sight in this American World,—a world which yields him no true self-consciousness, but only lets him see himself through the revelation of the other world. (p. 102)

Du Bois employs spiritual language here. The veil of which he speaks is the *birth caul.* In some births, the inner fetal membrane tissue doesn't rupture and it covers the head at delivery. This caul appears in about 1 in 1,000 births. Due to its rarity, some traditional cultures consider such a birth spiritually significant, and the caul is kept for good luck. The same is true of the seventh son reference. The seventh son is considered to have special powers, and references to such are to be found in many folk and blues songs as well as in the Bible. The "second-sight" is a reference to clairvoyant or prophetic vision.

Thus, Du Bois is saying that because of their experiential position, African Americans are gifted with special insight, a prophetic vision, into the American world. They see themselves not simply as they are; they also see their position from the perspective of the other world. In other words, blacks and other oppressed groups have a particular point of view of society that allows them to see certain truths about the social system that escape others. This idea of critical consciousness goes back to Marx. Marxian philosophy argues that only those on the outside of a system can understand its true workings; it is difficult to critically and reflexively understand a system if you accept its legitimation. In other words, capitalists and those who benefit from capitalism by definition believe in capitalism. It is difficult for a capitalist to understand the oppressive workings of capitalism because in doing so the person would be condemning her- or himself.

Contemporary feminists argue that it is the same with patriarchy: Men have a vested interest in the patriarchal system and will thus have a tendency to believe the ideology and have difficulty critiquing their own position. Dorothy Smith, a contemporary feminist, refers to this as *standpoint theory.* In general, this refers to a theory that is produced from the point of view of an oppressed subject:

> an inquiry into a totality of social relations beginning from a site outside and prior to textual discourses. Women's standpoint [is seen] here as specifically

subversive of the standpoint of a knowledge of ourselves and our society vested in relations of ruling. (D. E. Smith, 1987, p. 212)

Du Bois is thus arguing that African Americans have, by virtue of their position, a critical awareness of the American social system. This awareness can be a cultural resource that facilitates structural change. As Du Bois (1903/1996d) puts it, "This, then, is the end of his striving: to be a co-worker in the kingdom of culture, to escape both death and isolation, to husband and use his best powers and his latent genius" (p. 102).

One of the things that oppression in modernity has done is deny the voice of the Other. In the Durkheim chapter, we saw that one of the characteristics of modern society is the grand narrative. Nation-states provide all-encompassing stories about history and national identity. The purpose of such narratives is to offer a kind of Durkheimian rallying point for social solidarity. This sort of solidarity is necessary for nations to carry out large-scale programs, especially such things as colonization and war. The problem with such a narrative is that it hides inequities. For example, the grand narrative of equality in the United States was actually a story about white Anglo-Saxon Protestant males. Hidden in the national narrative and history was (and still is) the subjugation of Native Americans, African Americans, women, Mexican Americans, homosexuals, and so forth.

Contemporary theories of difference, then, focus on the subjective experience of the disenfranchised in contrast to this grand narrative. One of the things that is important in the fight for equality is allowing multiple voices to be heard; thus, we have recently moved from the cultural picture of the "melting pot" to that of the "salad bowl" in the United States. Du Bois's perspective is quite in keeping with this emphasis. In his own work, he uses the subjective mode to express the experience of the oppressed. He becomes a representative figure through which we might understand the plight of black people in America. Both Martineau and Gilman (Chapter 7) did use multiple approaches in expressing their ideas, and some of those used the first person, but not to the degree of Du Bois. Part of what this multiple-voiced approach entails is valuing the outsider's point of view. And, interestingly, Du Bois is much more in tune with this feminist idea of standpoint theory than either Martineau or Gilman, neither of whom privileges outsider knowledge. In that sense, Du Bois's work contains a more critical edge, again in keeping with much of contemporary analysis.

Du Bois was a prolific writer, so I have chosen to focus on particular points of his discourse about race. In my opinion, Du Bois's lasting contribution to social theory is his understanding of cultural oppression. In the following section on cultural oppression, we will see that it is just as necessary as structural oppression in the suppression of a social group. Du Bois's understanding of this process is quite good. He argues that cultural oppression involves exclusion from history, specific kinds of symbolic representations, and the use of stereotypes and their cultural logic of default assumptions. This cultural work results in a kind of "double consciousness," wherein the disenfranchised see themselves from two contradictory points of view. However, Du Bois isn't only interested in cultural oppression; he also gives us a race-based theory of world capitalism. In the section on the dark nations and world capitalism, we will see that it isn't only the elite capitalists that benefit from the exploitation of blacks and other people of color; the middle class benefits as well.

Concepts and Theory: Cultural Oppression

If we can get a sense of the subjectivity that Du Bois is trying to convey, we might also get a sense of what horrid weight comes with cultural oppression. Undergirding every oppressive structure is cultural exclusion. While the relative importance of structure and culture in social change can be argued, it is generally the case that structural oppression is legitimated and facilitated by specific cultural moves—historical cultural exclusion in particular. In this case, African Americans have been systematically excluded from American history, and they have been deprived of their own African history.

Exclusion From History

History plays an important part in legitimating our social structures; this is known as history as ideology. No one living has a personal memory of why we created the institutions that we have. So, for example, why does the government function the way it does in the United States? No one personally knows; instead, we have a historical account or story of how and why it came about. Because we weren't there, this history takes on objective qualities and feels like a fact, and this facticity legitimates our institutions and social arrangements unquestionably. But, Du Bois tells us, the current history is written from a politicized point of view: Because women and people of color were not seen as having the same status and rights as white men, our history did not see them. We have been blind to their contributions and place in society. The fact that we now have Black and Women's History Months underscores this historical blindness. Du Bois calls this kind of ideological history "lies agreed upon."

Du Bois, however, holds out the possibility of a *scientific* history. This kind of history would be guided by ethical standards in research and interpretation, and the record of human action would be written with accuracy and faithfulness of detail. Du Bois envisions this history acting as a guidepost and measuring rod for national conduct. Du Bois presents this formulation of history as a choice: We can either use history "for our pleasure and amusement, for inflating our national ego," or we can use it as a moral guide and handbook for future generations (Du Bois, 1935/1996c, p. 440).

It is important for us to note here that Du Bois is foreshadowing the contemporary emphasis on culture in studies of inequality. Marx argued that it is class and class alone that matters. Weber noted that cultural groups—that is, status positions—add a complexity to issues of stratification and inequality. But it was not until the work of the Frankfurt School (see the chapter on Marx) during the 1930s that a critical view of culture itself became important, and it was not until the work of postmodernists and the Birmingham School in the 1970s and 1980s that representation became a focus of attention. Yet Du Bois is explicating the role of culture and representation in oppression in his 1903 book, *The Souls of Black Folk*.

Representation

Representation is a term that has become extremely important in contemporary cultural analysis. Stuart Hall, for example, argues that images and objects by themselves don't mean anything. We see this idea in Mead's theory as well. The meaning has to be constructed, and we use representational systems of concepts and ideas. Representation, then, is the symbolic practice through which meaning is given to the world around us. It involves the production and consumption of cultural items and is a major site of conflict, negotiation, and potential oppression.

Let me give you an illustration from Du Bois. Cultural domination through representation implies that the predominantly white media do not truly represent people of color. As Du Bois (1920/1996b) says, "The whites obviously seldom picture brown and yellow folk, but for five hundred centuries they have exhausted every ingenuity of trick, of ridicule and caricature on black folk" (pp. 59–60). The effect of such representation is cultural and psychological: The disenfranchised read the representations and may become ashamed of their own image. Du Bois gives an example from his own work at *The Crisis* (the official publication of the NAACP). *The Crisis* put a picture of a black person on the cover of their magazine. When the (predominantly black) readers saw the representation, they perceived it (or consumed it) as "the caricature that white folks intend when they make a black face." Du Bois queried some of his office staff about the reaction. They said the problem wasn't that the person was black; the problem was that the person was *too* black. To this, Du Bois replied, "Nonsense! Do white people complain because their pictures are too white?" (p. 60).

While Du Bois never phrased it quite this way, Roland Barthes (1964/1967), a contemporary semiologist (someone who studies signs), explains that cultural signs, symbols, and images can have both denotative and connotative functions. Denotative functions are the direct meanings. They are the kind of thing you can look up in an ordinary dictionary. Cultural signs and images can also have secondary, or connotative, meanings. These meanings get attached to the original word and create other, wider fields of meaning. At times, these wider fields of meaning can act like myths, creating hidden meanings behind the apparent. Thus, systems of connotation can link ideological messages to more primary, denotative meanings. In cultural oppression, then, the dominant group represents those who are subjugated in such a way that negative connotative meanings and myths are produced. This kind of complex layering of ideological meanings is why members of a disenfranchised group can simultaneously be proud and ashamed of their heritage. As a case in point, the black office colleagues to whom Du Bois refers can be proud of being black but at the same time offer an explanation that an image is *too* black.

Stereotypes and Slippery Slopes

In addition to history and misrepresentation, the cultural representation of oppression consists of being defined as a problem: "Between me and the other world this is ever an unasked question. . . . How does it feel to be a problem?" (Du Bois, 1903/1996d, p. 101). Representations of the group thus focus on its shortcomings, and these images come to dominate the general culture as stereotypes:

> While sociologists gleefully count his bastards and his prostitutes, the very soul of the toiling, sweating black man is darkened by the shadow of a vast despair. Men call the shadow prejudice, and learnedly explain it as the natural defense of culture against barbarism, learning against ignorance, purity against crime, the "higher" against the "lower" races. To which the Negro cries Amen!" (p. 105)

I want to point out that last bit of the quote from Du Bois. He says that the black person *agrees* with this cultural justification of oppression. Here we can see one of the insidious ways in which cultural legitimation works. It presents us with an apparent truth that, once we agree to it, can reflexively destroy us. Here's how this bit of cultural logic works: The learned person says that discrimination and prejudice are necessary. Why? They are needed to demarcate the boundaries between civilized and uncivilized, learning and ignorance, morality and sin, right and wrong. We agree that we should be prejudiced against sin and evil, and against uncivilized and barbarous behavior, and we do so in a very concrete manner. For example, we are prejudiced against allowing a criminal in our home. We thus agree that prejudice is a good thing. Once we agree with the general thesis, it can then be more easily turned specifically against us.

Remember I spoke of Hofstadter's "slippery slope of sexism" in the section on Douglass? Du Bois has a similar slope in mind, but obviously one that entails race. In this section from *The Souls of Black Folk* from which we have been quoting, Du Bois says that the Negro stands "helpless, dismayed, and well-nigh speechless" before the "nameless prejudice" that becomes expressed in "the all-pervading desire to inculcate disdain for everything black" (Du Bois, 1903/1996d, p. 103). In *Darkwater* (1920/1996a), Du Bois refers to this slippery slope as a "theory of human culture" (p. 505). This work of culture has "worked itself through warp and woof of our daily thought." We use the term *white* to analogously refer to everything that is good, pure, and decent. The term *black* is likewise reserved for things despicable, ignorant, and fearful. There is thus a moral, default assumption in back of these terms that automatically, though not necessarily or even deliberately, includes the cultural identities of white and black.

In our cultural language, we also perceive these two categories as mutually exclusive. For example, we will use the phrase "this issue isn't black or white" to refer to something that is undecided, that can't fit in simple, clear, and mutually exclusive categories. The area in between is a gray, no-person's land. It is culturally logical, then, to perceive unchangeable differences between the black and white races, which is the cultural logic behind the "one drop rule" (the slavery-era determination that one drop of black blood makes someone black). Again, keep in mind that this movement between the specific and general is unthinkingly applied. People don't have to intentionally use these terms as ways to racially discriminate. The cultural default is simply there, waiting to swallow up the identities and individuals that lay in its path.

Double Consciousness

Du Bois attunes us to yet another insidious cultural mechanism of oppression: the internalization of the *double consciousness*. With this idea, Du Bois is drawing on his knowledge of early pragmatic theories of self. He knew William James and undoubtedly came in contact with the works of Mead and of Charles Horton Cooley. We can think of Mead's theory of role taking and Cooley's looking glass self (explained subsequently) and see how the double consciousness is formed. According to Du Bois, African Americans have another subjective awareness that comes from their particular group status (being black and all that that entails), and they have an awareness based on their general status group (being American). What this means is that people in disenfranchised groups see themselves from the position of two perspectives.

Definition

Double consciousness is part of Du Bois's understanding of the black experience in the United States. It generally refers to the experience of one's identity being fragmented into several, contradictory facets. These facets are at war with and negate one another so that the disenfranchised is left with no true consciousness. As Du Bois (1903/1996d) explains, "One ever feels his twoness,—an American, a Negro; two souls, two thoughts, two unreconciled strivings; two warring ideals in one dark body, whose dogged strength alone keeps it from being torn asunder" (p. 102).

Let's think about this issue using Charles Horton Cooley's notion of the looking glass self. In Cooley's theory, the sense of one's self is derived from the perceptions of others. There are three phases in Cooley's scheme. First, we imagine what we look like to other people, we then imagine their judgment of that appearance, and we then react emotionally with either pride or shame to that judgment. An important point for us to see is that Cooley didn't say that we actually perceive how others see and judge us; rather, we *imagine* their perceptions and judgments. However, this imagination is not based on pure speculation; it is based on social concepts of ways to look (cultural images), ways to behave (scripts), and ways we anticipate others will behave, based on their social category (expectations). In this way, Du Bois argues that African Americans internalize the cultural images produced through the ruling, white culture. They also internalize their own group-specific culture.

Blacks thus see themselves from at least two different and at times contradictory perspectives.

In several places, Du Bois relates experiences through which his own double consciousness was formed. In this particular one, we can see Cooley's looking glass at work:

> In a wee wooden schoolhouse, something put it into the boys' and girls' heads to buy gorgeous visiting-cards—ten cents a package—and exchange. The exchange was merry, till one girl, a tall newcomer, refused my card,—refused it peremptorily, with a glance. Then it dawned upon me with a certain suddenness that I was different from the others; or like, mayhap, in heart and life and longing, but shut out from their world by a vast veil. (Du Bois, 1903/1996d, p. 101)

We can thus see in Du Bois's work a general theory of cultural oppression. That is, every group that is oppressed structurally (economically and politically) will be oppressed culturally. The basic method is as he outlines it: Define and label the group in general as a problem, emphasize and stereotype the group's shortcomings and define them as intrinsic to the group, employ cultural mechanisms (like misrepresentation and default assumptions) so the negative attributes are taken for granted, and systematically exclude the group from the grand narrative histories of the larger collective.

Concepts and Theory: The Dark Nations and World Capitalism

Du Bois also sees the social system working in a very Marxian way. Like Marx, Du Bois argues that the economic system is vitally important to society and understanding history. Du Bois (1920/1996a) says that "history is economic history; living is earning a living" (p. 500). He also understands the expansion to global capitalism in Marxian, world systems terms. Capitalism is inherently expansive. Once limited to a few national borders, it has now spread to a worldwide economic system. One of the things this implies is that exploitation can be exported.

Capitalism requires a ready workforce that can be exploited. Goods must be made for less than their market value, and one of the prime ways to decrease costs and increase profits is by cutting labor costs. In national capitalism, this level of exploitation is limited to the confines of the state itself. Consequently, as the national economy expands and the living wage increases, the level of profit goes down. In other words, the average real wage of workers within a country increases over time—the average worker in the United States makes more today than she or he did 100 years ago. This proportional increase in income would lower profit margins. However, with global capitalism, exploitation can be exported. It costs less for a textile company to make clothing in China than in the United States, yet the market value of the garment remains the same. Profits thus go up.

Du Bois (1920/1996a) argues that this expansion of exploitation is due to increasing levels of "education, political power, and increased knowledge of the technique and meaning of the industrial process" (p. 504). There are, according to Marx and Du Bois, certain inevitable "trickle-down" effects of capitalism. As industrial technologies become more and more sophisticated, better-educated workers are needed. Increasing levels of education attune workers not only to the methods of production, but also to the abusive values and practices behind the system. This moment is a case in point: Most of you reading this text will work in the economy, yet here you are reading about scholarship that exposes the oppressive side of capitalism. Industrialization also brings workers closer together in factories and neighborhoods where they can communicate and become politically active.

The Need for Color

As a result, the economic system inexorably pushes against its national boundaries and seeks a labor force ready for exploitation. This notion of expanding systems and exporting exploitation is part of current world systems theory. *World systems theory* argues that core nations exploit periphery nations so places like the United States can have high wages and relatively cheap goods. In that theory, the issue is simply the economic standing of a nation, but Du Bois tells us that the reality of those nations is color. "There is a chance for exploitation on an immense scale for inordinate profit, not simply to the very rich, but to the middle class and to the laborers. This chance lies in the exploitation of darker peoples" (Du Bois, 1920/1996a, pp. 504–505). These are "dark lands" ripe for exploitation "with only one test of success,—dividends!" (p. 505). In other words, middle-class wages in advanced industrialized nations are based on there being racial groups for exploitation.

Du Bois thus sees race as a tool of capitalism. While Du Bois perceives distinct differences between blacks and whites that can be characterized in spiritual terms (the souls of black and white folk), he also argues that the color line is socially constructed and politically meaningful. The really interesting feature of Du Bois's perspective here is that he understands that both black and white social identities are constructed. Again, Du Bois beat contemporary social theory to the punch: We didn't begin to seriously think of white as a construct until the 1970s, and it didn't become an important piece in our theorizing until the late 1980s.

Du Bois argues that the idea of "personal whiteness" is a very modern thing, coming into being only in the nineteenth and twentieth centuries. Humans have apparently always made distinctions, but not along racial lines. Prior to modernity, people created group boundaries of exclusion by marking civilized and uncivilized cultures, religions, and territorial identities. As William Roy (2001) notes, these boundaries lack the essential features of race—they were not seen as biologically rooted or immutable, and people could thus change. Race, on the other hand, is perceived as immutable and is thus a much more powerful way of oppressing people.

As capitalism grew in power and its need for cheap labor increased, indentured and captive slavery moved to chattel slavery. Slavery has existed for much of human history, but it was used primarily as a tool for controlling and punishing a conquered people or criminal behavior, or as a method for people to pay off debt or get ahead. The latter is referred to as indentured servitude. People would contract themselves into slavery, typically for 7 years, in return for a specific service, like passage to America, or to pay off debt. In most of these forms of slavery, there were obligations that the master had to the slave, but not so with chattel slavery. Under capitalism, people could be defined as property—the word *chattel* itself means property—and there are no obligations of owner to property. In this move to chattel slavery, black became not simply *a* race but *the* race of distinction. The existence of race, then, immutably determined who could be owned and who was free, who had rights and who did not.

> Till the devil's strength be shorn,
>
> This some dim, darker David, a-hoeing of his corn,
>
> And married maiden, mother of God,
>
> Bid the black Christ be born!
>
> Then shall our burden be manhood,
>
> Be it yellow or black or white
>
> (Du Bois, 1920/1996a, p. 510)

Summary

- Du Bois's perspective is that of a black man. His subjective experience of race is central in his work. He often uses himself as a representational character, and he is vastly interested in drawing the reader into the experience of race. Du Bois also believes that being a member of a disenfranchised group, specifically African American, gives one a privileged point of view. People who benefit from the system cannot truly see the system. Most of the effects of active social oppression are simply taken for granted by the majority. They cannot see it.
- All structural oppression must be accompanied by cultural suppression. There are several mechanisms of cultural suppression: denial of history, controlling representation, and the use of stereotypes and default assumptions. As a result of cultural suppression, minorities have a double consciousness. They are aware of themselves from their group's point of view, which is positive, and they are conscious of their identity from the oppressor's position, which is negative. They are aware of themselves as black (in the case of African Americans), and they are aware of themselves as American—two potentially opposing viewpoints.
- Capitalism is based on exploitation: Owners pay workers less than the value of their work. Therefore, capitalism must always have a group to exploit. In global capitalism, where the capitalist economy overreaches the boundaries of the state, it is the same: There must be a group to exploit. Global capitalism finds such a group in the "dark nations." Capitalists thus export their

exploitation, and both capitalists and white workers in the core nations benefit, primarily because goods produced on the backs of sweatshop labor are cheaper.

Gender and Race: Thinking About Modernity and Postmodernity

Modernity and Identity

Martineau, Gilman, and Du Bois are clearly modernists. Du Bois believes in the hope of a scientific approach to history, Martineau champions the cause of positivism and the hope of the Enlightenment, and Gilman is a social evolutionist—social evolution being a theory born and bred for modernity. Yet, they are modern in a more fundamental way, as well. The notions of difference, identity, and social movements are themselves modern, as Douglass's work exemplifies.

Human identity and self are built upon difference. According to psychologists, one of the fundamental things an infant must learn is that she or he is a separate and distinct entity, different from the rest of the world and from significant others. This type of self is discovered through the gradual unfolding of the person's psychological makeup, which is the inner core of the individual. In this way of thinking about things, the individual is always identical to the inner self and is thus essentially different from others. Most of us, to one degree or another, have bought into this way of thinking about the self, unless we have been thoroughly brainwashed by sociology (the fervent hope of all your sociology professors). The self we are talking about here is the one with which you are born and the one that constitutes you all through your life. Your self is inside you, and the self inside the other person is different by definition—that is what makes us essentially different.

On the other hand, sociology understands identity and self in relation to society and culture. A person doesn't *discover* the self through one's lifespan; rather, she or he *constructs* the self in interaction with society and culture. This is an entirely different perspective. Here, the process of identity involves identifying one's self with an external element of culture, such as gender or race, not an internal characteristic. In other words, people are different not because they are *born* different, but because they relate to different social and cultural elements. There are two important points that we can glean from the sociological scheme: The construction of identity and self is carried out with reference to social structures and culture, and the construction of identity and self is always to some degree exclusionary. Let's expand a bit on the former point first.

According to sociologists, identity isn't a given; it isn't something we are born with, and it isn't defined by the boundaries of our body. For the psychologist, self and identity are embodied, and it is a simple matter to see where the self ends and the *alter* (the other) begins. Infants have to discover this, but it is nonetheless a given. Sociologists, however, argue that identities based on race, gender, religion, politics, music, work, and so forth are social constructions with which the individual identifies and through which the individual understands and gives meaning to his or her subjective experiences—in other words, one's self. That being the case, the boundaries upon which identities and selves are created are themselves cultural and social and not defined intrinsically by one's body. For psychologists, difference isn't a big deal because it is an intrinsic and natural part of our psychological makeup that is contained within a physical body; for sociologists, difference is important precisely because it is a function of culture. We can't take difference for granted as a psychologist can; we have to understand how it is created.

Part of what we have in the works of gender and race are theories about how these differences are produced. Du Bois, for example, gives us a cultural understanding of how difference is produced and maintained by the ruling class (through history, representation, stereotypes, cultural logic, and so on). Douglass points out how assumptions of human nature can be used to construct identities of difference. Gilman's theory has cultural components, but it is basically an evolutionary theory of difference: Men and women interacting with different environments produce difference. From this perspective, culture is produced by social structures and functions only to reinforce culture, although Gilman does seem to indicate that culture can have independent effects. Martineau doesn't give us a strong theory of difference; nevertheless, her ideas of courage and chastity fit in nicely with Du Bois's insights into cultural logic and Gilman's separation of spheres.

In terms of creating difference, race and gender are among the strongest social distinctions we make. There are obvious biological features, but as we saw with Mead, it's the meaning that matters. Even Gilman's theory with its genetic features (male/female energy, women's bodily changes) recognizes that the meaning we place on those characteristics is vitally important. The meaning is our reality, not the genetic or biological attributes. What makes race and gender our strongest social distinctions are the *legitimations* we can attach to them. We make them *appear* natural and unquestionable.

The Postmodern Twist

The idea of postmodernism gives the construction of difference an interesting twist. Postmodernists buy into the sociological idea of the relationship between the self and culture. Identities and selves are clearly created in relation to culture. Accordingly, as long as the culture of a society is fairly cohesive and stable, then the individual's self and identities will be, too. In fact, modernity requires such an identity and self. We have to see ourselves as part of some larger, abstract scheme of things, some imagined community called the nation. We have to have such an identity to participate in national projects, like going to war and paying taxes. If you want me to lay down my life for something, you'd better make sure that I think we are all on the same page about everything. I have to emotionally and psychically relate to the whole of society and its meaning.

More than that, modernity requires a stable, core identity in order to fulfill its goal of equality. As I've pointed out, included in the project of modernity are the ideas of democracy and equal treatment before the law. Ironically, these ideas initially pushed society to make distinctions between those who could vote and receive equal rights and those who could not. It's paradoxical, but there are some theoretical reasons for it. The modern movements for democracy and equality were social movements, and social movements require people to have a strong sense of identity.

One of the fundamental ways in which identity and difference are constructed is through exclusion. In psychology, this can by and large be taken for granted: I am by definition excluded from you because I am in my own body. For sociologists, exclusion is a cultural and social practice; it's something we *do,* not something we *are.* This fundamental point may sound elementary in the extreme, but it's important for us to understand it. In order for me to be me, I can't be you; in order for me to be male, I can't be female; in order for me to be white, I can't be black; ad infinitum. Cultural identity is defined over and against something else. Identity and self are based on exclusionary practices. The stronger the practices of exclusion are, the stronger will be the identity; and the stronger my identities are, the stronger will be my sense of self. Further, the greater the exclusionary practices, the more real will be my experience of identities and self.

Hence, the early social movements for equality and democracy had clear practices of exclusion. It gave the people the needed level of identity to make the sacrifices necessary to fight. The distinctions between royalty and commoner were torn down, and the differences between the recipients of "equal" rights were created. Let's note that no matter how limited they were in who should be included, they took the idea of equality further than it had ever been taken before. And let's also note that no matter how upset we might be at the limitations of these early social movements, the people in the world's democracies are *still* making distinctions about who can have civil rights and who cannot (the debate over gay marriage is a case in point).

The twist that postmodernism gives to all this is found in the ideas of cultural fragmentation and de-centered selves and identities. We've talked about all this before, particularly in the chapters on Durkheim and Weber, so I don't want to belabor the point. But, the entire idea of gender or race is modernist. It collapses all individualities into an all-encompassing identity. However, a woman isn't simply female; she also has many identities that crosscut that particular cultural interest and may shift her perceptions of self and other in one direction or another. For us to claim one or more of these identities—to claim to be female or black (or male and white)—is really for us to put ourselves under the umbrella of a grand narrative, which includes very strong exclusionary tactics.

More to the point, the construction of centered identities is becoming increasingly difficult in postmodernity. Remember the connection between identity and culture? As we've seen, postmodernists argue that culture in postindustrial societies is fragmented. If the culture is fragmented, then so are our identities. The idea of postmodernism, then, makes the issues of gender and race very complex. Racial and gender identities may thus be increasingly difficult to maintain. The culture has become more multifaceted, and so have our identities. This means that we have greater freedom of choice (culture and structure are not as determinative in postmodernity), which we all think we enjoy, but freedom of choice implies that the distinctions between gender and racial identities aren't as clear or as real as they once were (remember that claiming and maintaining a racial identity is a cultural practice). Thus, social movements around race and gender are difficult to create nowadays. If our identities are fragmented, then emancipatory politics with its social movements is difficult to pull off. We looked at this idea specifically in the chapter on Weber. The end result here is that the culture of race and gender becomes more vital than the structure.

Moreover, this multiplicity takes Du Bois's idea of double consciousness to a new level. For example, of her mixed racial heritage, Gloria Anzaldúa (2007) says,

> Cradled in one culture, sandwiched between two cultures, straddling all three cultures and their value systems, *la mestiza* undergoes a struggle of flesh, a struggle of borders, an inner war. . . . These numerous possibilities leave *la mestiza* floundering in uncharted seas. In perceiving conflicting information and points of view, she is subjected to a swamping of her psychological borders. (p. 387)

As I said, gender and race become more complex under postmodern considerations: The cultural boundaries aren't as clear, social movements become increasingly difficult to create, and life politics rather than emancipatory politics becomes the standard. I've heard it said, "Just when feminism made women's subjective experience important, postmodernism de-centered the subject." Yet, postmodernism also contains a value about identities. It may be a utopian vision, but it may also contain something we can hope for. Modern identities are strong and clear. We can talk about women's rights under modernity. But in doing so, we create a grand narrative that simultaneously excludes other identities, such as sexuality, religion, race, ethnicity, tastes, and so on. In postmodernity, the severity of structural and cultural differentiation may make identities difficult to maintain, but we should value complex identities, multiple voices, and local rather than grand identities.

BUILDING YOUR THEORY TOOLBOX

Learning More—Primary and Secondary Sources

For Douglass, I recommend that you start with his first biography and then move on to the selected readings:

- Douglass, F. (1882). *The life and times of Frederick Douglass, from 1817 to 1882, written by himself.* London: Christian Age.

(Continued)

(Continued)

- Douglass, F. (1999). *Selected speeches and writings,* P. S. Foner & Y. Taylor (Eds.). Chicago: Lawrence Hill Books.

For Du Bois, I recommend that you pick up *The Oxford W. E. B. Du Bois Reader,* edited by Eric Sundquist. It contains two of Du Bois's books in their entirety (*The Souls of Black Folk* and *Darkwater*) as well as myriad other important writings.

- Du Bois, W. E. B. (1996a). Darkwater. In E. J. Sundquist (Ed.), *The Oxford W. E. B. Du Bois reader.* New York: Oxford University Press.

In general, the best edited volume on classic African American thought is this one:

- Brotz, H. (Ed.). (2009). *African-American social & political thought: 1850–1920.* New Brunswick, NJ: Transaction Publishers.

Seeing the Social World (knowing the theory)

- Write a 250-word synopsis of the theoretical perspective of race theory.
- After reading and understanding this chapter, you should be able to define the following ideas in reference to Douglass's work: universalism/particularism, the Other, ruling species, chattel slavery, ontological wounds, emigration, democracy as an ongoing project.
- After reading and understanding this chapter, you should be able to define the following terms theoretically and explain their theoretical importance to Du Bois's theory of racial oppression: centered subject, seventh son, veil, grand narrative, history as ideology, representation, denotation, connotation, stereotypes, default assumptions, looking glass self, double consciousness, exploitation, dark nations, personal whiteness.
- Explain Douglass's argument about the democratic potential contained in the founding documents of the United States.
- Explain how democracy is an ongoing project.
- Describe how history and representation can be used as tools in oppression and in silencing the voices of disenfranchised groups.
- Explain how race is used in capitalism.

Engaging the Social World (using the theory)

- Go home and watch TV. Intentionally watch programs that focus on African Americans, and pay particular attention to commercials that feature people of color. Using Du Bois's understanding of representation and Barthes's ideas of denotation and connotation, analyze the images that you've seen. What are some of the underlying connotations of the representations of blacks, Chicanos, Asians, and other minorities? How do you think this influences the consciousness of members of these groups?

- According to Du Bois, one of the ways cultural oppression works is by excluding the voices and contributions of a specific group in the history of a society. The Anti-Defamation League has a group exercise called "name five." The challenge is to name five prominent individuals in each category: Americans; male Americans; female Americans; African Americans; Hispanic Americans; Asian or Pacific Islander Americans; Native Americans; Jewish Americans; Catholic Americans; pagan Americans; self-identified gay, lesbian, or bisexual Americans; Americans with disabilities; and Americans over the age of 65. For which categories can you name five prominent people? For which can't you name the five? What does this imply about the way we have constructed history in this country? In addition to trying this activity on yourself or a friend, go to the Biography Channel's webpage (www.biography.com). There you will find a searchable database of over 25,000 people whose lives are deemed important. Try each of the categories mentioned above. What did you find?

CHAPTER 9

Seeing Ahead

Defining Moments in Twentieth Century Theory: Talcott Parsons and the Frankfurt School

Part of what a classical theory text should do is prepare you for contemporary theory. Marx, Durkheim, and Weber are essential, but we've done a little bit more than is usual by considering how their ideas help us to understand our contemporary late- or postmodern society. Race and gender have become significant contemporary issues, and you now have significant background in those as well. And to really grasp contemporary sociological social psychology, Mead and Simmel are essential. In addition, there are two major "events" in sociology's history that occurred around the middle of the twentieth century that dramatically impact contemporary theory: Talcott Parsons and the Frankfurt School. One of the problems with these two is that they aren't classical and they aren't contemporary, so in terms of theory texts, it's difficult to know where to put them. I offer them here as a prelude to contemporary theory.

In many ways, these two forces took the social disciplines in opposite directions: Parsons saw himself building on the ideals of modern knowledge; the Frankfurt School did just the opposite and argued that rather than leading to social justice, social science destroys the possibility of freedom and equality. Both were centrally concerned with culture: Parsons saw culture as the most important factor leading to social cohesion and harmony, while the Frankfurt School of theorists saw culture, especially popular culture, as producing false consciousness. Both have influenced contemporary theory beyond their specific ideas.

Historically, the Frankfurt School developed first. However, I'm going to start with Parsons because he extends and systematizes the things we learned about modern knowledge and society in Chapter 1. In addition, Parsons's influence is in many ways more central, as many contemporary theorists continue to see themselves arguing for or against his work. There are good reasons for giving this much influence to Parsons, but there's also a sense in which I'm making him an ideal type, similar to what I did with modernity. Both Parsons and modernity exist and are clearly important, but I'm not presenting a well-reasoned and documented case for either. I'm using them as heuristics to further our discussion about theory and what sociology is up to—they represent ways for us to discover questions, ideas, and theories that come with each of our theorists.

Talcott Parsons: Defining Sociology

Talcott Parsons (1902–1979) was born in Colorado Springs, Colorado. Parsons began his university studies at Amherst (Massachusetts), planning to become a physician, but he later changed his major to economics. After receiving his BA in 1924, Parsons studied in Europe, completing his PhD work in sociology and economics at the University of Heidelberg, Germany. After teaching a short while at Amherst, Parsons obtained a lecturing position in 1931 at Harvard and was one of the first instructors in the new sociology department.

Parsons's Vision for the Social Sciences

Parsons was a man with a grand vision. He wanted to unite the social and behavioral disciplines into a single social science and to create a single theoretical perspective. Parsons worked at this not only theoretically, but also organizationally. In 1942, Parsons became department chair of the sociology department at Harvard University. One of the first things he did was to combine sociology, anthropology, and psychology into one department, the Department of Social Relations. The reason he did this was to break down the barriers between disciplines in order to create a general science of human action. His desire, then, wasn't simply to understand a portion of human action (as in sociology); he wanted, rather, to comprehend the totality of the human context and to offer a full and complete explanation of social action. The department existed from 1945 to 1972 and formed the basis of other interdisciplinary programs across the United States.

After 10 years of work, Parsons's first book was published in 1937: *The Structure of Social Action*. This book is characterized by Lewis Coser (1977) as a

> watershed in the development of American sociology in general and sociological theory in particular . . . [which] set a new course—the course of functional analysis—that was to dominate theoretical developments from the early 1940s until the middle of the 1960s. (p. 562)

More than any other single book, it introduced European thinkers to American sociologists and gave birth to structural functionalism, which Desmond Ellis described in 1971 as "the major theoretical orientation in sociology today" (p. 692). Parsons's other prominent works include *The Social System, Toward a General Theory of Action, Economy and Society, Structure and Process in Modern Societies*, and *The American University*. For much of the twentieth century, Parsons was "the major theoretical figure in English-speaking sociology, if not in world sociology" (Marshall, 1998, p. 480). As Victor Lidz (2000) notes, "Talcott Parsons . . . was, and remains, the pre-eminent American sociologist" (p. 388).

Parsons's Theoretical Project

There are at least three ways in which Parsons helped shape the center of sociological discourse in the twentieth century: the way he theorized, the problem he addressed, and the theory itself. We'll start with his theorizing. Recall that science is built upon positivism and empiricism. As such, science assumes that the universe is empirical, it operates according to law-like principles, and humans can discover those laws through rigorous investigation. Science also has very specific goals, as do most knowledge systems. Through discovery, scientists want to explain, predict, and control phenomena. In addition, there are two other important issues in positivistic theory, which we find in the following quotes from prominent contemporary theorists:

> The essence of science is precisely theory . . . as a *generalized* and coherent body of ideas, which explain the range of variations in the empirical world in terms of general principles. . . . [I]t is explicitly *cumulative and integrating*. (R. Collins, 1986a, p. 1345, emphasis added)

> A true science *incorporates the ideas* of its early founders in introductory texts and moves on, giving over the analysis of its founders to history and philosophy. (Turner, 1993, p. ix, emphasis added)

The first thing I want us to glean from the preceding quotes is that scientific theory is *generalized*. To make an idea or concept general means to make it applicable to an entire group of similar things or people. As you'll see when we consider Parsons's theory, his concepts are very general (and thus fairly dry—but, then, all scientific theory is that way). Scientific knowledge also involves both theory synthesis and cumulation. Synthesis involves bringing together two or more elements in order to form a new whole. For example, water is the synthesis of hydrogen and oxygen. *Theoretical synthesis,* then, involves bringing together elements from diverse theorists so as to form a theory that robustly explains a broader range of phenomena. Cumulation refers to the gradual building up of something, such as the cumulative effects of drinking alcohol. *Theory cumulation* specifically involves the building up of explanations over time. This incremental building is captured by Isaac Newton's famous dictum, "If I have seen further it is by standing on the shoulders of giants." Yet, what isn't clear in Newton's quote is that the ultimate goal of theory cumulation is to forget its predecessors.

To make this clear, let's compare the writings of two authors, Edgar Allan Poe and Albert Einstein. Here's one of Poe's famous stanzas, from "The Raven":

> Once upon a midnight dreary, while I pondered, weak and weary,
>
> over many a quaint and curious volume of forgotten lore,
>
> While I nodded, nearly napping, suddenly there came a tapping,
>
> As of someone gently rapping, rapping at my chamber door.
>
> "Tis some visitor," I muttered, "tapping at my chamber door;
>
> Only this, and nothing more."

Here's one of Einstein's famous quotes:

$$E = mc^2$$

There are some obvious differences between these two quotes: One is poetry and the other a mathematical equation. But I want you to see a bit more. Does it matter who wrote "Once upon a midnight dreary"? Yes, it does. A large part of understanding poetry is knowing who wrote it—who they were, how they lived, what their other works are like, what style they wrote in, and so on. These issues are part of what makes reading Poe different from reading Emily Dickinson. Now, does it matter who wrote $E = mc^2$? Not really. You can understand everything you need to know about $E = mc^2$ simply by understanding the equation. The author in this sense is immaterial.

One of the previous quotes is from Jonathan H. Turner's book *Classical Sociological Theory: A Positivist's Perspective.* Turner's (1993) goal in that book was "to codify the wisdom of the masters so that we can move on and *make books on classical theory unnecessary*" (p. ix, emphasis added). That last highlighted section is the heart of theory cumulation: Cumulating theory implies that we do away with the individual authors and historical contexts and keep only the theoretical ideas that explain, predict, and control the social world. In that spirit, here's a theoretical statement from Turner's book:

> The degree of differentiation among a population of actors is a gradual s-function of the level of competition among these actors, with the latter variable being an additive function of:
>
> the size of this population of actors,
>
> the rate of growth in this population,

the extent of economical concentration of this population, and

the rate of mobility of actors in this population. (p. 80)

First notice how general the statement is: It can be applied to any group of people, living anywhere, at any time. Notice also that there's no mention of from whom these ideas originally came. Now, you and I might know from whom this proposition comes (Durkheim), but does it matter? No. Like Einstein's formula, it's immaterial. If we are doing social science, what matters is whether or not we can show this statement to be false through scientific testing. If we can't, then we can have a certain level of confidence that the proposition accurately reflects a general process in the social world. In science, authorship is superfluous; *it's the explanatory power of the theory that matters.* The cumulation of these general statements is one of the main goals of scientific theory.

Of his groundbreaking work, Parsons (1949) says, "*The Structure of Social Action* was intended to be primarily a contribution to systematic social science and not to history" (pp. A–B). His work is actually a synthesis of three theorists. Parsons (1961) notes how he used the work of each one:

for the conception of the social system and the bases of its integration, the work of Durkheim; for the comparative analysis of social structure and for the analysis of the borderline between social systems and culture, that of Max Weber; and for the articulation between social systems and personality, that of Freud. (p. 31)

Yet Parsons clearly wants us to forget the historical and personal origins of the theories—for science, it's the power of the synthesized theory to illuminate and delineate social factors and processes that matters. This approach to theory is also what led to the three sociological perspectives or paradigms you were taught in your introduction to sociology courses: structural-functionalism, conflict theory, and interactionism. To say that someone is a functionalist, for example, is to pay more attention to the general features of the theory than what he or she contributes originally.

Parsons and the Problem of Social Order

Parsons saw himself as responding to *the problem of social order* posed by the philosopher Thomas Hobbes (1588–1679). Parsons's understanding of this Hobbesian problem of social order begins with the fact that all humans are ruled by passions. Moreover, all people are motivated to fulfill these passions, and, more important, they have the *right* to fulfill them because "there is 'no common rule of good and evil to be taken from the nature of the objects themselves'" (Hobbes, as cited in Parsons, 1949, p. 89). In other words, things aren't good or bad in themselves, and people have different desires for diverse things—thus, there is no basis for a rule. In the absence of any rule, people will use the most efficient means possible to acquire their goals. As Parsons says, "These means are found in the last analysis to be force and fraud" (p. 90). Thus, the most natural state of humanity is the war of all against all. The question, then is, how is social order achieved? Parsons's basic response is the normative order—social order achieved through norms. While some of the language might be new to you, most of Parsons's response will probably feel familiar. The reason that's probably the case is that Parsons's answer to the problem of social order has become for many sociologists the basic answer given in introduction to sociology classes.

Voluntaristic Action

In thinking about humans, it's sometimes convenient to make the distinction between behavior and action: All living things behave; only humans have the potential to act. Action implies choice and decision, whereas behavior occurs without thought, as when a plant's leaves reach for the sun. Of course, humans behave as well as act; there are a lot of things we do on autopilot, but we have the potential for action. Theory that focuses on this issue is referred to as *action theory,* which has been an interest of philosophy since the time of Aristotle. Parsons's work on action theory draws from Max Weber, who argued that action takes place when a person's behavior is meaningfully oriented toward other social actors, usually in terms of meaningful values or rational exchange. *Voluntaristic action,* then, is never purely individualistic: People choose to act voluntarily within a context of culture and social situations in order to meet individual goals. And because human needs are met socially, people develop shortcuts to action by creating norms and by patterning action through sets of ends and means.

Parsons calls this context of action *the unit act.* There are a variety of factors in the unit act. The first, and in some ways the most important, are the conditions of action. There are of course occasions when we have some control over the initial context—for example, you may decide to go to the movies on Saturday or study for a test. But once the choice is made, the actor has little immediate agency or choice over the conditions under which action takes place. Parsons has in mind such things as the presence of social institutions or organizations, as well as elements that might be specific to the situation, such as the social influence of particular people or physical constraints of the environment.

The second set of factors under which people act concerns the means and ends of action. Here we can see a fundamental difference between action and reaction: Action is goal oriented and involves choice. But for people to make choices among goals and means, the choices themselves must have different meanings. According to Parsons, the meanings of and relationships between means and ends are formed through shared value hierarchy. Cultural *values* are shared ideas and emotions regarding the worth of something, and values are always understood within a hierarchy, with some goals and means being more highly valued than others; otherwise it would be difficult to choose between one thing and another because you wouldn't care. Choices among means and ends are also guided by norms. *Norms* are actions that have sanctions (rewards or punishments) attached to them. Taken together, Parsons is arguing that human action is distinctly cultural and thus meaningful action.

Patterning Voluntaristic Action

While Parsons has now outlined the context wherein action takes place, the problem of social order isn't adequately answered. Two things need to be specifically addressed: First, Hobbes talked about a person's inner passions driving him or her; second, social order needs patterned behavior. Parsons argues that patterning action occurs on two levels: the structuring of patterned behaviors and individual internalization or socialization. Parsons understands internalization in Freudian terms. Freud's theory works like this: People are motivated by internal energies surrounding different need dispositions. As these different psychic motives encounter the social world, they have to conform in order to be satisfied. Conformity may be successful (well-adjusted) or unsuccessful (repressed), but the point to notice here is that the structure of the individual's personality changes as a result of this encounter between psychic energy and the social world. The superego is formed through these encounters. For Parsons, the important point is that cultural traditions become meaningful to and part of the need dispositions of individuals. The way we sense and fulfill our needs is structured internally by culture—notice how Parsons reconceptualized Hobbes's concern for passions; in Parsons's scheme they are social rather than individual. For Parsons, then, the motivation to conform comes principally from within the individual through Freudian internalization patterns of value orientation and meaning.

Action is also structured socially through modes of orientation and types of action that have become institutionalized. Modes of orientation simply refer to the way we come into a social situation with specific motives and values. These motives and values come together and form three types of action: instrumental, expressive, and moral. *Instrumental action* is composed of the need for information and evaluation by objective criteria. *Expressive action* is motivated by the need for emotional attachment and the desire to be evaluated by artistic standards. *Moral action* is motivated by the need for assessment by ultimate notions of right and wrong. People will tend to be in contact with others who are interested in the same type of action.

As we interact over time with people who are likewise oriented, we produce patterns of interaction and a corresponding system of status positions, roles, and norms. Status positions tell us where we fit in the social hierarchy of esteem or honor; roles are sets of expected behaviors that generally correspond to a given status position (for example, a professor is expected to teach); and norms are expected behaviors that have positive and/or negative sanctions attached to them. Generally, these cluster together in institutions. For functionalists such as Parsons, institutions are enduring sets of roles, norms, status positions, and value patterns that are recognized as collectively meeting some societal need. In this context, then, institutionalization refers to the process through which behaviors, cognitions, and emotions become part of the taken-for-granted way of doing things in a society ("the way things are").

These clusters of institutions meet certain needs that society has—the institutions *function* to meet those needs. The most important set of institutions for Parsons are the ones that produce *latent pattern maintenance*. If something is latent, it's hidden and not noticed. Social patterns are maintained, then, through indirect management. For this task, society uses culture and socialization. The chief socializing agents in society are the structures that meet the requirement of latent pattern maintenance—structures such as religion, education, and family.

In addition to latent pattern maintenance, Parsons gives us three other requisite functions for a system: adaptation, goal attainment, and integration. The *adaptation* function is fulfilled by those structures that help a system to adapt to its environment. Adaptation draws in resources from the environment, converts them to usable elements, and distributes them throughout the system (the economy). *Goal attainment* is the subsystem that activates and guides all the other elements toward a specific goal (government). In Parsons's scheme, *integration* refers to the subsystems and structures that work to blend together and coordinate the various actions of other structures. In society, the structure most responsible for this overt coordination is the legal system. Together, these four functions are referred to as AGIL: Adaptation, Goal attainment, Integration, and Latent pattern maintenance. All of these functions are embedded within one another and form a bounded system that tends toward equilibrium or balance; in other words, they form society.

The Frankfurt School: The Problem With Sociology

Karl Marx spawned two distinct theoretical approaches. One approach focuses on conflict and class as general features of society. The intent with this more sociological approach is to analytically describe and explain conflict. Conflict and power here are understood as fundamental to society. And, again, this approach is based on the same question as for Parsons: How is social order possible? The short version of Parsons's theory is that social order is achieved through commonly held norms, values, and beliefs. A norm, as we said above, "is a cultural rule that associates people's behavior or appearance with rewards or punishments" (Johnson, 2000, p. 209). This approach to understanding social order is sometimes called the "equilibrium model" because it's based on people internalizing and believing in the collective conscious. People do have selfish motivations, but they are offset by the collective conscious, thus creating a balance between individual desires and social needs.

Conflict theorists, however, would point out that there is an element of power underlying norms. Notice in the definition that norms are founded upon an ability to reward or punish behavior, both of which are based on power. Conflict theorists also point out that the values and beliefs commonly held in society can be explained in terms of the interests of the elite. Thus, for conflict theorists, social order is the result of constraint rather than consensus, and power is thus an essential element of society. Conflict theorists take the same basic scientific approach as Parsons, seeking generalizable processes and building theory cumulatively.

The other theoretical approach that is inspired by and draws from Marx is critical theory. In general, *critical theory* "aims to dig beneath the surface of social life and uncover the assumptions and masks that keep us from a full and true understanding of how the world works" (Johnson, 2000, p. 67). Critical theory doesn't simply explain how society operates. Rather, it uncovers the unseen or misrecognized ways in which society operates to oppress certain groups while maintaining the interests of others. Critical theory has one more defining feature: It is decidedly anti-positivistic. Thus, critical theory has a very clear agenda that stands in opposition to scientific sociology. This perspective was brought together by the Frankfurt School.

Historical Roots

Briefly, the Frankfurt School (also known as the Institute of Social Research) began in the early 1920s at the University of Frankfurt in Germany. It was formed by a tight group of radical intellectuals and, ironically, financed by Felix Weil, the son of a wealthy German merchant. Weil's goal was to create "an institutionalization of Marxist discussion beyond the confines both of middle-class academia and the ideological narrow-mindedness of the Communist Party" (Wiggershaus, 1986/1995, p. 16). As the Nazis gained control in Germany, the Frankfurt School was forced into exile in 1933, first to Switzerland, then New York, and eventually California. In 1953, the school was able to move back to its home university in Frankfurt. The various leaders and scholars associated with the school include Theodor Adorno, Max Horkheimer, Herbert Marcuse, Eric Fromm, and Jürgen Habermas.

Though there were many reasons why the Frankfurt School and its critical theory approach came into existence, one of the early problems that these theorists dealt with was the influence of Nazism in Germany. The two decades surrounding World War II were a watershed period for many disciplines. The propaganda machine in back of Nazism and the subsequent human atrocities left a world stunned at the capacity of humanity's inhumanity. The actions of the Holocaust and the beliefs that lay at their foundation spurred a large cross-disciplinary movement to understand human behavior and beliefs. One of the best-known attempts at understanding these issues is Stanley Milgram's (1974) psychological experiments on authority.

The sociological attempt at understanding this horror demanded that culture be studied as an independent entity, something Marx didn't do. It was clear that what happened in Germany was rooted in culture that was used to intentionally control people's attitudes and actions—this use of culture was formalized in 1933 with the Reich Ministry for Popular Enlightenment and Propaganda. In addition, ideology became seen as something different, something more insidious, than perhaps Marx first suspected. The Frankfurt School asks us to see ideology as more diffuse, as not simply a direct tool of the elite, but rather, as a part of the cultural atmosphere that we breathe.

In general, the Frankfurt School elaborates and synthesizes ideas from Karl Marx, Max Weber, and Sigmund Freud, and focuses on the social production of knowledge and its relationship to human consciousness. This kind of Marxism focuses on Marx's indebtedness to Georg Wilhelm Hegel. Marx basically inverted Hegel's argument from an emphasis on ideas to material relations in the economy. The Frankfurt School reintroduced Hegel's concern with ideas and culture but kept Marx's critical evaluation of capitalism and the state. Thus, like Marx, the Frankfurt School focuses on ideology; but, unlike Marx, critical theory sees ideological production as linked to culture and knowledge rather than simply class and the material relations of production. Ideology, according to critical theorists, is more broadly based and insidious than Marx supposed.

The Problem With Positivism:
Max Horkheimer and Theodor Adorno

The clearest expression of early critical theory is found in Max Horkheimer and Theodor Adorno's book *Dialectic of Enlightenment* (1972), first published in 1944.

Adorno was born in Germany on September 11, 1903. His father was a well-to-do wine merchant and musician. Adorno himself studied music composition in Vienna for 3 years beginning in 1925, after studying sociology and philosophy at the University of Frankfurt. He finished his advanced degree in philosophy under Paul Tillich (a Christian socialist) in 1931, and started an informal association with the institute. In two short years, Adorno was removed by the Nazis due to his Jewish heritage. He moved to Merton College, Oxford, in 1934 and to New York City in 1938, where he fully affiliated with the Frankfurt School in exile. When the school returned to Germany, Adorno became assistant director under Max Horkheimer, who had served as director since 1930. Horkheimer was German, also Jewish, and born into a wealthy family on February 14, 1895. After World War I, Horkheimer studied psychology and philosophy, finishing his doctorate in philosophy in 1925 at the University of Frankfurt, where he became a lecturer and eventually met Adorno. Horkheimer became the second director of the Institute of Social Research in 1930 and continued in that position until 1958, when Adorno took the directorship.

In the *Dialectic of Enlightenment,* Horkheimer and Adorno (1972) argue that the contradictions Marx saw in capitalism are eclipsed by the ones found in the Enlightenment. The Enlightenment promises freedom through the use of reason, rationality, and the scientific method. But in the end it brings a new kind of oppression, not one that is linked to the externalities of life (such as class), but one that extinguishes the spirit and breath of human nature. As we've seen, positivism is based on reason and assumes the universe is empirical. Reason is employed to discover the laws of nature in order to predict and control it. Horkheimer and Adorno argue that the very definition of scientific knowledge devalues the human questions of ethics, aesthetics, beauty, emotion, and the good life, all of which are written off by science as concerns only for literature, which under positivism isn't valued as knowledge at all.

Science is based on the Cartesian dichotomy of subject and object, but the human sciences (sociology, psychology, psychiatry, anthropology, history, and the like) have, in applying the scientific method, objectified the human being through and through. In objectifying, controlling, and opposing nature, the scientific method opposes human nature—human nature is treated like physical nature, objectified and controlled. It results in such things as seeing all human sensualities and sensibilities as evil forces that must be controlled (the Freudian id), and the human psyche, emotions, sensualities, and body must all be managed and brought under the regime of science. People in positivistic social sciences become statistics in a population that must be controlled for the interests of the state (current examples include sexual practices, emotion management, body weight, child rearing, smoking, drinking, and so on—all of which are seen as weaknesses or problems within the individual).

Through the Enlightenment and science, rationality has been enthroned as the supreme human trait. Yet Horkheimer and Adorno trace this ascendency back to fear of the unknown. Rationality began in religion, as magicians, shamans, and priests began to organize and write doctrine, and this impetus toward safety and control accelerated as society relieved such people from the burden of daily work; spirit guides became professional. Their work systematized and provided control over rituals and capricious spirits. Eventually, God was rendered predictable through the ideas of sin and ritualized redemption, and the idea of direct cause and effect was established. Religious issues became universal, with one version of reality, one explanation of the cosmos and humankind's place in it. The hierarchy of gods and individual responses were thus replaced by instrumental reason. The same fear of uncertainty was the motive behind science as well. The technical control of the physical world promised to relieve threats from disease, hunger, and pestilence. And to one degree or another it has done that. Yet science, like religion before it, takes on mythic form and reifies its own method: There is only one form of knowledge, only one way of knowing that is valid.

This unstoppable engine of rationality also extends to the control of everyday life (and we can see how much we've "progressed" in this since the time of Horkheimer and Adorno). The modern life is an *administered life.* Every aspect is open to experts and analysis and is cut off from real social contact and dependency. People are isolated through technology, whether it's the technologies of travel (cars and planes), technologies of communication (such as phones and computers), or the technology of management (bureaucracies). People rationally manage time, space, and relationships, as well as their own self. Self-help is the prescription of the day, guided by experts of every kind. But what's lost is self-*actualization*—there are only remnants of a self that hasn't been administered, only small portions to self-actualize, and even those are squashed in the name of the administered life.

Originally, the Enlightenment had an element of critical thought, where the process of thinking was examined. But it soon became as mechanical as the technologies that it creates, and it denies other ways of knowing and being. Horkheimer and Adorno's story of the Enlightenment is, then, "an account of how humankind, in its efforts to free itself from subjugation to nature, has created new and more all-encompassing forms of domination and repression" (Alway, 1995, p. 33).

The irony and problem is that the Enlightenment was to free humankind. Yet it has created a new kind of *unfreedom,* a binding of the mind that prevents it from perceiving its own chains of bondage. This of course is what Marx meant when he spoke of false consciousness, but for Horkheimer and Adorno, the blindness is ever more insidious. The very tools of thought that were to bring enlightenment instead bring the administered life. How is it possible to get out of this conundrum? This is precisely where critical theory comes in.

First, it's important to understand that there isn't a specific goal or program. Gone are the lofty goals of the Enlightenment, and the method of reason and rationality are useless as well. In back of this negation is a caution and realization. The caution is about being derailed again by believing that we've found *the way*—unlike the capitalists in Marx's scheme, the philosophers of the Enlightenment may be seen as having the best of intentions. We should, then, be cautious of any single answer. The realization is that human beings are social beings; we reflect and express the spirit of the age and the position we hold in society. Knowledge is therefore never pure.

Knowing this implies a second feature of critical theory: The way to freedom is through continual process and critical thinking, the goal of which is to unearth. In his introduction to Adorno's *The Culture Industry*, Bernstein (1991) notes,

> In reading Adorno, especially his writings on the culture industry, it is important to keep firmly in mind the thought that he is not attempting an objective, sociological analysis of the phenomena in question. Rather, the *question* of the culture industry is raised from the perspective of its relation to the possibilities for social transformation. The culture industry is to be understood from the perspective of its potentialities for promoting or blocking "integral freedom." (p. 2)

Notice that Adorno's reading of the culture industry is not intended as a systematic, objective, sociological study. Rather, it is an act of interrogation. But it is questioning with a purpose—in Adorno's case, to assess the potentials for integral freedom that pop culture provides. The image of integral freedom calls up many images, but among the most important are that freedom must be understood with reference to the whole human being—including social relations, economic achievement, spirituality, sensuality, aesthetics, and so on—and that its chief value is the dignity of the person. "Reason can realize its reasonableness only through reflecting on the disease of the world as produced and reproduced by man" (Horkheimer, 2004, p. 120).

BUILDING YOUR THEORY TOOLBOX

- After reading and understanding this chapter, you should be able to define the following terms. Make your definitions as theoretically robust as possible (and don't be afraid to consult other sources): generalized theory, theoretical synthesis, theory cumulation, the problem of social order, voluntaristic action, the unit act, values, norms, socialization, types of action, status positions, latent pattern maintenance, adaptation, goal attainment, integration, conflict theory, critical theory, administered life, unfreedom.

- After reading and understanding this chapter, you should be able to answer the following questions:

  1. Explain Parsons's vision for the social sciences.

  2. Describe the way in which Parsons theorized. Here I want you to write what we might think of as the ideal definition for social science.

  3. Explain how Parsons solves the problem of social order. In other words, according to Parsons, how are social actions patterned across time and space?

  4. Describe how and why the Frankfurt School came into existence.

  5. Explain critical theory's assessment of positivistic theory.

References

Agger, B. (2000). *Public sociology: From social facts to literary acts.* Lanham, MD: Rowman & Littlefield.

Alexander, J. C. (Ed.). (1985). *Neofunctionalism.* Beverly Hills, CA: Sage.

Alexander, J. C. (1988). *Durkheimian sociology: Cultural studies.* Cambridge, UK: Cambridge University Press.

Alexander, J. C. (2006). *The civil sphere.* New York: Oxford University Press.

Allan, K. (1998). *The meaning of culture: Pushing the postmodern critique forward.* Westport, CT: Praeger.

Allan, K., & Turner, J. H. (2000). A formalization of postmodern theory. *Sociological Perspectives, 43*(3), 363–385.

Alway, J. (1995). *Critical theory and political possibilities: Conceptions of emancipatory politics in the works of Horkheimer, Adorno, Marcuse, and Habermas.* Westport, CT: Greenwood Press.

Anweiler, O., Arnove, R. F., Bowen, J., Browning, R., Chambliss, J. J., Chen, T. H., et al. (2011). Education. *Encyclopædia Britannica.* Retrieved August 25, 2011, from http://www .britannica.com/EBchecked/topic/179408/ education

Anzaldúa, G. (2007). La conciencia de la Mestiza: Toward a new consciousness. In E. B. Freedman (Ed.), *The essential feminist reader* (pp. 385–390). New York: The Modern Library.

Barbalet, J. (2010). Citizenship in Max Weber. *Journal of Classical Sociology, 10*(3), 201–216.

Barthes, R. (1967). *Elements of semiology* (A. Lavers & C. Smith, Trans.). London: Jonathan Cape. (Original work published 1964)

Baudrillard, J. (1994). *Simulacra and simulation* (S. F. Blaser, Trans.). Ann Arbor: University of Michigan Press. (Original work published 1981)

Bauman, Z. (1992). *Imitations of postmodernity.* New York: Routledge.

Beauvoir, S. de (1989). *The second sex* (H. M. Parshley, Ed., Trans.). New York: Vintage. (Original work published 1949)

Bellah, R. N. (1970). *Beyond belief: Essays on religion in a post-traditional world.* New York: Harper & Row.

Bellah, R. N. (1975). *The broken covenant: American civil religion in time of trial.* New York: Seabury Press.

Bellah, R. N. (1980). *Varieties of civil religion.* San Francisco: Harper & Row.

Bellah, R. N., Madsen, R., Sullivan, W. M., Swidler, A., & Tipton, S. M. (1985). *Habits of the heart: Individualism and commitment in American life.* New York: Harper & Row.

Bellah, R. N., Madsen, R., Sullivan, W. M., Swidler, A., & Tipton, S. M. (1991). *The good society.* New York: Knopf.

Bendix, R. (1960). *Max Weber: An intellectual portrait.* New York: Doubleday.

Berger, P. (1963). *Invitation to sociology: A humanistic perspective.* New York: Anchor Books.

Berger, P. L. (1967). *The sacred canopy: Elements of a sociological theory of religion.* New York: Doubleday.

Berger, P. L., & Luckmann, T. (1966). *The social construction of reality: A treatise in the sociology of knowledge.* Garden City, NY: Doubleday.

Bernstein, J. M. (1991). Introduction. In T. Adorno, *The Culture Industry* (pp. 1–28). London: Routledge.

Blackstone, W. (1893). *Commentaries on the laws of England in four books.* Philadelphia: J. B. Lippincott. (Original work published 1765–1769)

Blau, P. M., & Meyer, M. W. (1987). *Bureaucracy in modern society* (3rd ed.). New York: McGraw-Hill.

Blumer, H. (1969). *Symbolic interactionism: Perspective and method.* Berkeley: University of California Press.

Bourdieu, P. (1984). *Distinction: A social critique of the judgment of taste* (R. Nice, Trans.). Cambridge, MA: Harvard University Press. (Original work published 1979)

Brotz, H. (Ed.). (2009). *African-American social & political thought: 1850–1920.* New Brunswick, NJ: Transaction Publishers.

Burawoy, M. (2005). 2004 presidential address: For public sociology. *American Sociological Review, 70,* 4–28.

Bureau of Justice Statistics Selected Findings. (2009). *Female victims of violence.* Retrieved October 21, 2011, from http://bjs.ojp.usdoj.gov/content/pub/pdf/fvv.pdf

Burns, T. (1998a). The social construction of consciousness, Part 1: Collective consciousness and its socio-cultural foundations. *Journal of Consciousness Studies, 5*(1), 67–85.

Burns, T. (1998b). The social construction of consciousness, Part 2: Individual selves, self-awareness, and reflexivity. *Journal of Consciousness Studies, 5*(2), 166–184.

Cassirer, E. (1944). *An essay on man.* New Haven, CT: Yale University Press.

Chalmers, D. J. (1995). Facing up to the problem of consciousness. *Journal of Consciousness Studies, 2*(3), 200–219.

Collier, J. (1904). Personal reminiscences by James Collier, for nine years the secretary and for ten years the Amanuensis of Spencer. In J. Royce & J. Collier, *Herbert Spencer: An estimate and review* (pp. 187–234). New York: Fox, Duffield, & Company.

Collins, P. H. (2000). *Black feminist thought: Knowledge, consciousness, and the politics of empowerment* (2nd ed.). New York: Routledge.

Collins, R. (1986a). Is 1980s sociology in the doldrums? *American Journal of Sociology, 91,* 1336–1355.

Collins, R. (1986b). *Max Weber: A skeleton key.* Newbury Park, CA: Sage.

Collins, R. (1988). *Theoretical sociology.* Orlando, FL: Harcourt Brace Jovanovich.

Collins, R. (1998). *The sociological of philosophies: A global theory of intellectual change.* Cambridge, MA: Harvard University Press.

Collins, R. (2009). *Conflict sociology: A sociological classic updated.* Boulder, CO: Paradigm Publishers.

Comte, A. (1957). *A general view of positivism* (J. H. Bridges, Trans.). New York: Robert Speller & Sons. (Original work published 1848)

Comte, A. (1975a). Plan of the scientific operations necessary for reorganizing society. In G. Lenzer (Ed.), *Auguste Comte and positivism: The essential writings* (pp. 9–67). New York: Harper Torchbooks. (Original work published 1822)

Comte, A. (1975b). The positive philosophy. In G. Lenzer (Ed.), *Auguste Comte and positivism: The essential writings* (pp. 71–194). New York: Harper Torchbooks. (Original work published 1830–1842)

Comte, A. (1975c). The second system. In G. Lenzer (Ed.), *Auguste Comte and positivism: The essential writings* (pp. 309–447). New York: Harper Torchbooks. (Original work published 1851–1854)

Condorcet, N. (1795). *Outlines of an historical view of the progress of the human mind.* Philadelphia: Lang and Ustick.

Coser, L. (1956). *The functions of social conflict.* Glencoe, IL: The Free Press.

Coser, L. A. (1977). *Masters of sociological thought: Ideas in historical and social context.* Prospect Heights, IL: Waveland Press.

Coser, L. A. (2003). *Masters of sociological thought: Ideas in historical and social context* (2nd ed.). Prospect Heights, IL: Waveland Press.

Csikszentmihalyi, M. (1990). *Flow: The psychology of optimal experience.* New York: Harper & Row.

Damasio, A. (1999). *The feeling of what happens: Body and emotion in the making of consciousness.* San Diego, CA: Harvest Books.

DeNavas-Walt, C., Proctor, B. D., & Smith, J. C. (2008). *Income, poverty, and health insurance coverage in the United States: 2008.* Retrieved February 18, 2012, from http://www.census.gov/prod/2009pubs/p60-236.pdf

Derksen, L. (2010). Micro/macro translations: The production of new social structures in the case of DNA profiling. *Sociological inquiry, 80*(2), 214–240.

Dewey, J. (1931). George Herbert Mead. *The Journal of Philosophy, 28*(12), 309–314.

Dewey, J. (2009). *Democracy and education: An introduction to the philosophy of education.* WLC Books. (Original work published 1916)

Dicker, R. (2008). *A history of U.S. feminisms.* Berkeley, CA: Seal Press.

Dilthey, W. (1991). *Selected works volume I: Introduction to the human sciences* (R. A. Makkreel & F. Rodi, Eds.). Princeton, NJ: Princeton University Press. (Original work published 1883)

Douglass, F. (1847, December 3). Our paper and its prospects. *The North Star.* Retrieved October 20, 2011, from http://docsouth.unc.edu/neh/douglass/support15.html

Douglass, F. (1882). *The life and times of Frederick Douglass, from 1817 to 1882, written by himself.* London: Christian Age.

Douglass, F. (2009a). An address to the colored people of the United States. In H. Brotz (Ed.), *African-American social & political thought: 1850–1920* (pp. 208–213). New Brunswick, NJ: Transaction Publishers. (Original work published 1848)

Douglass, F. (2009b). The nature of slavery. In H. Brotz (Ed.), *African-American social & political thought: 1850–1920* (pp. 215–220) New Brunswick, NJ: Transaction Publishers. (Original work published 1850)

Douglass, F. (2009c). The present and future of the colored race in America. In H. Brotz (Ed.,) *African-American social & political thought: 1850–1920* (pp. 267–277). New Brunswick, NJ: Transaction Publishers. (Original work published 1863)

Douglass, F. (2009d). Speech on the Dred Scott Decision. In H. Brotz (Ed.), *African-American social & political thought: 1850–1920* (pp. 247–262). New Brunswick, NJ: Transaction Publishers. (Original date 1857)

Dred Scott v. John F. A. Sandford, 60 U.S. 393 (1857).

Du Bois, W. E. B. (1996a). Darkwater. In E. J. Sundquist (Ed.), *The Oxford W. E. B. Du Bois reader* (pp. 481–623). New York: Oxford University Press. (Original work published 1920)

Du Bois, W. E. B. (1996b). In black. In E. J. Sundquist (Ed.), *The Oxford W. E. B. Du Bois reader* (pp. 59–60). New York: Oxford University Press. (Original work published 1920)

Du Bois, W. E. B. (1996c). The propaganda of history. In E. J. Sundquist (Ed.), *The Oxford W. E. B. Du Bois reader* (pp. 438–454.) New York: Oxford University Press. (Original work published 1935)

Du Bois, W. E. B. (1996d). The souls of black folk. In E. J. Sundquist (Ed.), *The Oxford W. E. B. Du Bois reader* (pp. 97–240). New York: Oxford University Press. (Original work published 1903)

Durkheim, É. (1938). *The rules of sociological method* (G. E. G. Catlin, Ed.; S. A. Solovay & J. H. Mueller, Trans.). Glencoe, IL: The Free Press. (Original work published 1895)

Durkheim, É. (1957). *Professional ethics and civic morals* (C. Brookfield, Trans.). London: Routledge.

Durkheim. É. (1960a). Montesquieu's contribution to the rise of social science. In *Montesquieu & Rousseau: Forerunners of sociology* (R. Manheim, Trans.; pp. 1–64). Ann Arbor: The University of Michigan Press. (Original work published 1892)

Durkheim. É. (1960b). Rousseau's social contract. In *Montesquieu & Rousseau: Forerunners of sociology* (R. Manheim, Trans.; pp. 65–138). Ann Arbor: The University of Michigan Press. (Original work published 1918)

Durkheim, É. (1961). *Moral education: A study in the theory and application of the sociology of education* (E. K. Wilson, Trans.). New York: The Free Press. (Original work published 1903)

Durkheim, É. (1984). *The division of labor in society* (W. D. Halls, Trans.). New York: The Free Press. (Original work published 1893)

Durkheim, É. (1993). *Ethics and the sociology of morals* (R. T. Hall, Trans.). Buffalo, NY: Prometheus. (Original work published 1887)

Durkheim, É. (1995). *The elementary forms of the religious life* (K. E. Fields, Trans.). New York: The Free Press. (Original work published 1912)

Eberly, D. E. (2000). The meaning, origins, and applications of civil society. In D. E. Eberly (Ed.), *The essential civil society reader: The classic essays* (pp. 3–29). Lanham, MD: Rowman & Littlefield.

Ellis, D. P. (1971). The Hobbesian problem of order: A critical appraisal of the normative solution. *American Sociological Review, 36,* 692–703.

Engels, F. (1978a). The origin of the family, private property, and the state. In R. C. Tucker (Ed.), *The Marx-Engels reader* (pp. 734–759.) New York: Norton. (Original work published 1884)

Engels, F. (1978b). Socialism: Utopian and scientific. In R. C. Tucker (Ed.), *The Marx-Engels* reader (pp. 683–717). New York: Norton. (Original work published 1880)

Engels, F. (1978c). Speech at the graveside of Karl Marx. In R. C. Tucker (Ed.), *The Marx-Engels reader* (pp. 681–682). New York: Norton. (Original work published 1883)

Engels, F. (2009). *The condition of the working class in England* (D. McLellan, Ed.). Oxford, UK: Oxford University Press. (Original work published 1844)

Fanon, F. (2004). *The wretched of the earth* (R. Philcox, Trans.). New York: Grove Press. (Original work published 1961)

Forbes.com. (2003). *What the boss makes.* Retrieved February 7, 2012, from http://www.forbes.com/finance/lists/12/2003/LIR.jhtml?passListId=12&passYear=2003&passListType=Person&uniqueId=1FSZ&datatype=Person

Foucault, M. (1994). *The order of things: An archaeology of the human sciences.* New York: Vintage. (Original work published 1966)

Friedan, B. (1963). *The feminine mystique.* New York: Norton.

Fromm, E. (1961). *Marx's concept of man.* New York: Continuum.

Fulbrook, M. (2004). *A concise history of Germany* (2nd ed.). Cambridge, UK: Cambridge University Press.

Gans, H. J. (1989). Sociology in America: The discipline and the public; American Sociological Association, 1988 presidential address. *American Sociological Review, 54,* 1–16.

Garfinkel, H. (1967). *Studies in ethnomethodology.* Cambridge, UK: Polity Press.

Garrison, W. L. (n.d.). *Public broadcasting service: Africans in America: Judgment Day.* Retrieved October 20, 2011, from http://www.pbs.org/wgbh/aia/part4/4p1561.html

Gergen, K. J. (1991). *The saturated self: Dilemmas of identity in contemporary life.* New York: Basic Books.

Gerth, H. H., & Mills, C. W. (Eds. & Trans.). (1948). *From Max Weber: Essays in sociology.* London: Routledge.

Giddens, A. (1990). *The consequences of modernity.* Stanford, CA: Stanford University Press.

Giddens, A. (1991). *Modernity and self-identity: Self and society in the late modern age.* Stanford, CA: Stanford University Press.

Giddings, F. H. (1922). *Studies in the theory of human society.* New York: Macmillan.

Gilman, C. P. (1975). *Women and economics: A study of the economic relation between men and women as a factor in social evolution.* New York: Gordon Press. (Original work published 1899)

Gilman, C. P. (2001). *The man-made world.* New York: Humanity Books. (Original work published 1911)

Hall, J. R., & Neitz, M. J. (1993). *Culture: Sociological perspectives.* Englewood Cliffs, NJ: Prentice Hall.

Hall, J. R., Neitz, M., & Barrani, M. (2003). *Sociology on culture.* London: Routledge.

Hall, S. (1996a). The question of cultural identity. In S. Hall, D. Held, D. Hubert, & K. Thompson (Eds.), *Modernity: An introduction to modern societies* (pp. 595–634). Malden, MA: Blackwell.

Hall, S. (1996b). The West and the Rest: Discourse and power. In S. Hall, D. Held, D. Hubert, & K. Thompson (Eds.), *Modernity: An introduction to modern societies* (pp. 184–227). Malden, MA: Blackwell.

Hanna, K. (2007). Riot Grrrl manifesto. In E. B. Freedman (Ed.) *The essential feminist reader* (pp. 394–396). New York: The Modern Library. (Original work published 1992)

Harrison, F. (1913). Introduction. In H. Martineau (Trans.), *The positive philosophy of Auguste Comte.* London: G. Bell and Sons.

Havens, J. J., & Schervish, P. G. (2003). *Why the $41 trillion wealth transfer estimate is still valid: A review of challenges and questions.* Retrieved February 2, 2012, from http://www.bc.edu/dam/files/research_sites/cwp/pdf/41trillion review.pdf

Hawking, S., & Mlodinow, L. (2010). *The grand design.* New York: Bantam Books.

Hegel, G. W. F. (1975). *Lectures on the philosophy of world history: Introduction* (H. B. Nisbet, Trans.). Cambridge, UK: Cambridge University Press. (Original work published 1830)

Heilbroner, R. L. (1986). The man and his times. In R. L. Heilbroner (Ed.), *The essential Adam Smith* (pp. 1–11). New York: Norton.

Hofstadter, D. (1985). *Metamagical themas: Questing for the essence of mind and pattern.* New York: Basic Books.

Horkheimer, M. (2004). *Eclipse of reason.* London: Continuum.

Horkheimer, M., & Adorno, T. W. (1972). *Dialectic of enlightenment.* New York: Herder & Herder.

James, W. (1904). Herbert Spencer. *The Atlantic Monthly, 94,* 99–108.

Jameson, F. (1984a). *The postmodern condition.* Minneapolis: University of Minnesota Press.

Jameson, F. (1984b). Postmodernism, or the cultural logic of late capitalism. *New Left Review, 146,* 53–92.

Johnson, A. G. (2000). *The Blackwell dictionary of sociology: A user's guide to sociological language* (2nd ed.). Malden, MA: Blackwell.

Kant, I. (1999). An answer to the question: What is Enlightenment? In M. J. Gregor (Ed.), *Practical philosophy* (pp. 11–22). Cambridge, UK: Cambridge University Press. (Original work published 1784)

Kitchen, M. (2006). *A history of modern Germany, 1800–2000.* Malden, MA: Blackwell.

Lash, S., & Urry, J. (1987). *The end of organized capitalism.* Madison: University of Wisconsin Press.

LeDoux, J. (1996). *The emotional brain: The mysterious underpinning of emotional life.* New York: Touchstone.

Leibniz, G. W. (1992). *Discourse on metaphysics and the monadology* (G. R. Montgomery, Trans.). Buffalo, NY: Prometheus Books. (Original works published 1686 and 1714)

Lengermann, P. M., & Niebrugge-Brantley, J. (1998). *The women founders: Sociology and social theory, 1830–1930.* Boston: McGraw-Hill.

Lidz, V. (2000). *Talcott Parsons.* In G. Ritzer (Ed.), *The Blackwell Companion to Major Social Theorists* (pp. 388–432). Malden, MA: Blackwell.

Lipset, S. M. (1962). Harriet Martineau's America. In H. Martineau, *Society in America* (pp. 5–42). New Brunswick, NJ: Transaction Publishers.

Luckmann, T. (1991). The new and the old in religion. In P. Bourdieu & J. S. Coleman (Eds.), *Social theory for a changing society* (pp. 167–182). Boulder, CO: Westview Press.

Lukács, G. (1971). *History and class consciousness: Studies in Marxist dialectics* (R. Livingstone, Trans.). London: Merlin Press. (Original work published 1922)

Lundberg, G. A. (1961). *Can science save us?* (2nd ed.). New York: David McKay.

MacIver, A. M. (1922). Saint-Simon and his influence on Karl Marx. *Economica, 6,* 238–245.

Mannheim, K. (1936). *Ideology and utopia: An introduction to the sociology of knowledge* (L. Wirth & E. Shils, Trans.). San Diego, CA: Harvest.

Marshall, G. (1998). *A dictionary of sociology* (2nd ed.). Oxford, UK: Oxford University Press.

Martineau, H. (1966). *Society in America.* New York: AMS. (Original work published 1837)

Martineau, H. (2003). *How to observe morals and manners.* New Brunswick, NJ: Transaction Publishers. (Original work published 1838)

Marx, K. (1977). *Capital: A critique of political economy* (Vol. 1; E. Mandel, Trans.). New York: Random House. (Original work published 1867)

Marx, K. (1978a). The civil war in France. In R. C. Tucker (Ed.), *The Marx-Engels reader* (pp. 618–652). New York: Norton. (Original work published 1891)

Marx, K. (1978b). Contribution to the critique: Introduction. In R. C. Tucker (Ed.), *The Marx-Engels reader* (pp. 53–65). New York: Norton. (Original work published 1844)

Marx, K. (1978c). Discovering Hegel. In R. C. Tucker (Ed.), *The Marx-Engels reader* (pp. 7–8). New York: Norton.

Marx, K. (1978d). Economic and philosophic manuscripts of 1844. In R. C. Tucker (Ed.), *The Marx-Engels reader* (pp. 66–125). New York: Norton. (Original work published 1932)

Marx, K. (1978e). *The Grundrisse.* In R. C. Tucker (Ed.), *The Marx-Engels reader* (pp. 221–293). New York: Norton. (Original work published by the Institute of Marx-Engels-Lenin, 1939–1941)

Marx, K. (1978f). Inaugural address of the working men's international association. In R. C. Tucker (Ed.), *The Marx-Engels reader* (pp. 512–519). New York: Norton. (Original work published 1864)

Marx, K. (1978g). Marx on the history of his opinions. In R. C. Tucker (Ed.), *The Marx-Engels reader* (pp. 3–6). New York: Norton. (Original work published 1859)

Marx, K. (1978h). Theses of Feuerbach. In R. C. Tucker (Ed.), *The Marx-Engels reader* (pp. 143–145). New York: Norton. (Original work published 1888)

Marx, K. (1995). Economic and philosophic manuscripts of 1844. In E. Fromm (Trans.), *Marx's concept of man* (pp. 90–196). New York: Continuum. (Original work published 1932)

Marx, K., & Engels, F. (1978). Manifesto of the Communist Party. In R. C. Tucker (Ed.), *The Marx-Engels reader* (pp. 469–500). New York: Norton. (Original work published 1848)

McMahon, D. M. (2006). *Happiness: A history.* New York: Grove Press.

Mead, G. H. (1934). *Mind, self, and society* (C. W. Morris, Ed.). Chicago: University of Chicago Press.

Menand, L. (2001). *The metaphysical club: A story of ideas in America*. New York: Farrar, Straus & Giroux.

Merriam-Webster. (2002). *Webster's third new international dictionary, unabridged*. Retrieved October 18, 2011, from http://unabridged.merriam-webster.com

Merton, R. K. (1949). *On theoretical sociology*. New York: The Free Press.

Merton, R. K. (1973). *The sociology of science: Theoretical and empirical investigations*. Chicago: University of Chicago Press.

Milgram, S. (1974). *Obedience to authority: An experimental view*. New York: Harper & Row.

Mills, C. W. (1956). *The power elite*. New York: Oxford University Press.

Mills, C. W. (1959). *The sociological imagination*. New York: Oxford University Press.

Moore, M. (Producer & Director). (1989). *Roger and me*. Burbank, CA: Warner Brothers.

Morris, C. W. (Ed.). (1962). *Mind, self, and society from the standpoint of a social behaviorist*. Chicago: University of Chicago Press.

National Bureau of Economic Research. (n.d.). *US business cycle expansions and contractions*. Retrieved February 20, 2012, from http://www.nber.org/cycles/cyclesmain.html#announcements

National Urban League. (2009). *State of black America*. Retrieved February 22, 2012, from http://www.nul.org/newsroom/publications/soba

Normano, J. F. (1932). Saint-Simon and America. *Social Forces, 11*, 8–14.

Nöth, W. (1995). *Handbook of semiotics*. Bloomington and Indianapolis: Indiana University Press.

Oakes, G. (Ed.). (1984). Introduction. In *Georg Simmel on women, sexuality, and love* (pp. 3–62). New Haven, CT: Yale University Press.

Obama, B. (2011). *Transcript: Obama's State of the Union address*. Retrieved June 1, 2011, from http://www.npr.org/2011/01/26/133224933/transcript-obamas-state-of-union-address

Orwell, G. (1946). *Shooting an elephant and other essays*. New York: Harcourt.

Parsons, T. (1937). *The structure of social action: A study in social theory with special reference to a group of recent European writers*. New York: McGraw-Hill.

Parsons, T. (1942). Max Weber and the contemporary political crisis II. *The Review of Politics, 4*(2), 155–172.

Parsons, T. (1949). *The structure of social action* (2nd ed.). New York: The Free Press.

Parsons, T. (1961). Culture and the social system. In T. Parsons (Ed.), *Theories of society: Foundations of modern sociological theory* (pp. 963–993). New York: The Free Press.

Price, R. (2005). *A concise history of France* (2nd ed.). Cambridge, UK: Cambridge University Press.

Ritzer, G. (2004). *The McDonaldization of society* (Rev. ed.). Thousand Oaks, CA: Pine Forge Press.

Roth, G. (1990). Marianne Weber and her circle. *Society, 27*(2), 63–69.

Rousseau, J. (1960). *Discourse on the origin of inequality* (F. Philip, Trans.). New York: Oxford University Press. (Original work published 1755)

Roy, W. (2001). *Making societies*. Thousand Oaks, CA: Pine Forge Press.

Schutz, A. (1967). *The phenomenology of the social world* (G. Walsh & F. Lehnert, Trans.). Evanston, IL: Northwestern University Press. (Original work published 1932)

Shils, E. A. (1949). Forward. In *The methodology of the social sciences* (E. A. Shils & H. A. Finch, Eds. & Trans.; pp. iii–x). New York: The Free Press.

Simmel, G. (1950). *The sociology of Georg Simmel* (K. H. Wolff, Ed. & Trans.). Glencoe, IL: The Free Press.

Simmel, G. (1971). *Georg Simmel: On individuality and social forms* (D. N. Levine, Ed.). Chicago: University of Chicago Press.

Smith, A. (1937). *An inquiry into the nature and causes of the wealth of nations* (E. Cannan, Ed.). New York: The Modern Library. (Original work published 1776)

Smith, D. E. (1987). *The everyday world as problematic: A feminist sociology.* Boston: Northeastern University Press.

Spencer, H. (1876). *Principles of sociology, Vol. I.* London: Williams and Norgate.

Spencer, H. (1954). *Social statics: The conditions essential to human happiness specified, and the first of them developed.* New York: Robert Schalkenbach Foundation. (Original work published 1882)

Spencer, H. (1968). *Reasons for dissenting from the philosophy of M. Comte.* Berkeley: Glendessary Press. (Original work published 1864)

Spencer, H. (1975a). *The principles of sociology* (Vol. 1). Westport, CT: Greenwood Press. (Original work published 1876–1896)

Spencer, H. (1975b). *The principles of sociology* (Vol. 3). Westport, CT: Greenwood Press. (Original work published 1876–1896)

Stanton, E. C. (2007). Declaration of sentiments and resolutions. In E. B. Freedman (Ed.), *The essential feminist reader.* New York: The Modern Library. (Original work published 1848)

Thatcher, M. (1987). Interview for *Woman's Own.* Retrieved February 20, 2012, from http://www.margaretthatcher.org/speeches/displaydocument.asp?docid=106689

Time. (1957, October 14). The Administration: Exit Charlie, grinning. Retrieved August 15, 2011, from http://www.time.com/time/magazine/article/0,9171,862753,00.html

Tocqueville, A. de (2002). *Democracy in America* (H. C. Mansfield & D. Winthrop, Eds. & Trans.). Chicago: University of Chicago Press. (Original work published 1835–1840)

Todd, J. (Ed.). (1993). *A vindication of the rights of woman and a vindication of the rights of men.* Oxford, UK: Oxford University Press.

Tucker, R. C. (1978). Preface to second edition. In R. C. Tucker (Ed.), *The Marx-Engels reader* (2nd ed., pp. ix–xiii). New York: Norton.

Turner, J. H. (1981). Émile Durkheim's theory of integration in differentiated social systems. *Pacific Sociological Review, 24,* 379–391.

Turner, J. H. (1985). In defense of positivism. *Sociological Theory, 3,* 24–30.

Turner, J. H. (1993). *Classical sociological theory: A positivist's perspective.* Chicago: Nelson-Hall.

Turner, J. H. (2000). *On the origins of human emotions: A sociological inquiry into the evolution of human affect.* Stanford, CA: Stanford University Press.

UC Santa Cruz Student Health Center. (n.d.). *Sexual assault prevention & education.* Retrieved September 2, 2009, from http://www2.ucsc.edu/rape-prevention/statistics.html

U.S. Bureau of Labor Statistics. (2009). *Highlights of women's earnings in 2008.* Retrieved September 2, 2009, from http://www.bls.gov/cps/cpswom2008.pdf

U.S. Department of Justice. (2008). *Growth in prison and jail populations slowing.* Retrieved February 22, 2012, from http://www.ojp.usdoj.gov/newsroom/pressreleases/2009/BJS090331.htm

University of Washington, Department of Sociology. (n.d.). *Guide to the University of Washington Department of Sociology records, 1920–2003.* Retrieved June 3, 2011, from http://www.lib.washington.edu/specialcoll/findaids/docs/uarchives/UA19_24_00UWDeptSociology.xml

Urry, J. (2006). Society. In J. Scott (Ed.), *Sociology: The key concepts* (pp. 167–170). London: Routledge.

Walker, R. (2007). Becoming third wave. In E. B. Freedman (Ed.), *The essential feminist reader* (pp. 397–401). New York: The Modern Library. (Original work published 1992)

Weber, M. (1975). *Max Weber: A biography* (H. Zohn, Ed. & Trans.). New York: Wiley. (Original work published 1926)

Weber, M. (1948a). Politics as a vocation. In H. H. Gerth & C. Wright Mills (Eds. & Trans.), *From Max Weber: Essays in sociology* (pp. 77–128). London: Routledge. (Original work published 1921)

Weber, M. (1948b). Science as a vocation. In H. H. Gerth & C. Wright Mills (Eds. & Trans.), *From Max Weber: Essays in sociology* (pp. 129–156). London: Routledge. (Original work published 1919)

Weber, M. (1948c). The social psychology of the world religions. In H. H. Gerth & C. Wright Mills (Eds. & Trans.), *From Max Weber: Essays in sociology* (pp. 267–301). London: Routledge. (Original work published 1915)

Weber, M. (1949). "Objectivity" in social science and social polity. In E. A. Shils & H. A. Finch (Eds. & Trans.), *The methodology of the social sciences* (pp. 50–112). New York: The Free Press. (Original work published 1904)

Weber, M. (1968). *Economy and society* (C. Wittich, Ed., & G. Roth, Trans.). New York: Bedminster. (Original work published 1922)

Weber, M. (1993). *The sociology of religion* (E. Fischoff, Trans.). Boston: Beacon Press. (Original work published 1922)

Weber, M. (1994). Suffrage and democracy in Germany. In P. Lassman & R. Speirs, (Eds.), *Weber: Political writings* (pp. 80–129). Cambridge, UK: Cambridge University Press. (Original work published 1917)

Weber, M. (2002). *The Protestant ethic and the spirit of capitalism* (S. Kalberg, Trans.). Los Angeles: Roxbury. (Original work published 1904–1905)

Weinstein, D., & Weinstein, M. A. (1993). *Postmodern(ized) Simmel.* London and New York: Routledge.

West, C. (1999). *The Cornel West reader.* New York: Civitas Books.

West, C. (2001). *Race matters* (2nd ed.). New York: Vintage Books.

Whatever happened to polio? (n.d.). Smithsonian Institution. Retrieved May 26, 2011, from http://americanhistory .si.edu/polio/americanepi/communities .htm

Wiggershaus, R. (1995). *The Frankfurt School: Its history, theories, and political significance* (M. Robertson, Trans.). Cambridge, MA: MIT Press. (Original work published 1986)

Williams, R. (1983). *Keywords: A vocabulary of culture and society* (Rev. ed.). New York: Oxford University Press.

Wills, G. (1978). *Inventing America: Jefferson's Declaration of Independence.* New York: Doubleday.

Wollstonecraft, M. (1993). A vindication of the rights of woman. In J. Todd (Ed.), *A vindication of the rights of woman and a vindication of the rights of men.* Oxford, UK: Oxford University Press. (Original work published 1792)

Wright, K. (2009, March). The ascent of Darwin. *Discover,* 34–38.

Wuthnow, R. (1987). *Meaning and moral order: Explorations in cultural analysis.* Berkeley: University of California Press.

Photo Credits

Chapter 2, Herbert Spencer: Wikimedia Commons.

Chapter 3, Karl Marx: Library of Congress.

Chapter 4, Émile Durkheim: Wikimedia Commons.

Chapter 5, Max Weber: Wikimedia Commons.

Chapter 6, George Herbert Mead: Wikimedia Commons.

Chapter 6, Georg Simmel: Wikimedia Commons.

Chapter 7, Harriet Martineau: Wikimedia Commons.

Chapter 7, Charlotte Perkins Gilman: Library of Congress. Copyright by C. F. Lummis.

Chapter 8, Frederick Douglass: Wikimedia Commons.

Chapter 8, W. E. B. Du Bois: Library of Congress. Photo by Cornelius M. Battey, 1918.

Chapter 9, Talcott Parsons: The Granger Collection, New York.

Index

overproduction and, 89–90
surplus labor and, 82–83
worker concentration and, 100–101
Inequality. *See also* Equality; Oppression
ascription and, 144
discrimination and, 310, 317–318
ideologies and, 95–99
institutions and, 48
Karl Marx and, 103
marriage and, 274–277
social solidarity and, 314
women and, 275–277
Inheritance, 144
Institutional differentiation, 151
Institutionalization of doubt, 149
Institutions
ceremonial, 46–47, 53–54
class consciousness and, 102
definition of, 43–44, 47
domestic, 44–46
ecclesiastical, 47–49
George Herbert Mead and, 234
Harriet Martineau and, 269, 271, 273
Herbert Spencer and, 43–49
history and, 315
legitimations, 43
marriage and, 45–46. *See also* Marriage
militaristic/industrial societies and, 40
moral culture and, 150
norms, 43
religion as, 273. *See also* Religion
requisite needs and, 33
social evolution and, 35
unequal treatment of women and, 293
Instrumental action, 335
Instrumental-rational action, 180
Integration
division of labor and, 139
Herbert Spencer and, 36
structural differentiation and, 139
Talcott Parsons and, 336
Intentionality, 3
Interactions. *See also* Action
culture and, 141
definition of, 229
George Herbert Mead and, 221–235
norms and, 142–143
social forms and, 240
Intermediary groups, 141–142
International corporations, 208
Intersectionality, 198
Iowa School, of symbolic
interaction, 237
Iron cage of bureaucracy, 206, 211

James, William, 27
Jameson, F.
machines and, 103–104
Marxian perspective and, 105
modern capitalism and, 103

Kant, I., 5, 64–67, 69, 169, 239–240
Keynes, John Maynard, 11–12
Kinship
bureaucracies and, 177
ceremonial institutions and, 47
Herbert Spencer and, 45–46
monogamy and, 45–46
religion and, 123, 211
sexuality and, 289
social evolution and, 35
totems and, 128
Knowledge, 5, 105. *See also* Education
bureaucracy and, 204–206, 211
capitalism and, 104
critical theory and, 10
definition of, 91
evolution of, 10–12
industrial societies and, 40
metaphysical stage of, 10–11
modernism and, 207
positive stage of, 11
postmodernism and, 209
power and, 280
religion and, 149, 279
Rousseau's views on, 117
sociology of, 91, 171
species-being and, 76
theological stage of, 10
theory and, 6–7
workers and, 100
Kuhn, M., 237

Labor. *See also* Division of labor; Work
capitalism and, 81, 91–95, 102, 192, 211
class and, 195
commodities and, 80–83
division of, 137–138
exploitation and, 81–83. *See also*
Exploitation
false consciousness/ideology and, 96
global capitalism and, 319–321
globalized division of, 102
Herbert Spencer and, 30
postmodernism and, 208
price controls and, 143
surplus, 82–83
value and, 80–83
Labor theory of value, 80–83

About the Author

Kenneth Allan received his PhD in sociology from the University of California, Riverside (1995), and is currently Professor of Sociology at the University of North Carolina at Greensboro (UNCG). He has authored several books in the area of theory, including *The Meaning of Culture: Moving the Postmodern Critique Forward* and his most recent book, *A Primer in Social and Sociological Theory: Toward a Sociology of Citizenship.* His current research interests are American individualism and civil society.

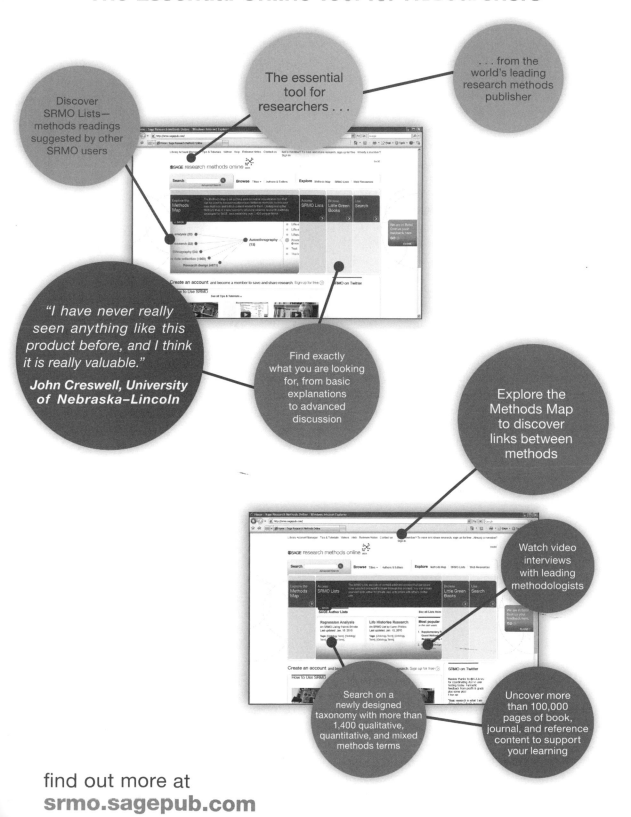